GENETIC ENGINEERING

Recent Developments in Applications

GENETIC ENGINEERING
Recent Developments in Applications

Dana M. Santos, PhD

Professor, Department of Anthropology,
Binghamton University, New York, U.S.A.

Apple Academic Press

Genetic Engineering: Recent Developments in Applications

© Copyright 2011*
Apple Academic Press Inc.

First Published in the Canada, 2011
Apple Academic Press Inc.
3333 Mistwell Crescent
Oakville, ON L6L 0A2
Tel. : (888) 241-2035
Fax: (866) 222-9549
E-mail: info@appleacademicpress.com
www.appleacademicpress.com

The full-color tables, figures, diagrams, and images in this book may be viewed at www.appleacademicpress.com

First issued in paperback 2021

ISBN 13: 978-1-77463-220-8 (pbk)
ISBN 13: 978-1-926692-67-8 (hbk)

Dana M. Santos, PhD

Cover Design: Psqua

Library and Archives Canada Cataloguing in Publication Data
CIP Data on file with the Library and Archives Canada

CONTENTS

INTRODUCTION

Genetic engineering involves the direct manipulation of genes. It is a common tool in both research and agriculture. A genetic engineer does research in medicine, biotechnology, agriculture, and related fields.

Important areas of research in medical genetic engineering include vaccine production, pharmaceutical development, and the treatment of disease, most especially cancer.

Genetic engineering in agriculture is involved in the genetic modification of crops and domestic animals to increase their yield, ease of production, and nutrition. Genetically modified food ranges from seedless vegetables to increasing the nutritional value of crops, particularly for nutrition-poor populations.

There are ethical issues that need to be addressed as the field develops, such as a concern that the genetic manipulation of human disease genes can turn into manipulation of appearance, intelligence, and so forth. There is also a major concern about genetically modified organisms (GMOs) and their effect on the environment and possibly human health.

Genetic engineering is important in disease research and the creation of new vaccines or treatments. Genetically modified organisms (GMOs) in agriculture are gaining popularity. Foods are created that are resistant to disease, have higher nutritional value, and increased production yield. The concern is that these GMOs may be negatively impacting indigenous organisms and harming the ecosystem as a whole.

Genetic engineering will continue to create new ways to advance disease research. This includes the manipulation of genes in genetic disorders. The agriculture industry continues to examine ways to produce disease-free, nutrition-high crops.

— **Professor Dana M. Santos**

Increased Hydrogen Production by Genetic Engineering of *Escherichia coli*

Zhanmin Fan, Ling Yuan and Ranjini Chatterjee

ABSTRACT

Escherichia coli is capable of producing hydrogen under anaerobic growth conditions. Formate is converted to hydrogen in the fermenting cell by the formate hydrogenlyase enzyme system. The specific hydrogen yield from glucose was improved by the modification of transcriptional regulators and metabolic enzymes involved in the dissimilation of pyruvate and formate. The engineered E. coli strains ZF1 (ΔfocA; disrupted in a formate transporter gene) and ZF3 (ΔnarL; disrupted in a global transcriptional regulator gene) produced 14.9, and 14.4 μmols of hydrogen/mg of dry cell weight, respectively, compared to 9.8 μmols of hydrogen/mg of dry cell weight generated by wild-type E. coli strain W3110. The molar yield of hydrogen for strain ZF3 was 0.96 mols of hydrogen/mol of glucose, compared to 0.54 mols of hydrogen/mol of glucose for the wild-type E. coli strain. The expression of the global transcriptional regulator protein FNR at levels above natural abundance had a synergistic effect on increasing the hydrogen yield in the ΔfocA genetic background. The modification of global transcriptional regulators to modulate the expression of multiple operons required for the biosynthesis of formate hydrogenlyase represents a practical approach to improve hydrogen production.

Introduction

The present emphasis on the diversification of global energy sources, while reducing the consumption of fossil fuels, has placed renewed interest on hydrogen as an alternative fuel and its synthesis through microbe-mediated bioconversions. Microbes have evolved unique mechanisms for hydrogen generation, and some of these mechanisms are being explored for biotechnological applications including nitrogenase-mediated and fermentative hydrogen production [1]. Efforts to improve hydrogen production have been focused on pathway redirection, identification and engineering of oxygen-tolerant hydrogenases, improvement in hydrogen molar yields and development of efficient hydrogen separation techniques from bioreactor headspace [1]–[5].

The production of hydrogen by facultative anaerobic organisms such as E. coli is a characteristic of mixed acid fermentation. Under anaerobic conditions, the fermentation products comprise a mixture of ethanol, succinate, lactate, acetate, and formate (Fig. 1). Succinate is produced via the succinate pathway. A key reaction in this pathway is carboxylation of phosphoenolpyruvate (PEP) to the four-carbon compound oxaloacetate by phosphoenolpyruvate carboxylase (PEPC) [6]. PEP is also converted to pyruvate which is subsequently converted to acetyl-CoA and formate by pyruvate formate lyase (PFL), and to lactate by lactate dehydrogenase (LDH). Formate hydrogenlyase (FHL) catalyzes the conversion of formate to carbon dioxide and hydrogen [7], [8]. The intracellular level of formate is determined by the rates of biosynthesis and metabolism of formate, and also by formate transporter FocA which is a membrane protein involved in formate transport into and out of the cell [9]. Knockout of the focA gene results in intracellular accumulation of formate [10]. The FHL enzyme complex is comprised of formate dehydrogenase H (FDHH), and a hydrogen-evolving hydrogenase (hydrogenase 3) [11], [12]. The biosynthesis of PFL and FHL are up-regulated by the action of multiple transcriptional regulators including the global transcriptional factors Fnr, ArcAB and integration host factor (IHF) [13]–[15], and repressed by the dual transcriptional regulator NarL (Table 1)[16]. The transcription of the fhl regulon is regulated by the primary and secondary transcriptional activator FHLA and ModE [17], [18]. The expression of fhlA is activated by Fnr in response to the cellular redox state [19], [20]. The biosynthesis of FDHH also requires the expression of selC gene which encodes a tRNA for the incorporation of seleno-cysteine to FDHH (Table 1) [8], [11], [21].

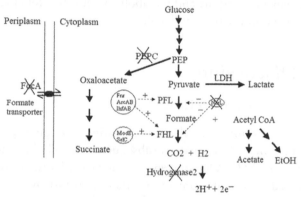

Figure 1. The genetic modification of metabolic pathways and regulatory components for hydrogen production in E. coli. The metabolic flows are indicated by solid arrows. Some key enzyme systems are labeled. The key global regulators and their regulatory targets are circled and indicated by dashed arrows, respectively. Pluses (+) represent activation, and minuses (–) represent repression. Crosses (X) indicate chromosomal gene disruptions.

Table 1. Genes engineered in this study, and their functions.

Gene	Function of gene product
focA	Transmembrane protein involved in formate transport.
hybC	Large catalytic subunit of hydrogenase 2; catalyzes dihydrogen oxidation.
narL	Global transcriptional regulator of anaerobic metabolism; represses the transcription of nik operon genes, fdhF and pflB genes; activates the transcription of formate dehydrogenase N, which couples formate oxidation to nitrate reduction.
ppc	Phosphoenolpyruvate carboxylase (PEPC) catalyzes the carboxylation of phosphoenolpyruvate (PEP) to oxaloacetate (OAA).
fnr	Global transcriptional regulator of anaerobic metabolism; activator of fhlA and pflB genes, and of the hyp, nik and moa operon genes.
arcAB	Global transcriptional regulator of aerobic and anaerobic metabolism; activator of pflB, and repressor of the hyb and hya operon genes.
ihfAB	Global transcriptional regulator activating the pflB and fhlA genes, and genes of the hyc and hyp operons.
modE	Secondary transcriptional activator of the fhl regulon.
selC	tRNA for selenocysteine Incorporation into FDH-F polypeptide of FHL.

Successful efforts to redirect pyruvate towards hydrogen in E. coli, have involved the up-regulation of the FHL complex and the disruption of pathways competing for pyruvate [4], [22]–[24]. It is estimated that over 50 genes distributed across at least 20 distinct genetic loci define fermentative hydrogen metabolism in E. coli [8]. In this study, our strategy involves the systematic modification of multiple, discrete, metabolic segments that include global regulatory, transport, and auxiliary components required for FHL biosynthesis, processing and assembly. The modification of key regulatory elements represents a practical strategy for the coordinated engineering of genes and operons that perform distinct biochemical functions related to the production of hydrogen. This approach has the potential to achieve the balance of cofactor and precursor supply pathways with the biosynthesis of structural polypeptides without placing an unnecessary metabolic burden on the cells [25]. Here, we describe increases in specific and molar yields of hydrogen achieved by the modification of the focA, ppc, narL, and fnr genes involved in precursor transport and metabolism, and the global regulation of fermentative hydrogen production by E. coli.

Results and Discussion

Strategy for Genetic Modification

The fermentation pathways initiate by the non-oxidative cleavage of pyruvate to acetyl-CoA and formate by PFL; formate is subsequently metabolized to hydrogen and carbon dioxide by FHL. Our strategy for improving fermentative hydrogen production involved the modification of precursor transport and metabolism, and of regulatory elements, to direct the flow of the two key precursor metabolites pyruvate and formate towards hydrogen, and achieve increased expression of both structural and auxiliary components of the FHL enzyme system. The biochemical reactions and regulatory elements that were modified by the inactivation of specific chromosomal genes, or by plasmid-directed gene over-expression are illustrated schematically in Figure 1; the functions of the corresponding gene products are summarized in Table 1.

The inactivation of the four selected genes in the chromosome of E. coli strain W3110 was hypothesized to increase the cellular concentrations of pyruvate and formate, reduce the re-oxidation of evolved hydrogen, and relieve the repression of genes required for PFL and FHL biosynthesis. The ppc gene encoding phosphoenolpyruvate carboxylase (PEPC) was disrupted to increase cellular pyruvate concentrations available for formate biosynthesis [6], [7]. The gene encoding the formate transporter, focA, was inactivated to elevate intra-cellular formate levels for FHL biosynthesis and conversion to hydrogen [10]. E. coli synthesizes two uptake hydrogenases, hydrogenase1 and 2, which oxidize hydrogen to protons with the release of electrons. In this study the hybC gene, which encodes the large catalytic subunit of hydrogenase 2, was inactivated to reduce the oxidation of hydrogen [26]. The global transcriptional regulator NarL represses transcription of the structural genes of FHL (fdhF), PFL (pflB), and of the genes of the nik operon, which are required for the transport and metabolism of nickel for Hydrogenase cofactor biosynthesis [16]. To eliminate the repression described, the narL gene was inactivated.

The global transcriptional regulators Fnr, ArcAB and IhfAB activate multiple genes of anaerobic pyruvate and formate metabolism. Targets for these global regulators include fhlA (the specific transcriptional activator of the fhl regulon), pflB, and the genes of the moa, nik and hyp operons required for the biosynthesis of the molybdopterin and Ni-Fe cofactors of FHL (Table 1) [15], [19], [27], [28]. The effects on hydrogen yield of multi-copy expression of the global regulators fnr, arcAB and lhfAB, as well as ModE (the secondary transcriptional activator of the fhl regulon), and the tRNA for selenocysteine incorporation (selC), were examined in wild-type W3110 and ΔfocA (ZF1) strains.

Increased Hydrogen Yields by Gene-Inactivated E. Coli Strains

The rate of hydrogen production and of growth by wild-type strain W3110, and four strains containing specific chromosomal gene disruptions were examined (Figures 2A and 2B). The maximum rate of hydrogen generation occurred between 9 and 16 hours following inoculation of M9 medium under the experimental conditions described. Strains ZF3 (W3110ΔnarL) and ZF4 (W3110Δppc) exhibited the highest rates of hydrogen production, and had accumulated about 20% to 30% more hydrogen in the culture headspace than W3110 in 25 hours (Figure 2A). Strain ZF1 (W3110ΔfocA) exhibited a slightly slower rate of hydrogen generation than W3110 and accumulated 33% less hydrogen by the end of the fermentation experiment. Strains ZF2 (W3110ΔhybC) and W3110 had comparable rates of hydrogen formation to each other. The slower rate of hydrogen generation observed in the later stages of fermentation correlated with an increased accumulation of biomass, possibly resulting from the hydrogen oxidizing activity of uptake hydrogenases 1, 2, and/or 4, generating protons and electrons for cellular ATP synthesis, or by inhibition of hydrogenase 3 by the evolved hydrogen that accumulated in culture headspace [8], [26]. The complete consumption of glucose accompanied by reduced glycolytic flux to pyruvate and subsequently to formate could account for the simultaneous reductions in growth and H2 production rates. Strains W3110, ZF3 (W3110ΔnarL) and ZF4 (W3110Δppc) exhibited similar generation times and reached similar final cell densities. Strain ZF2 (W3110ΔhybC) grew at a slightly slower rate than W3110, while strain ZF1 (W3110ΔfocA) grew at a slower rate than W3110 and reached a 50% lower final culture density (Figure 2B). The lower growth rates of strains ZF1 and ZF2 might result from the increased acidification of cellular cytoplasm from the disruption of formate export, and by the decrease in cellular ATP synthesis by the disruption of an uptake hydrogenase, respectively. The rate of anaerobic growth and of hydrogen production by the strains in TYP medium followed a profile similar to that seen in M9-glucose medium, with the highest rates of hydrogen production occurring between 4 and 7 hours of fermentation. The culture generation times were increased by about 2.5-fold in TYP medium, with A600 values reaching about 6.5 by 10 hours (data not shown).

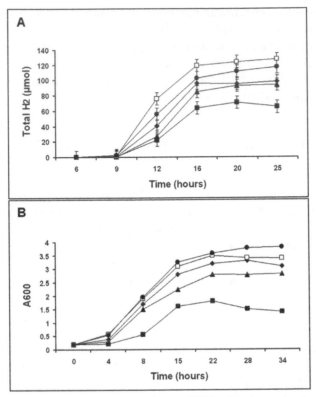

Figure 2. Hydrogen production and anaerobic growth rates of WT and single gene inactivated E. coli strains. Hydrogen production rate (A, μmol of H2) and anaerobic growth rate (B, A600) by strains W3110 (□), ZF1 (■), ZF2 (□), ZF3 (□), ZF4 (•) on M9 medium supplemented with 1% (w/v) glucose.

The specific hydrogen yields were determined for W3110 and the gene-inactivated strains at 17 hours following inoculation of M9-glucose medium (Table 2). All four gene-inactivated strains exhibited 14% to 50% higher specific hydrogen yields compared to W3110. Strains ZF1 (W3110ΔfocA) and ZF3 (W3110ΔnarL) had specific hydrogen yields of approximately 14 μmols/mg dry cell mass, representing an increase of about 46% over the yield by W3110. The increased hydrogen yield by strain ZF1 (W3110ΔfocA) is consistent with the hypothesis that an increase in the level of intracellular formate can either lead to enhanced synthesis of the FHL polypeptides via the transcriptional activator FhlA, or elevate the available substrate pool for FHL [29], [30]. The negative effect on growth rates by the focA deletion might be alleviated if the intracellular formate can be efficiently metabolized by FHL (further described below). The 50% increase in specific hydrogen yield by strain ZF3 (W3110ΔnarL) showed that the elimination of transcriptional repression of nik operon, and of fdhF and pflB genes by the NarL, results in the increased pyruvate metabolism towards hydrogen [8], [31].

Table 2. Hydrogen yields for strain W3110 with or without defined chromosomal gene disruptions.

Strains	Hydrogen yield (μmol/mg dry cell mass)	Hydrogen yield from glucose (mol/mol glucose)
W3110	9.80±0.72	0.54±0.04
ZF1 (W3310ΔfocA)	14.85±0.83	0.63±0.04
ZF2 (W3310ΔhybC)	12.14±0.88	0.70±0.05
ZF3 (W3310ΔnarL)	14.39±0.90	0.96±0.07
ZF4 (W3310ΔppC)	11.20±0.89	0.73±0.05
ZF5 (W3310ΔfocAΔhybC)	8.65±0.78	Not determined

The specific hydrogen yield of strain ZF2 (W3110ΔhybC) was 22% higher than W3110 (Table 2). Of the three E. coli uptake hydrogenases, hydrogenase 2 is believed to be the more active one based on in vitro kinetic studies [8]. The double mutant ZF5 (W3110ΔfocAΔhybC) did not produce more hydrogen than the single mutants ZF1 (W3110ΔfocA) and ZF2 (W3110ΔhybC) for reasons that remain unclear. Inactivation of the metabolic enzyme PEPC resulted in approximately 14% increase in specific hydrogen yields.

If all of the pyruvate is directed towards the PFL reaction the theoretical maximum yield of hydrogen from a hexose sugar is 2 mols of H2/mol glucose [7]. Strain ZF3 (W3110ΔnarL) also displayed the best molar yield of hydrogen (mols of H2/mol of glucose) of approximately 0.96 compared to 0.54 for W3110 (Table 2). The utility of eliminating transcriptional repression of multiple operons in anaerobic metabolism is highlighted by the effects of the narL gene deletion. This deletion confers a 1.5-fold increase in specific hydrogen yield and a 2-fold increase in molar hydrogen yield, while retaining the growth characteristics of the wild-type strain W3110.

To examine the partitioning of glucose in the two high hydrogen yielding strains the distribution of fermentation products of strains ZF1 (W3110ΔfocA) and ZF3 (W3110ΔnarL) were compared to that of W3110 (Table 3). These two strains carried out a balanced fermentation, with over 90% of the added glucose converted to products. No remaining glucose was detected in the medium by HPLC and spectrophotometric assays. Medium from strain ZF1 (W3110ΔfocA) contained 40% less formate compared to strains W3110 and ZF3 (W3110ΔnarL), consistent with less formate being secreted from the cell due to the disruption of the channel for formate. Lactate formation was increased by about 37% in strain ZF1 (W3110ΔfocA) compared to W3110, perhaps from the efficient induction of ldhA due to lower intra-cellular pH from formate accumulation, or by the inhibition of PFL activity and/or biosynthesis by formate. The formation of ethanol was lowered by 17% and 33% in strains ZF1 (W3110ΔfocA) and ZF3 (W3110ΔnarL), respectively, compared to that of W3110. These two strains accumulated succinate and acetate to levels similar to those of strain W3110.

Table 3. The fermentation products of strains W3110, ZF1 (W3110ΔfocA) and ZF3 (W3110ΔnarL).

	W3110	ZF1 (W3310Δ*focA*)	ZF3 (W3310Δ*narL*)
Hydrogen (μmol)	91.0±1.34	122.84±1.89	118.0±1.42
Formate (μmol)	114.24±1.32	66.86±0.87	112.83±1.26
Lactate (μmol)	158.92±1.47	218.40±1.88	159.66±1.21
Acetate (μmol)	106.36±1.17	107.96±1.15	109.14±1.15
Succinate (μmol)	41.48±0.76	42.85±0.72	43.46±0.81
Ethanol (μmol)	344.0±2.87	285.43±2.25	229.0±1.87

The Effect of Multi-Copy Gene Expression on Hydrogen Yields

Recent reports have described improvements in fermentative hydrogen yields in E. coli through the modification of the specific transcriptional activator, and repressor proteins of the FHL system, FhlA and HycA, respectively [22]. The three global regulators Fnr, ArcAB and IhfAB modulate the expression of multiple genes involved in the anaerobic metabolism of pyruvate in E. coli [9], [15], [19], [27], [28]. The impact on fermentative hydrogen yields by the over-production of the Fnr, ArcAB, and IhfAB proteins, and of ModE which is the secondary transcriptional activator of the hycoperon, was examined in strains W3110 and ZF1 (W3110ΔfocA). Strain ZF1 (W3110ΔfocA) was selected because an increased level of cellular formate, combined with the over-expression of formate-metabolizing genes might have a synergistic effect on hydrogen formation. The specific hydrogen yields following induction of the plasmid-bearing strains are shown in Table 4. Strain ZF7, over-expressing of the fnr gene in W3110 background, exhibited an approximately 5.5-fold increase in specific hydrogen yield compared to the control strain ZF6 (containing the plasmid without promoter-driven gene expression). Over-expression of the fnr gene in the ΔfocA genetic background (strain ZF13) resulted in the generation of 10 μmol of H_2/mg of dry cell mass compared to 6 μmol of H_2/mg of dry cell mass obtained with over-expression of the fnr gene in the wild-type W3110 background (strain ZF7) in TYP medium. When induced in M9-glucose medium for 8 hours, strains ZF7 and ZF13 accumulated approximately 24 μmol of H2/mg of dry cell mass and 34 μmol of H2/mg of dry cell mass, respectively (data not shown). Over-expression of the global regulators ArcAB and IhfAB, and of the secondary transcriptional activator ModE, had no discernible effects on hydrogen yields in either W3110 or ZF1 strains in TYP or in M9-glucose medium. The synergistic effect on increased hydrogen yield by the over-expression of Fnr in the ΔfocA genetic background suggests that genetic modifications to simultaneously increase both substrate and enzyme levels for hydrogen production are desirable to prevent the accumulation of formate inside the cell. These results show that specific genetic backgrounds provide additive effects on hydrogen yield when combined with increased gene dosage of particular metabolic enzymes or regulatory components.

Table 4. Hydrogen yields for strain W3110 and ZF1 (W3110ΔfocA) when expressing plasmid-borne genes.

Strain	Genotype	Hydrogen yield (μmol/mg dry cell mass)
ZF6	W3110/pACYC177L	1.07±0.03
ZF7	W3110/pfnr	6.23±0.45
ZF8	W3110/parcAB	1.11±0.04
ZF9	W3110/pihfAB	1.14±0.04
ZF10	W3110/pmodE	1.27±0.06
ZF11	W3110/pselC	1.81±0.06
ZF12	W3110ΔfocA/pACYC177L	0.00±0.00
ZF13	W3110ΔfocA/pfnr	10.32±0.58
ZF14	W3110ΔfocA/parcAB	0.19±0.02
ZF15	W3110ΔfocA/pihfAB	0.61±0.03
ZF16	W3110ΔfocA/pmodE	0.63±0.03
ZF17	W3110ΔfocA/pselC	1.19±0.06

While the Fnr protein activates transcription of operons involved in nickel and molybdenum metabolism for FHL biosynthesis [32], it has not been shown to affect the transcription of genes involved in selenium metabolism. To determine whether selenium metabolism is limiting to hydrogen production, the selC gene, encoding the tRNA for selenocysteine incorporation into the FDHH polypeptide, was over-expressed in strains W3110 and ZF1 (W3110ΔfocA). An approximately 63% increase in hydrogen yield was observed in strain ZF11 which expresses multiple copies of the selC gene compared to the negative control, strain ZF6 (Table 4). These results illustrate that hydrogen yields can be increased by improving the efficiency of a pathway for cofactor production for the FHL enzyme system.

One approach to obtaining sustained increases in fermentative hydrogen-yields will be to direct and improve the metabolism of pyruvate and formate towards hydrogen under anaerobic conditions. Genetic modifications that have led to increased hydrogen yields in E. coli have thus far focused on the disruption of key metabolic genes of the mixed acid fermentation pathways of E. coli, and of specific regulators of the FHL system. This report represents the first example of increased hydrogen production achieved through the modification of two global transcriptional regulators, a cofactor biosynthetic gene, and a transporter for formate. The construction of E. coli strains with specific chromosomal gene deletions, have revealed three unique genetic backgrounds that improve fermentative hydrogen production: inactivation of the focA, narL, and ppc genes lead to increased hydrogen yields over that of wild-type E. coli. The metabolism of selenium is one of several metabolic pathways required for the processing of metal ions for FHL biosynthesis. The improvement in hydrogen yield observed by the increased expression of the tRNA for selenocysteine suggests that cofactor biosynthesis pathways could be useful targets for modification. The increased hydrogen yields obtained by over-expression of the fnr gene, indicates that the anaerobic metabolism of pyruvate and formate can be coordinately up-regulated for improved flux towards hydrogen.

The results presented in this work illustrate the utility of modifying transport, regulatory, and auxiliary pathways related to FHL biosynthesis and function, to achieve increased hydrogen yields. The future application of genome engineering formats in combination with global gene expression analysis to E. coli will enable the identification of novel genes and metabolic functions underlying the improved hydrogen producing phenotype [25], [33]. Similar approaches might be extended to enteric microbes that contain the FHL enzyme system to improve hydrogen production from alternate carbon sources, and for the simultaneous production of value-added chemical entities derived from the anaerobic metabolism of pyruvate.

Materials and Methods

Genetic Manipulations

All strains and plasmids used in this study are listed in Table 5, and PCR primers used are shown in Table 6. The disruption of chromosomal genes in E. coli K-12 strain W3110 was performed by the method of Datsenko and Wanner [34]. Plasmid pKD13 was used as the template for amplification of the FRT-flanked Km^R gene with primers containing 44 to 50 nucleotide homology extensions. Plasmids pKD46 and pCP20 were used for the replacement of the target gene with the Km^R gene, and for subsequent Km^R gene removal, respectively. Primer pairs, focA-H2P4/focA-H1P1, hybC-H2P4/hybC-H1P1, nar-H2P4/nar-H1P1, and ppc-H2P4/ppc-H1P1 were used for the disruption of the focA, hybC, narL, and ppc genes, respectively. The entire coding sequences of the hybC, narL, and ppc genes were deleted, while only 589 nucleotides at the 5′ terminal sequence of the focA gene were deleted since the sequence downstream of it overlaps the pfl operon. The gene disruptions were verified by PCR analyses of genomic DNA using the primers k1 and k2 to the Km^R gene, and the FW and RV primers to the flanking sequences of the disrupted genes.

Table 5. E. coli strains and plasmids used in this study.

Strains and plasmids	Genotype or description	Reference or source
Strains		
W3110	F-I-IN(rrnD-rrnE)1 rph-1	CGSC# 4470
ZF1	W3110ΔfocA	This work
ZF2	W3110ΔhybC	This work
ZF3	W3110ΔnarL	This work
ZF4	W3110Δppc	This work
ZF5	W3110ΔfocAΔhybC	This work
ZF6	W3110 containing pACYC177L	This work
ZF7	W3110 containing pfnr	This work
ZF8	W3110 containing parcAB	This work
ZF9	W3110 containing pihfAB	This work
ZF10	W3110 containing pmodE	This work
ZF11	W3110 containing selC	This work
ZF12	ZF1containing pACYC177L	This work
ZF13	ZF1 containing pfnr	This work
ZF14	ZF1 containing parcAB	This work
ZF15	ZF1 containing pihfAB	This work
ZF16	ZF1 containing pmodE	This work
ZF17	ZF1 containing pselC	This work
Plasmids		
pfnr	pACYC with LacZ promoter and W3110 fnr gene	This work
parcAB	pACYC with LacZ promoter and W3110 arcA and arcB genes	This work
pihfAB	pACYC with LacZ promoter and W3110 ihfA and ihfB genes	This work
pmodE	pACYC with LacZ promoter and W3110 ihfA and ihfB genes	This work
pselC	pACYC with LacZ promoter and W3110 selC gene	This work

The plasmid pACYC177 (Invitrogen) was used for the multi-copy expression of the E. coli genes fnr, arcAB, ihfAB, modE, and selC. The genes were amplified by PCR, using Pfx50™ DNA polymerase (Invitrogen) to produce blunt ended PCR products, from genomic DNA of E. coli strain W3110. For the construction of pfnr, pmodE and pselC, the lacZ promoter sequence from the pGEM®-T easy vector (Promega) was transferred into pAYCY177 by amplification using primers Plac-BamHI-FW and Plac-PstI-RV, digestion with BamHI and PstI, and ligation into the corresponding sites of pACYC177. The modified pACYC177 vector was designated pACYC177L. The fnr, modE, and selC genes were amplified with primer pairs: fnr-FW/fnr-RV, modE-FW/modE-RV and selC-FW/selC-RV, respectively, digested with PstI and FspI, and ligated into the corresponding sites of pACYC177L. For the construction of parcAB and pihfAB, primer Plac-ScaI-RV was used instead of primer Plac-PstI-RV for the amplification of the lacZ promotor, and ScaI was used for digestion of the PCR fragment. This modified pACYC177 vector was designated as pACYC177L1. The arcB and ihfB genes were amplified using primer pairs arcB-XbaI-FW/arcB-BglI-RV and ihfB-XbaI-FW/ihfB-BglI-RV, respectively. The two forward primers contained a 40-nucleotide linker sequence from genomic DNA of W3110. The PCR products were digested with BglI, and ligated to the FspI/BglI site of pACYC177L1. Finally, the arcA and ihfA genes were amplified using primer pairs arcA-FW/arcA-XbaI-RV and ihfA-FW/ihfA-XbaI-RV, digested with XbaI, and ligated to ScaI/XbaI site of pACYC177L1.

Table 6. Primers used in this study.

Primer	Sequence	Source
focA-H2P4	5'-CTTTGTTAGTATCTCGTCGCCGACTTAATAAAGAGAGAGTTAGTATTCCGGGGATCCGTCGACC-3'	This work
focA-H1P1	5'-GATACTGTGCTCAAAACCGCTGGCAACAAACATCGCGACCGGCAGTGTAGGCTGGAGCTGCTTC-3'	This work
focA-FW	5'-GCCAGGCGAGATATGATCTA-3'	This work
focA-RV	5'-CTGCGTACTTCGACAACCAT-3'	This work
hybC-H2P4	5'-CCTTTAAAACAAAACGATCATAATCGTCATGAGGCGAGCAAAGCATGAGCATTCCGGGGATCCGTCGACC-3'	This work
hybC-H1P1	5'-ATCGGTCAGCAAAATATTGCCGACCCCTAAGACTAAAATACGCATTACAGGTGTAGGCTGGAGCTGCTTC-3'	This work
hybC-FW	5'-GTCAGCACTGAGCGCACTGT-3'	This work
hybC-RV	5'-CAACGCCGGAACTTCTTCAT-3'	This work
nar-H2P4	5'-CACATTCATTAAGGTTATTGCTCATTTAAAGCCTGAAGGAAGAGGTTTACATTCCGGGGATCCGTCGACC-3'	This work
nar-H1P1	5'-TTCGCAGCATCGGGTGATCGTCAATCAGCAGGATAGTAGCCGGTTCCTGAGTGTAGGCTGGAGCTGCTTC-3'	This work
nar-FW	5'-GCCAATAGCCGAGAATACTG-3'	This work
nar-RV	5'-GACTCCGCCAGTTCAATACC-3'	This work
ppc-H2P4	5'-CGTGAAGGATACAGGGCTATCAAACGATAAGATGGGGTGTCTGGGGTAATATTCCGGGGATCCGTCGACC-3'	This work
ppc-H1P1	5'-ATTTCAGAAAACCCTCGCGCAAAAGCACGAGGGTTTGCAGAAGAGGAAGAGTGTAGGCTGGAGCTGCTTC-3'	This work
ppc-FW	5'-CGCATCTTATCCGACCTACA-3'	This work
ppc-RV	5'-TGGCCTGTAGCAGAGTAGAG-3'	This work
Plac-BamH-FW	5'-ACATGGATCCGCGCCCAATACGCAAAC-3'	This work
Plac-PstI-RV	5'-ACATCTGCAGAGCTGTTTCCTGTGTGAAAT-3'	This work
Plac-ScaI-RV	5'-ACATAGTACTAGCTGTTTCCTGTGTGAAAT-3'	This work
Fnr-FW	5'-ACTACTGCAGATGATCCCGGAAAAGCGAATTA-3'	This work
Fnr-RV	5'-TCTATGCGCATCAGGCAACGTTACGCGTA-3'	This work
modE-FW	5'-ACTACTGCAGATGCAGGCCGAAATCCTTC-3'	This work
modE-RV	5'-TCTATGCGCATTAGCACAGCGTGGCGATA-3'	This work
selC-FW	5'-ACTACTGCAGATACGCTGCGTGTATTAGGC-3'	This work
selC-RV	5'-TCTATGCGCACTTAGACCAGGTATTCCTGAAG-3'	This work
arcB-XbaI-FW	5'-TCTAGATAAGGCCGCGGAGGATTACACTATGATGAAGCAAATTCGTCTGCT-3'	This work
arcB-BglI-RV	5'-ATCTAGCCGGAAGGGCTCATTTTTTAGTGGCTTTTGCC-3'	This work
arcA-FW	5'-ATGCAGACCCCGCACATTCT-3'	This work
arcA-XbaI-RV	5'-ACTATCTAGATTAATCTTCCAGATCACCGCA-3'	This work
arcA-XbaI-RV	5'-ACTATCTAGATTAATCTTCCAGATCACCGCA-3'	This work
ihfB-XbaI-FW	5'-TCTAGATAAGGCCGCGGAGGATTACACTATGACCAAGTCAGAATTGAT-3'	This work
ihfB-BglI-RV	5'-ATCTAGCCGGAAGGGCTTAACCGTAAATATTGGCGC-3'	This work
ihfA-FW	5'-ATGGCGCTTACAAAAGCTGA-3'	This work
ihfA-XbaI-RV	5'-ACTATCTAGATTACTCGTCTTTGGGCGA-3'	This work

Bacterial Strains, Media, and Growth Conditions

All strains of E. coli used in this study are listed in Table 5. E. coli strains were routinely cultured at 37°C in either phosphate-buffered rich medium TYP (10 g/L tryptone, 5 g/L yeast extract, 12 g/L K_2HPO_4, 3 g/L KH_2PO_4; pH 7.6) or in M9 salts medium (Sigma-Aldrich) containing trace elements [35], [36]. When required, kanamycin was added at a concentration of 30 μg/ml. For studies requiring plasmid-directed gene expression, 1 mM IPTG was added to the cultures 1 hour following inoculation of the fermentation medium, and induction was performed for approximately 4 hours.

Fermentations were carried out in 37 ml serum vials (Wheaton) containing 15 ml of TYP or M9 medium. Following the addition of medium, the vials were sealed using silicone stoppers and aluminum crimp seals (Supelco), and autoclaved. Prior to inoculation, the headspace was evacuated and flushed with Argon (ultra high purity, Airgas). Supplements included in TYP or M9 medium were: 0.5% or 1% (w/v) glucose; 1 μM sodium molybdate; 1 μM sodium selenite; 50 μM nickel sulfate. Approximately 1.3% (v/v) of inoculum culture, prepared under aerobic conditions in TYP, was used to inoculate fermentation medium. Specific hydrogen yields were determined at approximately 5 hours, and 17 hours in TYP and M9 medium, respectively. Fermentations in 10 ml of M9 medium containing 0.5% glucose (w/v) were run for 17 hours to determine the molar yields of hydrogen. The concentrations of all products generated from glucose were determined for strains cultured anaerobically for 34 hours in 15 ml of M9 medium containing 1.0% glucose (w/v) to allow adequate time for the consumption of all of the glucose contained in the medium. All experiments were repeated six times with at least duplicate fermentations contained within each experiment. The A600 for strains cultured anaerobically in M9 medium was measured by removing samples anoxically with a syringe in intervals for 34 hours.

Analytical Methods

The gas mixture from culture headspace was withdrawn using gastight syringes (Hamilton), and monitored for hydrogen formation. Hydrogen was detected using a Shimadzu GC-17A gas chromatograph (GC) equipped with a thermal conductivity detector (TCD), a 60/80 Molecular Sieve 5A column (Supelco), with argon as the carrier gas. The Clarity software package (DataApex) was used to analyze the chromatograms.

The formation of fermentation products were detected by high performance liquid chromatography (HPLC) using an Aminex HPX-87H ion exchange column (Bio-Rad; 7.8×300 mm) and a Varian ProStar 410 liquid chromatography instrument equipped with UV absorbance and refractive index detectors. Culture supernatant was filtered through 0.45 μM cellulose acetate filters, and 40 μL of sample was injected. The column was eluted isocratically with 5 mM H2SO4 at a flow rate of 0.5 ml/minute. Data was analyzed using the Star Chromatography

Workstation software. Ethanol was quantified using the Quantichrome assay kit (BioAssay Systems). Glucose was quantified by both HPLC (described above), and by the glucose oxidase (GO) assay kit (Sigma).

Dry cell mass was determined by filtration of the entire culture through a 0.8 μM nitrocellulose membrane, drying the membrane at 60°C in a vacuum oven for about 14 hours, and determining the weight of the membrane.

Acknowledgements

We are grateful to Keith Kepler for technical assistance, and to Kenneth Zahn for insightful discussions.

Author Contributions

Conceived and designed the experiments: ZF LY RC. Performed the experiments: ZF RC. Analyzed the data: ZF LY RC. Contributed reagents/materials/analysis tools: ZF RC. Wrote the paper: RC. Revised manuscript drafts: ZF. Revised manuscript: LY.

References

1. Buckley MaW, J (2006) Microbial Energy Conversion. American Academy of Microbiology Report.

2. Nath K, Das D (2004) Improvement of fermentative hydrogen production: various approaches. Appl Microbiol Biotechnol 65: 520–529.

3. Maeda T, Sanchez-Torres V, Wood TK (2007) Enhanced hydrogen production from glucose by metabolically engineered Escherichia coli. Applied Microbiology and Biotechnology 77: 879–890.

4. Yoshida A, Nishimura T, Kawaguchi H, Inui M, Yukawa H (2006) Enhanced hydrogen production from glucose using ldh- and frd-inactivated Escherichia coli strains. Appl Microbiol Biotechnol 73: 67–72.

5. Samir Kumar Khanal, Wen-Hsing Chen, Ling Li, Sung S (2004) Biological hydrogen production: effects of pH and intermediate products International. Journal of Hydrogen Energy 29: 1123–1131.

6. Gokarn RR, Eiteman MA, Altman E (2000) Metabolic analysis of Escherichia coli in the presence and absence of the carboxylating enzymes phosphoenolpyruvate

carboxylase and pyruvate carboxylase. Applied and Environmental Microbiology 66: 1844–1850.

7. Ingraham JLaN FC (1987) Escherichia Coli and Salmonella Typhimurium: Cellular and Molecular Biology. ASM Press 262–278.

8. Sawers G (1994) The hydrogenases and formate dehydrogenases of Escherichia coli. Antonie Van Leeuwenhoek 66: 57–88.

9. Sawers RG (2005) Formate and its role in hydrogen production in Escherichia coli. Biochemical Society Transactions 33: 42–46.

10. Suppmann B, Sawers G (1994) Isolation and characterization of hypophosphite–resistant mutants of Escherichia coli: identification of the FocA protein, encoded by the pfl operon, as a putative formate transporter. Mol Microbiol 11: 965–982.

11. Vignais PAea, FEMS Microbiology Reviews, 2001. 25: p. 455–501. (2001).

12. Self WT, Hasona A, Shanmugam KT (2004) Expression and regulation of a silent operon, hyf, coding for hydrogenase 4 isoenzyme in Escherichia coli. J Bacteriol 186: 580–587.

13. Salmon K, Hung SP, Mekjian K, Baldi P, Hatfield GW, et al. (2003) Global gene expression profiling in Escherichia coli K12. The effects of oxygen availability and FNR. J Biol Chem 278: 29837–29855.

14. Perrenoud A, Sauer U (2005) Impact of global transcriptional regulation by ArcA, ArcB, Cra, Crp, Cya, Fnr, and Mlc on glucose catabolism in Escherichia coli. J Bacteriol 187: 3171–3179.

15. Constantinidou C, Hobman JL, Griffiths L, Patel MD, Penn CW, et al. (2006) A reassessment of the FNR regulon and transcriptomic analysis of the effects of nitrate, nitrite, NarXL, and NarQP as Escherichia coli K12 adapts from aerobic to anaerobic growth. Journal of Biological Chemistry 281: 4802–4815.

16. Overton TWGL, Patel MD, Hobman JL, Penn CW, Cole JA, Constantinidou C (2006) Microarray analysis of gene regulation by oxygen, nitrate, nitrite, FNR, NarL and NarP during anaerobic growth of Escherichia coli: new insights into microbial physiology. Biochem Soc Trans 34: 104–107.

17. Self WT, Hasona A, Shanmugam KT (2001) N-terminal truncations in the FhlA protein result in formate- and MoeA-independent expression of the hyc (formate hydrogenlyase) operon of Escherichia coli. Microbiology-Sgm 147: 3093–3104.

18. Self WT, Grunden AM, Hasona A, Shanmugam KT (1999) Transcriptional regulation of molybdoenzyme synthesis in Escherichia coli in response to molybdenum: ModE-molybdate, a repressor of the modABCD (molybdate transport)

operon is a secondary transcriptional activator for the hyc and nar operons. Microbiology 145(Pt 1): 41–55.

19. Sawers RG (2005) Formate and its role in hydrogen production in Escherichia coli. Biochem Soc Trans 33: 42–46.

20. Hopper S, Babst M, Schlensog V, Fischer HM, Hennecke H, et al. (1994) Regulated expression in vitro of genes coding for formate hydrogenlyase components of Escherichia coli. J Biol Chem 269: 19597–19604.

21. Blokesch M, Paschos A, Theodoratou E, Bauer A, Hube M, et al. (2002) Metal insertion into NiFe-hydrogenases. Biochem Soc Trans 30: 674–680.

22. Yoshida A, Nishimura T, Kawaguchi H, Inui M, Yukawa H (2005) Enhanced hydrogen production from formic acid by formate hydrogen lyase-overexpressing Escherichia coli strains. Appl Environ Microbiol 71: 6762–6768.

23. Redwood MD, Mikheenko IP, Sargent F, Macaskie LE (2008) Dissecting the roles of Escherichia coli hydrogenases in biohydrogen production. FEMS Microbiol Lett 278: 48–55.

24. Maeda TS-TV, Wood TK (2007) Enhanced Hydrogen production from glucose by metabolically engineered Escherichia coli. Appl Microbiol Biotechnol 77: 879–890.

25. Adrio JL, Demain AL (2006) Genetic improvement of processes yielding microbial products. FEMS Microbiol Rev 30: 187–214.

26. Paulette MVignais, Bernard Billoud, Meyer J (2006) Classification and phylogeny of hydrogenases. FEMS Microbiology Reviews 25: 455–501.

27. Perrenoud A, Sauer U (2005) Impact of global transcriptional regulation by ArcA, ArcB, Cra, Crp, Cya, Fnr, and Mlc on glucose catabolism in Escherichia coli. Journal of Bacteriology 187: 3171–3179.

28. Salmon K, Hung SP, et al. (2003) Global gene expression profiling in Escherichia coli K12. The effects of oxygen availability and FNR. J Biol Chem 278(32): 29837–55. 278: 29873–29855.

29. Maupin JA, Shanmugam KT (1990) Genetic regulation of formate hydrogenlyase of Escherichia coli: role of the fhlA gene product as a transcriptional activator for a new regulatory gene, fhlB. J Bacteriol 172: 4798–4806.

30. Schlensog V, Lutz S, Bock A (1994) Purification and DNA-binding properties of FHLA, the transcriptional activator of the formate hydrogenlyase system from Escherichia coli. J Biol Chem 269: 19590–19596.

31. http://biocyc.org/ECOLI/NEW-IMAGE?typeENZYME&objectNARL-MONOMER .

32. http://biocyc.org/ECOLI/NEW-IMAGE?typeENZYME&objectPD00197 .

33. Petri R, Schmidt-Dannert C (2004) Dealing with complexity: evolutionary engineering and genome shuffling. Curr Opin Biotechnol 15: 298–304.

34. Datsenko KA, Wanner BL (2000) One-step inactivation of chromosomal genes in Escherichia coli K-12 using PCR products. Proc Natl Acad Sci U S A 97: 6640–6645.

35. Miller JH (1972) Experiments in Molecular Genetics Cold Spring Harbor. NY: Cold Spring Harbor.

36. Jacobi A, Rossmann R, Bock A (1992) The hyp operon gene products are required for the maturation of catalytically active hydrogenase isoenzymes in Escherichia coli. Arch Microbiol 158: 444–451.

Use of the λ Red-Recombineering Method for Genetic Engineering

Joanna I. Katashkina, Yoshihiko Hara, Lyubov I. Golubeva, Irina G. Andreeva, Tatiana M. Kuvaeva and Sergey V. Mashko

ABSTRACT

Background

Pantoea ananatis, a member of the Enterobacteriacea family, is a new and promising subject for biotechnological research. Over recent years, impressive progress in its application to L-glutamate production has been achieved. Nevertheless, genetic and biotechnological studies of Pantoea ananatis have been impeded because of the absence of genetic tools for rapid construction of direct mutations in this bacterium. The λ Red-recombineering technique previously developed in E. coli and used for gene inactivation in several other bacteria is a high-performance tool for rapid construction of precise genome modifications.

Results

In this study, the expression of λ Red genes in P. ananatis was found to be highly toxic. A screening was performed to select mutants of P. ananatis that were resistant to the toxic affects of λ Red. A mutant strain, SC17(0) was identified that grew well under conditions of simultaneous expression of λ gam, bet, and exo genes. Using this strain, procedures for fast introduction of multiple rearrangements to the Pantoea ananatis genome based on the λ Red-dependent integration of the PCR-generated DNA fragments with as short as 40 bp flanking homologies have been demonstrated.

Conclusion

The λ Red-recombineering technology was successfully used for rapid generation of chromosomal modifications in the specially selected P. ananatis recipient strain. The procedure of electro-transformation with chromosomal DNA has been developed for transfer of the marked mutation between different P. ananatis strains. Combination of these techniques with λ Int/Xis-dependent excision of selective markers significantly accelerates basic research and construction of producing strains.

Background

Pantoea ananatis belongs to the Enterobacteriacea family. The P. ananatis strain AJ13355 (SC17) was isolated from soil in Iwata-shi (Shizuoka, Japan) as a bacterium able to grow at acidic pH and showing resistance to high concentrations of glutamic acid [1]. These physiological features made this organism an interesting object for biotechnological studies, and for this reason its genome has been sequenced by Ajinomoto Co. (unpublished results). Nevertheless, up to the recent past, the absence of efficient genetic tools has hampered manipulations of this bacterium and retarded both basic research and applied investigations.

Over the last decade [2-7], the most powerful method for generating a wide variety of DNA rearrangements in E. coli has been termed "recombineering" (recombination-mediated genetic engineering) [8]. The term generally refers to in vivo genetic engineering with DNA fragments carrying short homologies with a bacterial chromosome, using the proteins of a homologous recombination system of the bacteriophage λ (λ Red system). Lambda red operon includes only three genes encoding Exo, Beta and Gam proteins. Gam inhibits the host nucleases, RecBCD and SbcCD, thereby protecting the dsDNA substrate for recombination [9,10]; Exo degrades linear dsDNA from each end in a 5'→3' direction, creating dsDNA with 3' single-stranded DNA tails [11-14]; and Beta stably binds a ssDNA greater than 35 nucleotides in length and mediates pairing the one with a complementary target [15-17].

The λ Red-mediated recombineering technology developed initially for modification of the genome of Escherichia coli K12 [2,4-7], was later broadened to other E. coli strains including enteropathogenic ones [18], and to Salmonella [19,20], Shigella [21,22], Yersinia [23,24], Pseudomonas [25], as well. Probably, one of the factors impeding application of this system in other hosts is the toxicity of expression of the λ Red genes for the cells.

In the present study, to overcome this toxicity for Pantoea ananatis, the special strain resistant to simultaneous expression of the λ Red genes was selected. Using this mutant, construction of all types of chromosomal rearrangements previously obtained in E. coli by recombineering was reproduced. The approach described may be used for adjusting the technology to other hosts.

Results

Construction of the New Broad-Host-Range λ Red-Expressing Plasmid

To provide regulated expression of the λ Red genes in different bacteria, the plasmid pRSFRedTER [GenBank:FJ347161] based on the broad-host-range replicon of RSF1010 [26] has been constructed. This plasmid is useful for λ Red-mediated recombineering because: 1) the replicon is stably maintained in many Gram-negative [27] and some Gram-positive bacteria [28]; 2) the λ Red genes are placed under the control of the P_{lacUV5} promoter recognized by different bacterial RNA polymerases [29,30]; 3) the auto-regulated element P_{lacUV5}-lacI provides IPTG-inducible expression of the λ Red genes with low basal level [31]; 4) the plasmid contains the levansucrase gene from B. subtilis allowing rapid and efficient recovery of this plasmid from the cells in a medium containing sucrose.

To test recombineering efficiency, pRSFRedTER mediated disruption of galK gene in E. coli MG1655 chromosome by integration of the PCR-generated DNA fragment carrying the KmR gene from pUC4K flanked by attLλ and attRλ sites (attLλ-KmR -attRλ) was performed. The constructed plasmid provided about 300 transformants per 108 survivors following electroporation. A similar frequency was obtained when pKD46 plasmid [4] was used as λ Red-expressing plasmid. In each case a chromosome structure of ten KmR colonies was confirmed with PCR analysis.

Another plasmid carrying λ Red genes and named pRSFRedkan [GenBank:FJ347162], has been constructed via substitution of CmR and B. subtilis sacB genes by the KmR gene from pUC4K [32].

Concerted Expression of λ Red Genes is Highly Toxic for the P. Ananatis Wild-Type Cells

Clones of P. ananatis SC17 strain [1] obtained after electroporation by pRSFRedTER and plated on a solid LB-medium supplemented with Cm (50 μg/ml) were of very small size. Bacteria from these colonies grew very slowly in comparison with bacteria carrying pRSFsacB plasmid, which served as a vector for cloning of λ Red genes. Addition of IPTG (1 mM) for induction of expression of λ Red genes, led to complete cessation of the growth of pRSFRedTER containing cells. This effect was not detected for the cells carrying pRSFsacB and was based on the toxicity of the expression of λ Red genes.

To establish which component of the λ Red system caused the toxic effect, we constructed the pRSFGamBet [GenBank;FJ347163] plasmid lacking the exo gene encoding 5'→3' exonuclease.

It is well known that exonuclease activity is necessary for integration of dsDNAs only; integration of ssDNAs, such as chemically synthesized oligos, requires only recombinase (product of the bet gene, see [8]). To test the functional activity of the constructed pRSFGamBet, the plasmid was used to promote recombination between the artificial ss-oligos comprising two 36-nt homologies to the galK sequences and chromosome of the E. coli MG1655galK::(attLλ-KmR-attRλ) strain. As a result of recombination a native structure of galK gene was restored. Between (1.5–2.5) × 104 Gal+ integrants per 108 survivors following electroporation were obtained in three independent experiments.

When introduced into P. ananatis SC17 strain, the pRSFGamBet plasmid did not inhibit cell growth even under the induced conditions (in the presence of 1 mM IPTG). Hence, the detected toxicity of pRSFRedTER was apparently caused by exo expression, or by simultaneous expression of all λ Red genes in P. ananatis cells.

Selection of the Recipient Strain for λ Red-Mediated Recombineering in P. Ananatis

A mutant P. ananatis strain, SC17(0), resistant to concerted expression of all λ Red genes, and thus manifesting the properties of a suitable recipient strain for λ Red-mediated integration of the linear DNAs into the chromosome, was obtained as follows. About 10 clones from 10^6 transformants obtained after electroporation of P. ananatis SC17 strain by pRSFRedTER, were of larger size after being plated on LB-agar with Cm. In LB-broth, bacteria from the "large" clones had a growth rate similar to the control strain with the pRSFsacB plasmid, and induction of the λ Red genes by IPTG caused only slight retardation in the growth of these cells.

Several of the selected "large" clones were cured from the plasmid on LB-agar containing sucrose, and re-transformed with pRSFRedTER. All clones grown after this re-transformation were of large size, similar to the parental clones. Three of the pRSFRedTER transformants that grew well were used as recipient strains for λ Red-mediated disruption of hisD gene. A PCR substrate, containing (attLλ-KmR-attRλ)-marker flanked by 40-bp homologous to the hisD gene, was electroporated into these strains. From 100 to 150 KmRHis- clones per 108 survivors following electroporation were obtained for each tested recipient strain. The insertion of the marker in the desired point of hisD gene was confirmed by PCR-analysis of 10 independent KmRHis- clones in each case. The observed integration frequency was similar to that obtained in the corresponding experiments with E. coli [4-6]. One of the plasmid-less strain used as a recipient in this experiment was named as SC17(0), and the obtained hisD strain constructed on its basis – as SC17(0)hisD::(attLλ-KmR-attRλ).

We tried to determine the nature of the mutation/mutations that provide resistance to concerted expression of λ Red genes to P. ananatis. There were no auxotrophic properties for SC17(0) strain growing on the M9 minimal media, supplemented with different carbon sources, in comparison with initial SC17 strain [1]. One of the possible explanations for the SC17(0) resistance is the reduced level of accumulation of λ Red proteins in this strain. To test the level of accumulation of λ Red proteins in SC17 and SC17(0), we performed SDS-PAGE of extracts of both strains carrying pRSFRedTER plasmid. Unfortunately, bands of λ Red proteins were not detected among the total cellular proteins even in conditions of IPTG induction for the both plasmid-carrier strains. The reduction in level of accumulation of λ Red proteins in SC17(0) strain, also, can be caused by decreased copy-number of RSF1010-replicon carrying plasmids in this strain. However, no reliable change in copy-number of pRSFsacB plasmid extracted from the cells of P. ananatis SC17 or SC17(0) strains could be experimentally found. On the other hand, the toxicity of the expression of λ Red genes has been detected for the plasmid-carrier cells of SC17 strain grown even without IPTG addition to the medium. In this case the transcription of the operon mediated by auto-regulated PlacUV5-lacI genetic element has to be tightly repressed. According to Skorokhodova et al. addition of 1 mM IPTG to the culture medium provides an increase of transcriptional level up to 10–20 fold [31]. But, the transcription level of λ Red genes under such conditions is not toxic for SC17(0). Hence, it is unlikely, that the synthesis of λ Red proteins in SC17 under the repressed conditions could be significantly higher than in SC17(0) after induction.

Perhaps, mutation/mutations that are present in the genome of the SC17(0) strain does not affect most of the genes encoding factors of global cellular regulation. At least, the patterns of total cellular proteins separated by 2D-PAGE in such a fashion that about 350 of individual polypeptides could be quantitatively analyzed [33], were not different for the SC17 and SC17(0) strains (data not shown).

Thus, the molecular mechanism of the resistance of the mutant SC17(0) strain to concerted expression of all λ Red genes is unknown as yet. Nevertheless, as will be shown below, various λ Red-driven modifications of bacterial chromosome could be provided on the basis of this selected strain.

Use of the Combined λ Red-Int/Xis System for Introduction of Multiple Modifications

Earlier, a λ Int/Xis-driven system for removing the markers from E. coli chromosome was constructed, similar to one developed by Peredelchuk & Bennet [34,35]. It comprised of the plasmids carrying removable markers of KmR or CmR flanked by attL$_\lambda$ and attR$_\lambda$ sites and the plasmid pMW-intxis-ts with thermo-sensitive pSC101-like replicon. This plasmid provided thermo-inducible expression of the xis-int genes under the control of λ P$_R$ promoter regulated by the temperature sensitive λ CIts857 repressor. Even being partially induced at 37°C, this system provided a high frequency (about 30%) of marker excision in E. coli. The high frequency of marker eviction at +37°C was very important for use of the system in P. ananatis because this bacterium has a lower temperature optimum than E. coli and cannot grow at 42°C (the standard temperature for temperature sensitive λ CIts857 repressor inactivation).

The pMW-intxis-ts plasmid contains ApR gene as a selective marker that was not practicable for P. ananatis because of its high natural resistance to Ap. For this reason, we have substituted this marker with the CmR gene. The resulting plasmid (pMW-intxis-cat) was introduced to the SC17(0)hisD::(attLλ-KmR-attRλ) strain described above by electroporation. More than 30% of the transformants grown at 37°C on the plates containing LB-agar with Cm had lost the Km resistance. Loss of the KmR cassette in the kanamycin-sensitive colonies was verified by PCR. Thus, the KmR cassette can be used in the next step of the chromosomal modifications of this strain (Fig. 1A). Curing of the KmS clones from the pMW-intxis-cat plasmid (with a frequency of 10%) was performed by re-streaking bacteria on LB-agar without antibiotic followed by cultivation at 37°C.

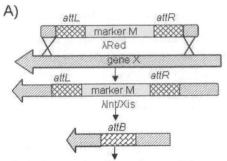

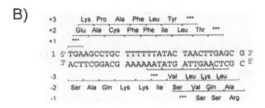

Figure 1. Scheme of a construction of multiple chromosomal modifications, using the combined λ Red-Int/Xis system. A) Selective marker M flanked by attL$_\lambda$/attR$_\lambda$ is used for introduction of an appropriate mutation into the chromosome by λ Red-dependent recombination. Then the marker is eliminated from the chromosome by λ Int/Xis site-specific recombination. As a result, only the 31 bp long attB$_\lambda$ sequence linked to the mutation remains in the chromosome. Selective marker M can then be used in the next step of the introduction of multiple chromosomal modifications. B) The sequence of the attB$_\lambda$ site. One of the six ORFs provided by this sequence does not contain stop codons. Hence, it is possible to design an "in-frame" deletion of a gene. Asterisks mark stop codons.

The attBλ site (31 bp in length), remaining in the chromosome after marker excision, contains six possible reading frames. One of these reading frames does not contain stop codons (Fig. 1B). Therefore, usage of the removable markers flanked with attLλ and attRλ sites allows design and construction of "in frame" deletions.

Using the λ Red-driven chromosomal modification followed by λ Int/Xis-mediated excision of the selective marker, it was possible to provide, step-by-step, the multiple chromosomal modifications in P. ananatis SC17(0) strain. This approach was repeatedly used for different modifications. Among them were 1) combinations of the simple or "in frame" deletions of several genes/operons; 2) integration of the marked heterologous genes into the chromosome of P. ananatis and 3) modification of the regulatory regions of the genes of interest [36]. Up to the present, several P. ananatis strains carrying more than 10 different modifications have been constructed using this strategy; the presence of multiple attBλ sites in their chromosomes did not hamper the repeated exploitation of this system.

Transfer of Marked Mutations by Electroporation of Chromosomal DNA

General transduction is the most efficient and popular method for transfer of mutations between different E. coli strains. Although P. ananatis and E. coli are close relatives, the known E. coli transducing phages cannot infect P. ananatis cells. Therefore, development of another method for transfer of mutations between P. ananatis strains was necessary.

The electroporation of genomic DNA has been described for the transfer of genetic markers between different backgrounds of E. coli and Pseudomonas [37,38]. We tried to apply this technique to P. ananatis. The SC17(0)hisD::(attLλ-KmR-attRλ) strain was used as a donor of KmR marker. Wild type strain SC17 was used as a recipient for electro-transformation of chromosomal DNA.

Previously, it was shown that special electroporation conditions are needed for the transformation of E. coli cells with large DNA molecules (see [39] for details). Different cultivation conditions of recipient strain and parameters of electroporation (electric field strength – E, time constant – τ) were tested. For P. ananatis, the highest yield of integrants (about 100 KmR His- integrants per 108 survivors following electroporation) was obtained under the following conditions. Recipient strain was grown up to absorbance of 0.8 – 1.0. Then electro-competent cells were prepared using 10 ml of culture as described in the "Plasmid electro-transformation" section (see "Methods"). Electroporation was performed at E = 12.5 kV/cm and τ = 10 msec (resistance of 400 Ω and capacity of 25 μF). Electro-transformation with chromosomal DNA is very fast method: all procedures, including DNA isolation and electroporation, can be performed in one day.

We found that marked chromosomal modification, obtained in P. ananatis SC17(0) strain via λ Red-recombineering method, could be easily transferred into the wild type SC17 strain by electroporation of chromosomal DNA. At the time of writing up to ten different mutations had been combined in the chromosome of the SC17 strain by repeated electro-transformation with chromosomal DNA followed by λ Int/Xis-driven excision of selective marker. The frequency of marked mutation transfer varied from several up to several hundreds of integrants per trial.

Two-Step λ Red-Mediated Introduction of Unmarked Mutations Into P. ananatis Chromosome

A two-step λ Red-mediated procedure for the introduction of unmarked mutations was elaborated for P. ananatis SC17(0). It comprises: 1) λ Red-driven insertion of dual selective/counter-selective marker into the desired point; 2) elimination of the marker via λ Red-mediated integration of the short dsDNA fragment containing the mutation of interest flanked with sites homologous to the appropriate target. Such an approach based on ET-recombination or λ Red-recombination was previously exploited for introduction of the unmarked mutations in E. coli [40,41]. One of the most popular counter-selective markers used for this purpose is the sacB gene from B. subtilis, whose introduction imparts sucrose sensitivity to the bacterium.

To create a template for the PCR amplification of the dual selective/counter-selective marker, cat/sacB, the pRSFPlacsacB plasmid was constructed. The cassette PlacUV5-sacB-cat was amplified with primers containing 36-nt homologies to the target on their 5'-ends, and integrated into SmaI recognition site located in hisD gene using SC17(0) harboring pRSFRedkan as a recipient and the cat gene in the cassette, as the selective marker. A short 170-bp long dsDNA fragment harboring an appropriate mutation and 82-bp long flanks homologous to the target region has been constructed (Fig. 2A). The desired modification of bacterial chromosome (substitution of the artificial XhoI for the native SmaI site) was finally achieved via the λ Red-mediated integration of the obtained DNA fragment accompanied by elimination of PlacUV5-sacB-cat, using sacB gene for counter-selection (Fig. 2B). The integrants of interest were found with a rather high frequency (25%) among the clones grown on LB-agar containing 30% sucrose.

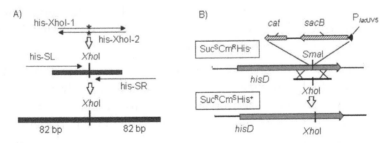

Figure 2. Construction of the unmarked nucleotide exchange in the P. ananatis hisD gene. A) Construction of a dsDNA fragment with appropriate mutation in the center. First, his-XhoI-1 and his-XhoI-2 oligos are annealed to each other. The nucleotide exchange of interest included in the sequence of his-XhoI-1/his-Xho2 olig is indicated by asterisks. The resulting dsDNA fragment is then used as DNA-template for PCR-amplification with his-SL and his-SR oligos. As a result a linear dsDNA fragment, harboring a XhoI restriction site and 82 bp long arms homologous to the target region of hisD gene, was obtained. B) The cassette containing dual selective/contra-selective marker is integrated into the target point of hisD gene. The constructed in vitro linear dsDNA fragment or ssDNA harboring appropriate mutation in the center is then integrated into the chromosome of this strain by the λ Red recombination system. As a result, the dual selective/contra-selective marker is eliminated from the chromosome with simultaneous introduction of the desired mutation into the hisD gene. This mutation leads to substitution of the native SmaI restriction site by the XhoI restriction site and restoration of the amino-acid sequence of HisD protein. Integrants are selected as colonies resistant to sucrose. Such colonies are subsequently tested for Cm sensitivity, ability for growth without histidine and presence of XhoI restriction site in the target chromosome point.

This procedure could be applied for introduction of unmarked mutations to the genome of SC17(0) strain. Transfer of the obtained unmarked mutations to other strains by electro-transformation with chromosomal DNA is difficult because of impossibility of direct selection of integrants. Hence, to provide construction of unmarked mutations in the genome of the P. ananatis strains, sensitive to expression of all λ Red genes, we performed the method of λ Beta-driven integration of single-stranded oligos into the chromosome of SC17 strain.

λ Beta-Driven Integration of Single-Stranded Oligos

As mentioned above, ssDNAs, e.g. oligos containing short flanks homologous to the target site, can be integrated into the E. coli chromosome using only the product of the λ bet gene [8]. As the expression of gam and bet genes from pRS-FGamBet plasmid was not toxic towards P. ananatis SC17 even under induced conditions, we tried to perform λ Beta-driven integration of oligos into the chromosome of this strain.

The hisD::PlacUV5-sacB-cat chromosomal modification was transferred into the SC17 strain by electro-transformation with genomic DNA isolated from SC17(0)hisD::PlacUV5-sacB-cat strain. The resulting strain, selected on LB medium supplemented with Cm, was named SC17hisD::PlacUV5-sacB-cat. The pRSFGamBet-kan [GenBank:FJ347164] plasmid was constructed by substitution of sacB and cat genes of pRSFGamBet for the KmR gene from pUC4K. This plasmid was introduced to the obtained SC17hisD::PlacUV5-sacB-cat strain. It is known from the literature [8,42] that the efficiency of λ Red-dependent integration of the oligos depends on a direction of replication through the recombination site. The direction of replication at the hisD locus in the P. ananatis chromosome is not known. Hence, the plasmid-carrier cells were independently electroporated with hisD-XhoI-1 and hisD-XhoI-2 ss-oligos complementary to each other. Cells were plated on LB medium containing 30% sucrose. No colonies were observed after 24-hour cultivation when hisD-XhoI-2 olig was used for electroporation. About 200 clones were obtained after electro-transformation with hisD-XhoI-1 olig. To test the phenotype of the obtained transformants, 100 clones were replicated on the following solid mediums: LB medium supplemented with 30% sucrose, LB medium supplemented with Cm and M9 minimal medium. Seven of the tested clones had SucRCmSHis+ phenotype. These colonies were further verified by PCR and restriction of the amplified product. The presence of the expected mutation (substitution of the SmaI by XhoI recognition site) was confirmed in all of the seven SucRCmSHis+ clones.

Although the frequency of selection of the desired mutation was not high, in principle, λ Beta-dependent integration of ssDNAs allowed the construction of the unmarked mutations in the P. ananatis.

Discussion

It is currently well established that high-frequency recombination between short homologies can be catalyzed in E. coli cells by the λ Red functions [4-7,43-45]. Unfortunately, up to the present, the range of bacteria for which this system has been utilized is limited. The main goal of the present study was to widen use of the λ Red-recombineering technology to P. ananatis, a bacterium of interest in the field of metabolic engineering. A broad-host-range λ Red-expressing plasmid useful for P. ananatis was constructed. The observed general toxicity of simultaneous expression of the λ gam, bet, and exo genes for P. ananatis SC17 cells was overcome by selection of the special recipient, SC17(0).

It is known that expression of λ Red functions in E. coli cells can lead to toxic effects [46,47]. Certainly, expression of λ Red genes has been investigated for E. coli in detail. In addition to its direct influence on traditional recombination pathways due to inhibition of RecBCD and SbcCD nucleases by λ Gam, the activity of λ Red proteins can interfere with the processes of replication and repair (see, [48,49] for reviews). For example, prolonged expression of gam gene could lead to formation of linear multimers of high, medium and low copy-number plasmids, and, even, of minichromosomes [50-53]. Expression of exo gene in addition to gam enhances this effect [50,52]. The plasmid linear multimers, also, may interfere with λ Red-recombination [18]. Murphy and co-workers showed that extended expression of λ Red-recombination functions could significantly induce a spontaneous mutagenesis, probably caused by interfering with mismatch repair UvrD-dependent pathway of E. coli [18].

Toxicity of expression of λ Red genes for the bacteria closely related to E. coli, but differing in the enzymes of replication, recombination and reparation, could not be predicted in advanced, as in the case of P. ananatis SC17. It is difficult, as well, to give an univocal explanation of how λ Red-mediated toxicity could be overcome. It seems that the decrease of this toxicity could be based on (i) lower intracellular level of λ Red proteins in the mutant cell, caused by reduced level in the synthesis of λ Red proteins or by increased efficiency of the specific proteolysis of λ Red proteins, (ii) decreased affinity of specific targets for interaction with λ Red proteins or increased level of biosynthesis of these targets. As for the P. ananatis mutant strain obtained, both variants of the explanation may be possible. As mentioned in the Results section, it is not very probable that SC17(0) strain possesses reduced level of the synthesis of λ Red proteins. But this possibility can not be completely rejected.

Nevertheless, even without information concerning the nature of the SC17(0) mutation, it is possible to use the corresponding strain for the desired λ Red-mediated rearrangements of P. ananatis chtomosome. All types of chromosome modifications constructed in E. coli by the λ Red-mediated recombineering, have been successfully reproduced in SC17(0) with pRSFRedTER as λ Red-expressing plasmid. Typically, the yield of recombinants varied from several tens to several hundreds per trial.

Using SC17(0) as the initial recipient for λ Red-promoting modifications it is subsequently possible to transfer the marked mutation to other P. ananatis strains of interest using the method of electro-transformation with chromosomal DNA. This method is a unique way to combine the set of marked mutations constructed in different P. ananatis strains into a single strain. In addition, potentially, co-transfer of rather closed mutations could be prevented by digestion of the chromosomal DNA by the appropriate restriction endonucleases, whose recognition sites are located between the mutations.

As the number of combined mutations of interest exceeds the number of available antibiotic resistance markers, curing of the intermediate strains from the used selective markers is necessary. Moreover, the presence of antibiotic resistance genes in the genomes of the industrial strains is rigorously restricted by the legislations of different countries. A wide variety of systems for marker curing based on the site-specific recombination systems (Cre/lox, Flp/FRT) are well-known [54-57]. These systems provide "symmetrical" recombination reaction between two identical sites flanking the removing marker, e.g. FRTxFRT = 2FRT. In this case, after removing the marker, the active site remains in the chromosome. Repeated action of such systems can lead to inversion or deletion of extended chromosomal fragments caused by site-specific recombination between the sites remaining at different points in the chromosome. Therefore, use of systems providing "asymmetrical" recombination reaction would be preferable. One such system is the site-specific recombination system of λ phage. It includes the Int and Xis proteins encoded by int and xis genes that, together with the host factors (IHF, RecA and Fis), provide the following reaction: attLλx attRλ = attPλ+attBλ. Peredelchuk & Bennet [34] were the first who used this system for removal of the selective markers flanked by attLλ and attRλ sites. The attBλ site remaining in the chromosome after marker excision cannot recombine with attLλor attRλ site in the next steps of strain construction. Thus, repeated action of the system would not influence the strain stability. Certainly, residual attBλ sites could recombine with each other via host general recombination system or provoke replication errors, especially if many left over scars (attBλ) are presented in chromosome and their positions were rather close to each other. Such events would lead to deletions of chromosomal fragments.

We adjusted the λ Int/Xis-dependent system for use in high-efficient marker excision in P. ananatis. Note that it is also possible, in particular, to design marker-less strains, carrying "in-frame" deletions.

Finally, the two-step procedure for introducing the unmarked mutations into the P. ananatis genome was demonstrated using B. subtilis sacB gene as a counter-selective marker. Desirable mutants were achieved at the second stage via λ Red-mediated integration of the short dsDNA in SC17(0) or ssDNA in any other P. ananatis strain carrying the dual selective/counter-selective marker in the target point of the genome.

Up to the present, the developed λ Red-mediated method has been used for deletion of more than 50 P. ananatis genes whose products were involved in central metabolism, respiration, transcription regulation, etc. Several marker-less P. ananatis strains carrying multiple (> 10) different chromosomal modifications including "in-frame" deletions, point mutations, rearrangements of regulatory regions, in particular, were constructed for basic research and applied purposes using combined application of the λ Red-Int/Xis systems and electro-transformation with chromosomal DNA.

Conclusion

The λ Red-mediated recombineering has been adjusted for rapid and efficient construction of genome rearrangements in P. ananatis. In combination with the established procedures of λ Int/Xis-dependent marker elimination and electro-transformation with chromosomal DNA, this method provides a simple route to obtaining marker-less strains carrying multiple mutations of different types (deletions, substitutions of regulatory regions, integration of heterologous genes, and point mutations). The described approach of selection of the recipient strain resistant to expression of λ Red genes could be useful in exploiting λ Red-recombineering in other bacteria.

Methods

Strains and Plasmids

Strains and plasmids used or generated in this study are listed in Table 1.

Table 1. Strains and plasmids used or generated in this study.

Name	Main characteristics/accession number	Source or reference
Strains		
Pantoea ananatis SC17 (AJ13355)	mutant with decreased secretion of mucus	[1]
Pantoea ananatis SC17(0)	Derivative of SC17 resistant to expression of λ Red genes	This work
E. coli K12 MG1655	Wild type	VKPM
Plasmids		
RSF1010	GenBank accession number NC_001740	[26]
pUC4K	GenBank accession number X06404	[32]
pMW-*attL$_\lambda$*-Km^R-*attR$_\lambda$*	Donor *attL$_\lambda$*-Km^R-*attR$_\lambda$* cassette; Ap^R; Km^R	[35]
pKD46	pINT-ts; λ *gam, bet,* and *exo* genes are under P$_{araB}$ promoter; Ap^R	[4]
pRSFRedTER	λ *gam, bet,* and *exo* genes are under control of P-element; *sacB* gene; Cm^R	This work
pRSFRedkan	λ *gam, bet,* and *exo* genes are under control of P-element; Km^R	This work
pRSFGamBet	λ *gam* and *bet* genes are under control of P- element; *sacB* gene; Cm^R	This work
pRSFGamBetkan	λ *gam* and *bet* genes are under control of P- element; Km^R	This work
pRSFPlacsacB	P$_{lacUV5}$-*sacB-cat* cassette; Cm^R	This work
pMW-intxis-cat	pSC101-ts; λ *xis-int* genes transcribed from λ P$_R$ promoter under CIts857 control; Cm^R	This work

A P-element marks the auto-regulated element P$_{lacUV5}$-*lacI* [see [31] for details].

Media and Growth Conditions

E. coli and P. ananatis strains were cultivated with aeration in LB medium at 37°C and 34°C, respectively. The following antibiotic concentrations were used to select transformants and to maintain the plasmids: Km – 40 mg/l, Cm – 50 mg/l. The M9 salt medium supplemented with galactose (1 g/l) or glucose (1 g/l) was used to select Gal$^+$ or His$^+$ cells.

Recombinant DNA Techniques

DNA manipulations were performed according to standard methods [58]. Restrictases were provided by "Fermentas" (Lithuania). T4-DNA ligase was from Promega (USA). All reactions were performed according to the manufacturer's instructions. PCR was carried out with Taq-polymerase ("Fermentas"). Primers were purchased from "Syntol" (Russia).

Construction of Integrative Cassettes

To provide cassettes for λ Red-dependent integration, the appropriate selective marker was amplified by PCR with oligos containing on their 5'-ends 36-nt sequences homologous to the target region. To disrupt E. coli galK and P. ananatis hisD genes, a removable Km^R marker flanked by attL$_\lambda$ and attR$_\lambda$ was amplified with galK-5/galK-3 and hisD-5/hisD-3 primers, respectively. The pMW-attL$_\lambda$-Km^R-attR$_\lambda$ plasmid was used as DNA template. To obtain insertion into the P. ananatis hisD gene, the Plac-sacB-cat cassette was amplified in PCR with his-Plac-5/his-cat-3 primers using pRSFPlacsacB plasmid as template.

Construction of the Linear dsDNA Fragment to Exchange Native SmaI Recognition Site in the P. Ananatis hisD Gene by XhoI Site

First, his-XhoI-1 and his-XhoI-2 oligos complementary to each other were annealed. Both oligos contained sequences corresponding to the XhoI restriction site in the center and arms homologous to the sequence surrounding native SmaI site of the P. ananatis hisD. As a result, the short dsDNA fragment containing the XhoI recognition site in its center and 33 bp long arms homologous to the target region, was obtained. The obtained fragment was amplified and extended by PCR with the primers his-SL and his-SR. The resultant DNA fragment generated by PCR contained XhoI recognition site in its center flanked with 82 bp arms homologous to the appropriate site in P. ananatis hisD gene.

Plasmid Electro-Transformation

An overnight culture of P. ananatis strain grown at 34°C with aeration was diluted with fresh LB broth 100 times and the cultivation was continued up to the OD_{600} = 0.5–0.8. Cells from ten milliliters were washed three times with an equal volume of deionized ice water followed by washing with 1 ml of 10% cold glycerol and resuspended in 35 µl of 10% cold glycerol. Just before electroporation, 10–100 ng of the plasmid DNA dissolved in 2 µl of deionized water was added to the cell suspension. The procedure of plasmid electro-transformation was performed using the GenePulser and Pulse Controller ("BioRad", USA). The applied pulse parameters were: electric field strength of 20 kV/cm, time constant of 5 msec. After electroporation, 1 ml of LB medium enriched with glucose (5 g/l) was immediately added to the cell suspension. Then the cells were cultivated under aeration at 34°C for 2 h and plated on LB-agar containing the appropriate antibiotic. This was followed by an overnight incubation at 34°C. A competence of P. ananatis cells, determined for RSF1010 plasmid, was 106 CFU per µg of DNA. Typically, 105–106 antibiotic resistant colonies were obtained per 108 survivors following electroporation.

Gene Rearrangement

Overnight cultures of P. ananatis or E. coli strains harbouring the plasmid, expressing appropriate λ Red genes, grown in LB broth with Cm (for pRS-FRedTER, pRSFGamBet plasmids) or Km (for pRSFRedkan, pRSFGamBetkan plasmids) were diluted 100 times with the same fresh medium supplemented with 1 mM IPTG for induction of the λ Red genes. At culture density of 0.5–0.6, electro-competent cells were prepared as described above. From 200 to 500 ng of a PCR-generated linear dsDNA or 100 ng of a ss-oligos were used for transformation. Electroporation was carried out at electric field strength of 25 kV/cm and time constant of 5 msec for both types of DNA substrates. The chromosome structure of the obtained transformants was verified in PCR with galK-t1/galK-t2 primers for E. coli galK gene disruption, hisD-t1/hisD-t2 for P. ananatis hisD gene disruption and for insertion of the double selective/contra-selective marker into the P. ananatis hisD.

Gene disruption provided with pKD46 plasmid was performed as described in (4).

P. Ananatis Electro-Transformation with Chromosomal DNA

Cells were grown in LB medium up to $OD_{600} = 0.8$–1.0. Electro-competent cells were prepared as described above. From 1 to 2 mg of a chromosomal DNA, isolated using a Genomic DNA Isolation Kit (Sigma), was used for transformation. Electroporation was carried out with an electric field strength of 12.5 kV/cm and time constant of 10 msec.

SDS-PAGE

SDS-PAGE of cell extracts was performed according to Laemmli [59] with polyacrylamide gel with linear gradient of concentrations from 10% to 15%.

Abbreviations

Ap: ampicillin; Ap^R: ampicillin resistance; Cm: chloramphenicol; Cm^R: chloramphenicol resistance; cat: chloramphenicol resistance gene; Km: kanamycin; Km^R: kanamycin resistance; kan: kanamycin resistance gene; sacB: gene encoding the levansucrase; attL$_\lambda$: attL site of phage λ; attR$_\lambda$: attR site of phage λ; attB$_\lambda$: attB site of phage λ; IPTG: isopropyl-β-D-thiogalactopyranoside; nt: nucleotide; bp: base pair(s); oligos: oligonucleotides; PCR: polymerase chain reaction; ssDNA: single-stranded DNA; dsDNA: double-stranded DNA.

Authors' Contributions

JIK and YH are the project leaders in AGRI and in Ajinomoto, respectively. JIK obtained SC17(0), designed the main experiments concerned with λ Red-driven modifications, and drafted the manuscript. YH developed the transfer of λ Red-driven mutations to P. ananatis strains differed from SC17(0). LIG constructed all recombinant plasmids, expressing λ Red genes, tested their function initially in E. coli, and edited the manuscript. TMK performed the λ Red-driven modifications of the chromosome of P. ananatis SC17(0) strain. IGA performed λ Red-driven oligos integration into chromosome of P. ananatis SC17 strain. SVM supervised and coordinated the work and edited the manuscript. All authors have read and approved the final version of the manuscript.

Acknowledgements

Authors acknowledge Prof. B.L. Wanner kindly provided us with BW25113/pKD46 strain. This study was carried out at the request of and with financial support from the Ajinomoto Co.

References

1. Izui H, Hara Y, Sato M, Akiyoshi N: Method for producing L-glutamic acid. United States Patent.

2. Murphy KC: Use of bacteriophage λ recombination functions to promote gene replacement in Escherichia coli. J Bacteriol 1998, 180:2063–2071.

3. Zhang Y, Buchholtz F, Muyrers JPP, Stewart AF: A new logic for DNA engineering using recombination in Escherichia coli. Nature Genetics 1998, 20:123–128.

4. Datsenko KA, Wanner BL: One-step inactivation of chromosomal genes in Escherichia coli K12 using PCR products. Proc Natl Acad Sci USA 2000, 97:6640–6645.

5. Murphy KC, Campellone KG, Poteete AR: PCR-mediated gene replacement in Escherichia coli. Gene 2000, 246:321–330.

6. Yu D, Ellis HM, Lee EC, Jenkins NA, Copeland NG, Court DL: An efficient recombination system for chromosome engineering in Escherichia coli. Proc Natl Acad Sci USA 2000, 97:5978–5983.

7. Zhang Y, Muyrers JP, Testa G, Stewart AF: DNA cloning by homologous recombination in Escherichia coli. Nat Biotechnol 2000, 18:1314–1317.

8. Ellis HM, Yu D, DiTizio T, Court LD: High efficiency mutagenesis, repair, and engineering of chromosomal DNA using single-stranded oligonucleotides. Proc Natl Acad Sci USA 2001, 98:6742–6746.

9. Karu AE, Sakaki Y, Echols H, Linn S: The gamma protein specified by bacteriophage λ. Structure and inhibitory activity for the RecBC enzyme of Escherichia coli. J Biol Chem 1975, 250:7377–7387.

10. Murphy KC: λ Gam protein inhibits the helicase and chi-stimulated recombination activities of Escherichia coli recBCD enzyme. J Bacteriol 1991, 173:5808–5821.

11. Little JW: An exonuclease induced by bacteriophage λ. II. Nature of the enzymatic reaction. J Biol Chem 1967, 242:679–686.

12. Carter DM, Radding CM: The role of exonuclease and β protein of phage λ in genetic recombination. II. Substrate specificity and the mode of action of lambda exonuclease. J Biol Chem 1971, 246:2502–2512.

13. Cassuto E, Lash T, Sriprakash KS, Radding CM: The role of exonuclease and β protein of phage λ in genetic recombination. V. Recombination of λ DNA in vitro. Proc Natl Acad Sci USA 1971, 68:1639–1643.

14. Hill SA, Stahl MM, Stahl FM: Single-strand DNA intermediates in phage λ's Red recombination pathway. Proc Natl Acad Sci USA 1997, 94:2951–2956.

15. Muniyappa K, Radding CM: The homologous recombination system of phage λ. Pairing activities of β protein. J Biol Chem 1986, 261:7472–7478.

16. Mythili E, Kumar KA, Muniyappa K: Characterization of the DNA-binding domain of β protein, a component of λ Red-pathway, by UV catalyzed cross-linking. Gene 1996, 182:81–87.

17. Li Z, Karakousis G, Chiu SK, Reddy G, Radding CM: The β protein of phage λ promotes strand exchange. J Mol Biol 1998, 276:733–744.

18. Murphy KC, Campellone KG: Lambda Red-mediated recombinogenic engineering of enterohemorrhagic and enteropathogenic E. coli. BMC Mol Biol 2003, 4:11.

19. Husseiny MI, Hensel M: Rapid method for the construction of Salmonella enterica serovar Typhimurium vaccine carrier strains. Infect Immun 2005, 73(3):1598–1605.

20. Karlinsey JE: Lambda-Red genetic engineering in Salmonella enterica serovar Typhimurium. Methods Enzymol 2007, 421:199–209.

21. Shi Z-X, Wang H-L, Hu K, Feng E-L, Yao X, Su G-F, Huang P-T, Huang L-Y: Identification of alkA gene related to virulence of Shigella flexneri 2a by mutational analysis. World J Gastroenterol 2003, 9:2720–2725.

22. Ranallo RT, Barnoy S, Thakkar S, Urick T, Venkatesan MM: Developing live Shigella vaccines using lambda red recombineering. FEMS Immunol Med Microbiol 2006, 47:462–469.

23. Derbise A, Lesic B, Dacheux D, Ghigo JM, Carniel E: A rapid and simple method for inactivating chromosomal genes in Yersinia. FEMS Immunol Med Microbiol 2003, 38:113–116.

24. Lesic B, Bach S, Ghigo JM, Dobrindt U, Hacker J, Carniel E: Excision of the high-pathogenicity island of Yersinia pseudotuberculosis requires the combined actions of its cognate integrase and Hef, a new recombination directionality factor. Mol Microbiol 2004, 52:1337–1348.

25. Lesic B, Rahme LG: Use of the lambda Red recombinase system to rapidly generate mutants in Pseudomonas aeruginosa. BMC Mol Biol 2008, 9:20.

26. Scholz P, Haring V, Wittmann-Liebold B, Ashman K, Bagdasarian M, Scherzinger E: Complete nucleotide sequence and gene organization of the broad-host-range plasmid RSF1010. Gene 1989, 75:271–288.

27. Frey J, Bagdasarian M: The molecular biology of IncQ plasmids. In Promiscuous plasmids of Gram-negative bacteria. Edited by: Thomas CM. New York: Academic Press, Inc; 1989:79–94.

28. Gormley EP, Davies J: Transfer of plasmid RSF1010 by conjugation from Escherichia coli to Streptomyces lividans and Mycabacterium smegmatis. J Bacteriol 1991, 173:6705–6708.

29. Brunschwig E, Darzins A: A two-component T7 system for the overexpression of genes in Pseudomonas aeruginosa. Gene 1992, 111:35–41.

30. Dehio M, Knorre A, Lanz C, Dehio C: Construction of versatile high-level expression vectors for Bartonella henselae and the use of green fluorescent protein as a new expression marker. Gene 1998, 215:223–229.

31. Skorokhodova AYu, Katashkina ZhI, Zimenkov DV, Smirnov SV, Gulevich AYu, Biriukova IV, Mashko SV: Design and study on characteristics of auto- and smoothly regulated genetic element O3/P lac UV5/O lac→lacI. Biotechnologiya (Russian) 2004, 5:3–21.

32. Taylor LA, Rose RE: A correction in the nucleotide sequence of the Tn903 kanamycin resistance determinant in pUC4K. Nucleic Acids Res 1988, 16:358.

33. Rabilloud T: Proteome research: two-dimensional gel electrophoresis and identification methods. Berlin: Springer; 2000.

34. Peredelchuk MY, Bennett GN: A method for construction of E. coli strains with multiple DNA insertion in the chromosome. Gene 1997, 187:231–238.

35. Minaeva NI, Gak ER, Zimenkov DV, Skorokhodova AYu, Biryukova IV, Mashko SV: Dual-In/Out strategy for genes integration into bacterial chromosome: a novel approach to step-by-step construction of plasmid-less marker-less

recombinant E. coli strains with predesigned genome structure. BMC Biotechnology 2008, 8:63.

36. Katashkina JI, Golubeva LI, Kuvaeva TM, Gaidenko TA, Gak ER, Mashko SV: Method for constructing recombinant bacterium belonging to the genus Pantoea and method for producing L-amino acids using bacterium belonging to the genus Pantoea. Russian Federation Patent application 2006134574.

37. Kilbane JJ 2nd, Bielaga BA: Instantaneous gene transfer from donor to recipient microorganism via electroporation. Biotechniques 1991, 10:354–365.

38. Choi KH, Kumar A, Schweizer HP: A 10-min method for preparation of highly electrocompetent Pseudomonas aeruginosa cells: application for DNA fragment transfer between chromosomes and plasmid transformation. J Microbiol Methods 2006, 64:391–397.

39. Sheng YuL, Mancino V, Birren B: Transformation of Escherichia coli with large DNA molecules by electroporation. Nucleic Acids Res 1995, 23:1990–1996.

40. Muyrers JJP, Zhang Y, Benes V, Testa G, Ansorge W, Stewart AF: Point mutation of bacterial artificial chromosomes by ET recombination. EMBO reports 2000, 1:239–243.

41. Heermann R, Zeppenfeld T, Jung K: Simple generation of site-directed point mutations in the Escherichia coli chromosome using Red®/ET® recombination. Microbial Cell Factories 2008, 7:14.

42. Li X, Costantino N, Lu L, Liu D, Watt RM, Cheah KSE, Court DL, Huang J-D: Identification of factors influencing strand bias in oligonucleotide-mediated recombination in Escherichia coli. Nucleic Acids Res 2003, 31:6674–6687.

43. Swaminathan S, Ellis HM, Waters LS, Yu D, Lee E-C, Court DL, Sharan SK: Rapid engineering of bacterial artificial chromosomes using oligonucleotides. Genesis 2001, 29:14–21.

44. Datta S, Costantino N, Court DL: A set of recombineering plasmids for gram-negative bacteria. Gene 2006, 379:109–115.

45. Thomason LC, Costantino N, Shaw DV, Court DL: Multicopy plasmid modification with phage λ Red recombineering. Plasmid 2007, 58:148–158.

46. Friedman SA, Hays BH: Selective inhibition of Escherichia coli activities by plasmid-encoded GamS function of phage lambda. Gene 1986, 43:255–263.

47. Sergueev K, Yu D, Austin S, Cuort D: Cell toxicity caused by products of the p(L) operon of bacteriophage lambda. Gene 2001, 227:227–235.

48. Court DL, Sawitzke JA, Thomason LC: Genetic engineering using homologous recombination. Annu Rev Genet 2002, 36:361–388.

49. Sawitzke JA, Thomason LC, Costantino N, Bubuenko M, Datta S, Court DL: Recombineering: In vivo genetic engineering in E. coli, S. enterica, and beyond. Methods Enzymol 2007, 421:171–199.

50. Enquist LW, Skalka A: Replication of bacteriophage λ DNA dependent on the function of host and viral genes. I. Interaction of red, gam and rec. J Mol Biol 1973, 75:185–212.

51. Silberstein Z, Cohen A: Synthesis of linear multimers of OriC and pBR322 derivatives in Escherichia coli K-12: role of recombination and replication functions. J Bacteriol 1987, 169:3131–3137.

52. Silberstein Z, Maor S, Berger I, Cohen A: Lambda Red-mediated synthesis of plasmid linear multimers in Escherichia coli K12. Mol Gen Genet 1990, 223:496–507.

53. Murphy K: λ Gam protein inhibits the helicase and χ-stimulated activities of Escherichia coli RecBCD enzyme. J Bacteriol 1991, 173:5808–5821.

54. Dale EC, Ow DW: Gene transfer with subsequent removal of the selection gene from the host genome. Proc Natl Acad Sci USA 1991, 88:10558–10562.

55. Posfai G, Koob M, Hradecna Z, Hasan N, Filutowich M, Szybalski W: In vivo excision and amplification of large segments of the Escherichia coli genome. Nucleic Acids Res 1994, 22:2392–2398.

56. Cherepanov PP, Wackernagel W: Gene disruption in Escherighia coli : TcR and KmR cassettes with the option of FLP-catalyzed excision of the antibiotic-resistance determinant. Gene 1995, 158:9–14.

57. Kristensen CS, Eberl L, Sanchez-Romero JM, Givskov M, Molin S, De Lorenzo V: Site-specific deletions of chromosomally located DNA segments with the multimer resolution system of broad-host-range plasmid RP4. J Bacteriol 1995, 177:52–58.

58. Sambrook J, Fitsch EF, Maniatis T: Molecular Cloning: A Laboratory Manual. Cold Spring Harbor: Cold Spring Harbor Press; 1989.

59. Laemmli UK: Cleavage of structural proteins during the assembly of the head of bacteriophage T4. Nature 1970, 227:680–685.

Recombination-Mediated Genetic Engineering of a Bacterial Artificial Chromosome Clone of Modified Vaccinia Virus Ankara (MVA)

Matthew G. Cottingham, Rikke F. Andersen,
Alexandra J. Spencer, Saroj Saurya, Julie Furze,
Adrian V. S. Hill and Sarah C. Gilbert

ABSTRACT

The production, manipulation and rescue of a bacterial artificial chromosome clone of Vaccinia virus (VAC-BAC) in order to expedite construction of expression vectors and mutagenesis of the genome has been described

(Domi & Moss, 2002, PNAS 99 12415–20). The genomic BAC clone was 'rescued' back to infectious virus using a Fowlpox virus helper to supply transcriptional machinery. We apply here a similar approach to the attenuated strain Modified Vaccinia virus Ankara (MVA), now widely used as a safe non-replicating recombinant vaccine vector in mammals, including humans. Four apparently full-length, rescuable clones were obtained, which had indistinguishable immunogenicity in mice. One clone was shotgun sequenced and found to be identical to the parent. We employed GalK recombination-mediated genetic engineering (recombineering) of MVA-BAC to delete five selected viral genes. Deletion of C12L, A44L, A46R or B7R did not significantly affect CD8⁺ T cell immunogenicity in BALB/c mice, but deletion of B15R enhanced specific CD8⁺ T cell responses to one of two endogenous viral epitopes (from the E2 and F2 proteins), in accordance with published work (Staib et al., 2005, J. Gen. Virol. 86, 1997–2006). In addition, we found a higher frequency of triple-positive IFN-γ, TNF-α and IL-2 secreting E3-specific CD8⁺ T-cells 8 weeks after vaccination with MVA lacking B15R. Furthermore, a recombinant vaccine capable of inducing CD8⁺ T cells against an epitope from Plasmodium berghei was created using GalK counterselection to insert an antigen expression cassette lacking a tandem marker gene into the traditional thymidine kinase locus of MVA-BAC. MVA continues to feature prominently in clinical trials of recombinant vaccines against diseases such as HIV-AIDS, malaria and tuberculosis. Here we demonstrate in proof-of-concept experiments that MVA-BAC recombineering is a viable route to more rapid and efficient generation of new candidate mutant and recombinant vaccines based on a clinically deployable viral vector.

Introduction

Modified Vaccinia virus Ankara (MVA) is a replication-deficient attenuated poxvirus that was derived from Vaccinia virus by Mayr et al. through more than 500 blind passages in chick embryo fibroblast (CEF) cell culture [1]. With the notable exception of the Syrian hamster cell line BHK-21 [2], MVA is unable to replicate productively in mammalian cells, though genome replication, late gene expression and immature virion formation usually occur [3], [4]. During its attenuation, MVA acquired large genomic deletions totalling about 30 kb, resulting in the complete loss of 26 open reading frames (ORFs), together with truncation or fragmentation of a further 21 ORFs and numerous smaller scale mutations [5]. MVA does not therefore express many of the known poxviral immune evasion and virulence factors [6]–[8], though how this determines its abortive phenotype in mammalian cells is not well understood [9]. Despite these features, MVA is at least as immunogenic as conventional replicating Vaccinia virus [10]–[13] and has the advantage of a considerably improved safety profile [14].

Attention has focussed on MVA not only as a possible 'next generation' small-pox vaccine [15]–[17], but also as a recombinant vaccine vector, with particular value in combination with other vectors (e.g. DNA, adenovirus, or an avian pox-virus) as a component of heterologous prime-boost regimens designed to elicit high frequencies of antigen-specific T cells [18]. This approach is undergoing development and clinical trial for diseases such as HIV-AIDS [19], [20], tuber-culosis [21], [22] and malaria [23], [24], where protective T cell responses are required for vaccine efficacy. In the case of liver-stage malaria vaccines, specific cytotoxic T cell responses correlate with the limited levels of protection achieved so far in humans [25], but current vectors and regimens are not immunogenic enough to achieve significant efficacy in malarious regions [26]. There is there-fore a need to identify recombinant vaccine vectors with greater immunogenicity, either by using novel vectors (e.g. simian adenoviruses [27]) or by improving existing platforms.

The routes available for attempts to improve MVA vector immunogenicity can be crudely divided into addition or removal of genes. Co-expression of co-stimulatory molecules, such as B7.1 [28] and 4-1BB ligand [29] and various cy-tokines, including IL-12 [30], IL-15 [31] and GM-CSF [32] have been reported to increase recombinant poxvirus vaccine immunogenicity. These interventions can achieve a two- to four-fold increase in peak murine IFN-γ T cell responses. The parallel strategy is to delete poxviral genes encoding immunomodulators that attenuate adaptive immune responses, but are not required for growth in vitro. During attenuation MVA lost many such factors, including soluble decoy recep-tors for IFN-γ [33], TNF-α [34], IFN-α/β [35] and various chemokines [36]; an inhibitor of IL-1β converting enzyme [37], the complement control protein [38] and an intracellular inhibitor of Toll-like receptor (TLR) and IL-1 receptor (IL-1R) signalling [39]. On the other hand, it retains genes encoding secreted inter-leukin- [40]–[42] and chemokine- [36], [43] binding proteins; a dehydrogenase involved in steroid synthesis [44], [45], a second inhibitor of TLR/IL-1 recep-tor signalling [46], and other genes implicated in poxviral virulence [47]–[50]. Evidence that deletion of such genes provides a route for improvement of MVA immunogenicity is provided by reports of augmented protective mouse CD8+ T cell responses elicited by MVA lacking B15R, encoding an IL-1β binding protein [51], and A41L, encoding a chemokine binding protein of unknown specificity [43].

In addition to a priori candidates, there are at least 30 MVA genes whose function remains unknown. Although many poxviral proteins retain sequence homology and functional similarity with host proteins, it has recently been shown in the case of Vaccinia virus protein N1 that preservation of structure and function can occur in the absence of significant sequence similarity [52]. An unbiased experimental approach is therefore required for identification of MVA genes that negatively affect its immunogenicity. In order to expedite this approach, we have constructed a bacterial artificial chromosome (BAC) clone of the MVA genome using the elegant method devised by A. Domi and B. Moss for Vaccinia virus [53], with slight modifications. Recombination-mediated genetic engineering (recombineering) of this construct [54], [55] permits more rapid generation of mutants for analysis than can be achieved by traditional methods, with the possible exception of host-range selection on rabbit cells using K1L [56], [57]. To test this concept, we deleted five MVA genes using recombineering of a fully sequenced BAC clone, rescued the constructs to infectious virus, and here demonstrate a modest but significant improvement in immunogenicity of MVA lacking B15R, supporting and extending published work [51]. For large-scale biomanufacturing of an MVA-based vaccine for clinical use, the ability to use clonal purifed BAC DNA as the starting point, rather than a conventional plaque-purified recombinant, is likely to be of considerable value. With this aim in mind, as well as pre-clinical studies, we demonstrate immunogenicity of a recombinant antigen inserted at the traditional thymidine kinase locus of MVA-BAC by recombineering.

Materials and Methods

Recombinant MVA Construction

Figure 1 shows the plasmid construct used to generate a recombinant virus (referred to as MVA-BAC-parent) containing the entire sequence of the BAC vector pBELO-BAC11 at the Deletion III locus of MVA (between the remnants of A51R and A56R) together with a GFP reporter gene driven by the p4B late promoter from Fowlpox virus. The fragments required were amplified by PCR with primers designed to introduce restriction sites for assembly by standard methods. The flanking regions for recombination correspond to positions 148412-149083 and 149340-149855 of MVA genome sequence U94848 [5]. The plasmid was linearised with PacI and transfected into MVA-infected primary chick embryo fibroblast (CEF) cells (Institute for Animal Health, Compton, UK). The cells were trypsinised and GFP-positive cells were sorted by flow cytometry on a Dako Cytomation MoFlo prior to plaque picking of the recombinant virus to purity, amplification and titration on CEFs. Cells were maintained in Dulbecco's modified Eagle medium (DMEM) supplemented with 10% FCS or with 2% FCS for viral growth. Identity and purity were confirmed by PCR. MVA expressing the recombinant antigen TIP was prepared as described (A. Spencer et al., manuscript in preparation).

Figure 1. Schematic of construct for recombination into deletion III locus of MVA, based on pBELO-BAC11. DIIIL and DIIIR = left and right flanks for homologous recombination into MVA genome; GFP = green fluorescent protein gene; p4B = Fowlpox virus late promoter; CmR = chloramphenicol acetyltransferase gene for selection in E. coli. Flanking FseI sites are absent from MVA and permit excision of the cassette. The PacI site is provided for linearisation of the plasmid prior to recombination with MVA in infected transfected cells.

Generation of MVA-BAC

CEF cells in 6-well plates were infected with MVA-BAC-parent at 5 pfu/cell and 2 h later the inoculum was replaced with growth medium containing 45 µM isatin-β-thiosemicarbazone (IβT). The compound was kindly synthesised by Dr J. Robertson (Chemistry Research Laboratory, Oxford University) and was dissolved to 5 mg/mL in acetone for storage at –20°C, then diluted to 1mg/mL in 0.25 M NaOH immediately prior to use. The cells in some wells were transfected with pCI-Cre (obtained from A. Domi, NIAID, NIH, Bethesda, MD) using Lipofectamine (Invitrogen). The DNA was phenol-extracted from the cultures 24h later as described [53] and resuspended in 20 µl of Tris-EDTA. Of this material, 3 µl was electroporated into DH10B E. coli (Invitrogen) prior to selection on LB plates with chloramphenicol (12.5 µg/mL) and miniprep by alkaline lysis from liquid LB cultures. Clones were screened by PCR for the presence of the genes MVA005, MVA010, MVA044, MVA086 and MVA188. For restriction mapping and pulsed-field gel electrophoresis, BAC DNA was prepped by QIAgen kit and 0.5–1 µg of DNA was digested overnight in a 15 µl reaction with the relevant enzymes (NEB). MVA-BAC clone 26 was shotgun sequenced at eightfold coverage by Lark Technologies (Cogenics), Inc.

BAC Rescue

BHK-21 cells (maintained in DMEM+10% FCS) were seeded into 6-well plates and infected with FP9-LacZ [58], an attenuated strain of Fowlpox virus expressing β-galactosidase, at 1 pfu/cell in Optimem (Gibco). After 1–2 hours, the cells were transfected with 4 µg of Qiagen-purified MVA-BAC DNA using Lipofectamine 2000 (Invitrogen) in Optimem. The BAC DNA was stored at 4°C, was never vortexed, and was pipetted only with cut-off tips. The cells were monitored for GFP expression by epifluorescence microscopy and were harvested by freeze-thaw 4–6 days later. The lysate was used to infect fresh BHK cell monolayers and the cultures were monitored for the appearance of virus growth. Rescued viruses were amplified in BHK cells using DMEM without serum for infection and growth, before purification over sucrose cushions and titration on CEFs.

BAC Recombineering

GalK-based recombineering was done exactly as described by S. Warming et al. [55], from whom the reagents and strains were obtained (see also http://recombineering.ncifcrf.gov/). Sequencing of PCR products and primer synthesis was performed by MWG-Biotech AG.

Mouse Immunogenicity

Female BALB/c and C57BL/6 mice aged 6 to 8 weeks were obtained from the Biomedical Services Unit, Oxford University and procedures were conducted according to the UK Animals (Scientific Procedures) Act 1986. Mice were anaesthetised with ketamine-dormitor prior to intradermal immunisation with 10^6 pfu of MVA divided into two 25 µl injections (one per ear). Splenocytes were harvested as described [58] seven days post-immunisation for flow cytometric analysis or fourteen days post-immunisation for ELIspot analysis. ELIspots were conducted using 18–20 h stimulation with 1 µg/mL synthetic peptide (ProImmune) in IVPH plates (Millipore) coated with anti-IFN-γ antibody AN18 (Mabtech). Spots were developed with R46A2-biotin (Mabtech) followed by streptavidin alkaline phosphatase (Mabtech) and substrate kit (BioRad) and enumerated on an automated reader (Autoimmun Diagnostika). For flow cytometry, cells were stimulated for 6h with 1 µg/mL peptide in the presence of 2 µl/mL Golgi-Plug (BD) and stained with anti-CD8 pacific blue conjugate and anti-CD4 APC-Alexa-750 conjugate prior to fixation in 10% neutral buffered formalin (Sigma). Intracellular cytokine staining was performed using FITC-conjugated anti TNF-α, PE-conjugated anti-IL-2 and Alexa-647-conjugated anti-IFN-γ diluted in Cytoperm (BD). All antibodies were obtained from eBiosciences. Data were acquired on a CyAn flow cytometer (Dako) and analysed using FlowJo (Treestar). PESTLE and SPICE software were obtained from Dr M. Roederer, Vaccine Research Centre, NIH. Other statistical analyses were performed using Prism (GraphPad Software, Inc.).

Results

Cloning of MVA Genome into BAC

A procedure very similar to that developed by Domi and Moss for VAC-BAC [53] was used to create BACs containing the genome of MVA. First, a cassette containing pBELO-BAC11, a LoxP recombinase site and a GFP reporter gene driven by a late poxviral promoter (see Figure 1) was inserted into the deletion III locus of MVA by conventional recombination in infected CEF cells. Cells infected with this recombinant virus, referred to at MVA-BAC-parent, were treated with isatin-β-thiosemicarbazone (IβT) to inhibit viral hairpin resolution and promote genome concatemerisation [59]. After transformation of E. coli with DNA from these cells, dozens of chloramphenicol-resistant colonies were observed regardless of treatment of infected cells with IβT. Unexpectedly, transfection of cells with pCI-Cre prior to infection in an attempt to induce recombination between LoxP sites to produce circular molecules from head-to-tail concatemers [53] dramatically decreased the number of colonies obtained, probably due to cytotoxicity of the transfection reagent. Alternatively, it is possible that cryptic LoxP sites such as those present in mammalian genomes [60] resulted in illegitimate recombination events.

Colonies were screened for potential full-length clones by PCR at five loci spaced along the genome, and candidates were further characterised by restriction mapping with HindIII and XhoI and pulsed-field gel electrophoresis after digestion with FseI to excise the BAC cassette (data not shown). No full-length clones were isolated from pCI-Cre transfected samples (of 5 colonies screened), but from untransfected cells, four apparently full-length clones were obtained, three derived from IβT-treated cells (of 24 colonies screened) and one from untreated cells (of 18 colonies screened). This frequency is very similar to that observed for VAC-BAC [53].

BAC Rescue

The next step was to "rescue" the BAC clones to infectious MVA using a Fowlpox virus helper to provide transcriptional machinery. The BHK cell line, which is the only mammalian cell line known to support MVA replication, is ideal for this purpose since it is non-permissive for the host-range restricted avipoxviruses. Following transfection into BHK cells infected with Fowlpox virus, all four of the MVA-BAC clones were converted to infectious MVA able to replicate and form plaques on BHK and CEF cells. We have not conducted formal growth rate analysis of the resulting viruses, but viral titres of 8.0×10^8 to 2.4×10^9 pfu/mL (final volume ~0.5 mL) were achieved following sucrose cushion purification from BHK cultures totalling 1500 cm^2. The titrations were performed on CEF cells, in which Fowlpox virus can replicate, and absence of significant residual Fowlpox virus was demonstrated by X-Gal staining for the LacZ marker gene present in the helper virus.

Immunogenicity of MVA-BACs

The principal utility of MVA is as a vaccine vector, particularly for eliciting T cell responses. The immunogenicity of the viruses derived from the four MVA-BAC clones was therefore assessed in mice by determining CD8+ T cell responses to viral antigens following intradermal inoculation. Tscharke et al. have described in detail the immunodominance hierarchy of CD8+ T cell "determinants" (i.e., epitopes) following infection of BALB/c and C57BL/6 mice with Vaccinia virus, MVA and other poxviruses [61], [62]. We used the described synthetic peptides to stimulate splenocytes from immunised BALB/c or C57BL/6 mice and quantified antigen-specific T cell responses by IFN-γ ELIspot assay (Figure 2).

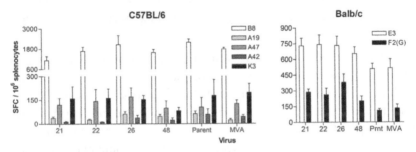

Figure 2. Immunogenicity of four MVA-BAC clones (21, 22, 26, 48) in C57BL/6 and BALB/c mice compared to parent virus (containing the BAC cassette) and MVA (lacking the BAC cassette). Mice were immunised with 106 pfu i.d. and the specific T cell responses to the indicated endogenous synthetic peptide epitopes [61], [62] were measured 14 days later by IFN-γ ELIspot assay. Data are means and SEM from groups of four mice.

Mice were immunised intradermally with the four MVA-BAC clones, the MVA-BAC-parent (from which the BAC clones were derived), or MVA (i.e. virus lacking the BAC cassette). In C57BL/6 mice, all were immunogenic and there was no statistically significant difference by one-way ANoVA between the T cell responses to any of these viruses for each of the five available determinants. In BALB/c mice, the same was true for the immunodominant determinant E3, but there were significant differences in the subdominant determinant specific F2(G) responses ($p<0.01$, one-way ANoVA), and a post-hoc test revealed that the statistically significant pairwise comparisons are clone #26 versus parent or MVA ($p<0.05$, Newman-Keuls multiple comparison test). Despite the lack of conventional significance for other clones, there is a trend towards higher responses to BAC-derived MVA in BALB/c mice, but this is not an effect of the presence of the BAC cassette at deletion III, since MVA-BAC-parent also contains this insertion.

The inability to distinguish the MVA-BAC clones by restriction map, viral growth yield or murine immunogenicity suggests that they are likely to be identical, but the possibility of some bias in the isolation procedure necessitates verification of clone fidelity by complete sequencing, especially prior to embarking on a programme of genetic manipulation.

Sequence of MVA-BAC

Three MVA genome sequences are available in GenBank, one published by Antoine et al. [5] (U94848), one deposited by Bavarian Nordic GmbH (DQ983236) and the "Acambis 3000" strain (AY603355) deposited by the CDC. The Bavarian Nordic sequence is identical to the Acambis sequence, and these differ from Antoine et al.'s sequence only by five single nucleotide substitutions (excluding the repeat regions of the inverted terminal repeats, which are not fully described in the Acambis and Bavarian Nordic sequences). All of these loci are very polymorphic amongst strains of Vaccinia virus (http://www.poxvirus.org/). The mutations and genes affected are shown in Table 1.

Table 1. Polymorphisms in MVA and MVA-BAC clone #26.

Position[a]	Gene affected	Mutation	Coding change	MVA strains possessing mutation[b]	Example Vaccinia virus strain possessing mutation[c]
68740	G6R	A>G	N>D	A/BN, MVA-BAC	Lister, WR, TAN
73886	L3L	C>T	R>K	A/BN, MVA-BAC	WR
108489	A4L	C>T	Silent	A/BN, MVA-BAC	WR
114308	A10L	T>C	Silent	A/BN, MVA-BAC	WR
114945	A10L	C>T	R>K	A/BN, MVA-BAC	Lister
30326	F7L	6 bp ins.	NK ins.	MVA-BAC	TAN

[a]Refers to residue numbering of MVA genome sequence U94848 [5].
[b]A/BN = Acambis/Bavarian Nordic (see text).
[c]WR = Western Reserve; COP = Copenhagen; TAN = TanTien.
doi:10.1371/journal.pone.0001638.t001

MVA-BAC clone #26 was selected for shotgun sequencing at eightfold coverage, resulting in a single contig (excluding the repeat regions), and was found to be identical to the Acambis/Bavarian Nordic strain, with the exception of a 6 bp insertion in the non-essential gene F7L [63]. The mutation codes for an extra Asn-Lys repeat and makes MVA-BAC identical to the TanTien strain of Vaccinia virus at this polymorphic locus, where other strains have two Asn-Lys repeats, with the exception of the Copenhagen strain which has eight (http://www.poxvirus.org/). In order to determine at what point in the genesis of MVA-BAC this mutation occurred, we sequenced the progenitor viruses at this locus by PCR amplification from genomic DNA. MVA-BAC-parent was found to possess the 6 bp insertion mutation, but its progenitor (from which the recombinant was derived) did not. Sequencing of PCR products revealed that a selection of other MVA recombinants from our laboratory and our original stock provided by Anton Mayr did not possess the mutation. It is not clear whether this 6 bp insertion arose during production of the MVA-BAC parent, or represents isolation of a rare genotype present at low levels in the source MVA. Evidence for genetic heterogeneity within poxvirus strains [64], [65](personal communication, R. Regnery) favours the latter possibility.

Deletion of MVA Genes by BAC Recombineering

Recombination mediated genetic engineering (recombineering) permits modification of BAC DNA by homologous recombination in E. coli. We used the system described by Warming et al. [55], which is based on stringent temperature-sensitive expression of the λ Red genes exo, bet and gam coupled with GalK-based positive and negative selection. A similar approach, not using GalK, has been described for VAC-BAC by Domi and Moss [54], who also verified the stability of their Vaccinia virus BAC clone after induction of the λ Red system. Five genes reported to affect MVA immunogenicity or Vaccinia virus immunogenicity or virulence were selected from the literature for deletion by insertion of GalK in place of the ORF (see Table 2). Long oligonucleotide primers were used to introduce 50 bp homology arms at the 5′ and 3′ ends of GalK by PCR and these products were electroporated into heat-induced SW102 cells carrying MVA-BAC prior to selection on galactose minimal medium, as described [55]. Before rescue by transfection into Fowlpox virus infected BHK cells, as above, the modified BACs were checked by restriction digest with HindIII and XhoI and by sequencing two PCR products spanning the junctions at the GalK insertion. In only one case (the A46R deletion), was any mutation detected: a primer-derived single nucleotide deletion in the intergenic region downstream of the insertion. This mutation was ignored as it did not fall within any predicted regulatory sequences (promoters or terminators). GalK deletion mutant BACs were also checked for absence of undeleted contaminants by PCR across the deleted locus. All five modified GalK-carrying MVA-BACs converted into infectious virus, and again, though we did not formally compare growth rates, all were amplified to titres of 2.8×10^8 to 2.6×10^9 pfu/mL (final volume ~0.5 mL) following sucrose cushion purification from 1500 cm^2 of BHKs.

Table 2. MVA-BAC genes deleted by GalK recombineering.

Gene	Function of encoded protein	Refs	Left homology arm (bp)[a]	Right homology arm (bp)[a]	Mutations in MVA[b]
C12L	IL-18 binding protein	[41,42]	13195-13224	13415-13464	6 bp in-frame deletion
A44L	Hydroxysteroid dehydrogenase	[44,45]	142608-142657	143549-143598	1 amino acid substitution
A46R	Inhibitor of Toll-like receptor signalling	[46]	143941-143990	144699-144748	-
B7R	Chemokine binding protein (SCP-3[c])	[36]	156989-157038	157564-157613	15 bp in-frame deletion
B15R	IL-1β binding protein	[51]	161971-162020	162999-163048	-

[a]Refers to residue numbering of MVA genome sequence U94848 [5].
[b]Coding mutations specific to MVA, i.e. not known in any other strain of Vaccinia virus.
[c]Smallpox virus encoded chemokine receptor (SECRET) domain containing protein 3.
doi:10.1371/journal.pone.0001638.t002

Immunogenicity of MVA-BAC Deletion Mutants

The effect of the above modifications on immunogenicity was assessed in intra-dermally immunised BALB/c mice. Peak specific T cell responses were analysed seven days later by flow cytometry with intracellular cytokine staining following stimulation with the F2(G) and E3 peptides (Figure 3). No statistically significant differences between groups were observed (one-way ANoVA for each epitope). Of the selected genes, only B15R, which encodes an IL-1β-binding protein, has been reported to affect MVA immunogenicity, rather than Vaccinia virus immunoge-nicity or virulence [51]. At the peak of the response, Staib et al. described a small non-significant effect of B15R deletion, very similar to that shown in Figure 3 for the E3 epitope, and with an identical p-value by t-test (p = 0.07); however, in the memory phase six months post-vaccination, they observed a highly significant ~4.5-fold increase in CD8+ T cell response to the B15R deletion mutant. We therefore assessed memory responses at 8 weeks post-vaccination, and observed a small, but statistically significant increase in CD8+ T cell responses to E3 (p = 0.039, t-test), but not to F2(G). As well as the readout time, variation in dose (108 vs. 106 pfu), route (i.p. vs. i.d.), antigen (H3 vs. E3 peptide) and mouse strain (HHD vs. BALB/c) are likely to underlie the suggested difference in the magnitude of the effect of B15R deletion on memory T cell responses.

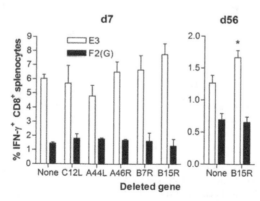

Figure 3. Peptide-specific splenic CD8+ T cell responses 7 days (d7) or 56 days (d56) after i.d. immunisation of BALB/c mice with 106 pfu of MVA-BAC derived viruses, either unmodified or with indicated genes deleted by insertion of GalK.* p = 0.039 (t-test). Data are means and SEM from groups of four mice.

In order to characterise the effect of B15R deletion further, we analysed E3-specific CD8+ T cell expression of TNF-α and IL-2 in addition to IFN-γ. SPICE software [66] was used to discriminate individual cells expressing the seven dif-ferent 'Boolean' combinations of cytokines from the IFN-γ+, TNF-α+ and IL-2+ gated populations (Figure 4). This approach reveals that at the peak of the response (d7), most of the cells are IFN-γ and TNF-α double-positives. Furthermore, although there is no conventionally significant difference, the increase in bulk IFN-γ responses observed with the B15R deletion mutant is primarily the result of an increased number of these double-positive cells, rather than of the IFN-γ single-positive or the IFN-γ, TNF-α and IL-2 triple-positive cells that account for the remainder of the CD8+ T cell response. By contrast, at day 56, the effect of B15R deletion is seen to reside mainly within the triple-positive population, and the difference is statistically significant (p = 0.018, Willcoxon-Rank test). Deletion of B15R therefore affects both the quality and magnitude of the E3-specific CD8+ T cell response to MVA. Enhanced production of the mitogenic cytokine IL-2 during the early memory phase (8 weeks) may underpin the large difference in T cell frequency observed in Staib et al.'s 6 month experiment [51].

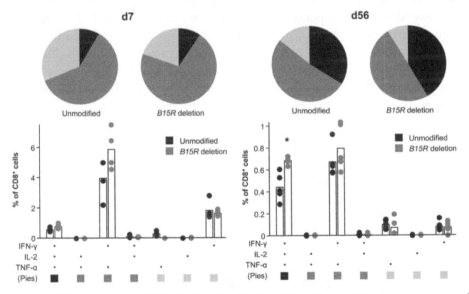

Figure 4. Multifunctionality of CD8+ T cells induced 7 days (d7) or 56 days (d56) after immunisation of BALB/c mice with unmodified MVA-BAC or B15R deletion mutant. Histograms show frequency of cells expressing each of the seven possible combinations of IFN-γ, IL-2 and TNF-α with bars showing the mean and circles showing the values for individual mice (n = 4). Pie charts show fraction of the total response comprised of cells expressing all three cytokines (black), any two cytokines (dark gray), or one cytokine only (light gray). * p = 0.018, Willcoxon-Rank test. Analysis performed using SPICE software [66].

Insertion of a Recombinant Antigen into MVA by BAC Recombineering

In addition to more rapid production of deletion mutants, the MVA-BAC system would also have value as a faster method for insertion of genes encoding exogenous protective antigens or candidate 'molecular adjuvants' with the potential to increase vector immunogenicity. The elegance of the GalK based recombineering system [55] relies on the ability to counterselect for clones lacking GalK using deoxygalactose, which is metabolised to a toxic intermediate. This enables insertion of genes without the need for a selectable marker. However, loss of GalK can occur both by recombination with an electroporated DNA, or by deletion of a segment of the BAC. Unlike GalK insertion, which has an efficiency close to 100% in MVA-BAC, the specific replacement of GalK with an exogenous sequence has been achieved to date at a frequency of only up to 1% compared to spontaneous deletions.

We used this technique to insert an antigen expression cassette carrying no tandem reporter gene or selectable marker into the traditional thymidine kinase (TK) locus of MVA-BAC. This construct comprised the p7.5 early/late promoter driving an 82 amino acid epitope string, named "TIP" (for Tuberculosis, Immunodeficiency virus and Plasmodium), which contains relevant epitopes from various pathogens for vaccine-induced protection in mouse models, including the protective H-2Kd-restricted Pb9 epitope from Plasmodium berghei circumsporozoite protein [67]. Figure 5 shows that Pb9-specific CD8+ T cell responses are elicited in MVA-BAC-TIP immunised mice to a level comparable with those elicited by conventional MVA-TIP. Although not statistically significant, there is a trend toward higher Pb9 responses (p = 0.07, t-test) and lower E3 responses (p = 0.11, t-test) in the conventionally derived MVA-TIP compared to the BAC-derived virus. This difference in immunodominance hierarchy may be attributable to differences in the nature of the cassette inserted at the TK locus (the lack of a late-promoter-driven GFP marker gene and opposite orientation of the TIP gene) or to the presence of the BAC cassette at deletion III.

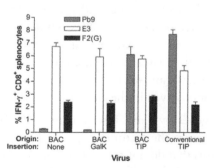

Figure 5. Immunogenicity of MVA-BAC expressing a recombinant antigen inserted at the standard thymidine kinase insertion site, in comparison to conventional recombinant MVA and recombineering precursors containing no insertion or GalK. The recombinant antigen, TIP, is an epitope string containing the Pb9 epitope from Plasmodium berghei circumsporozoite protein [67]. Bars show mean specific CD8+ T cell responses to the indicated peptides 7 days after i.d. vaccination of with 10^6 pfu, from groups of four BALB/c mice, with error bars showing SEM.

Discussion

Traditional methods for genetic manipulation of poxviruses, either for mutagenesis or expression of exogenous genes, rely on onerous plaque purification of rare recombinant viruses. Building on the results of Domi and Moss for VAC-BAC [53], [54], here we circumvent this requirement by creating an MVA-BAC that can be manipulated using recombineering and rescued to clonal viruses. With an efficiency similar to that observed in the case of VAC-BAC, we generated four MVA-BAC clones, which were identical by restriction map and had indistinguishable CD8$^+$ T cell immunogenicity in two strains of mice. We sequenced one clone, and found it to be identical to published genomic sequences of MVA, with the exception of a minor mutation that is polymorphic amongst Vaccinia virus strains and was present in the parent MVA recombinant. We did not conduct an investigation into stability of the construct, since good stability had been shown for the even larger VAC-BAC construct [53], [54].

Of five candidate genes selected from the literature for deletion by MVA-BAC recombineering (Table 2), only one had any statistically significant effect on responses to viral CD8+ T cell epitopes following intradermal immunisation of BALB/c mice. The effect of B15R deletion was more pronounced 8 weeks post-vaccination, though not as great as was previously reported at a 6 month timepoint in HHD mice [51]. Although we did not observe an effect of similar magnitude using an earlier readout, we show using SPICE polychromatic flow cytometry analysis that antigen-specific CD8+ T cells from mice immunised with MVA lacking B15R are composed of a higher proportion of triple-positive cells that express IL-2 in addition to IFN-γ and TNF-α. This 'multi-functional' phenotype may be responsible for the large differences in cell frequency at 6-months post-vaccination [51], since IL-2 is strongly implicated in memory T cell homeostasis [68] and its expression by virus-specific CD8+ T cells has been reported to promote antigen-specific proliferation of these cells even in the absence of CD4+ T cell help [69]. Whether responses to a recombinant antigen will be augmented in a B15R deletion background remains to be tested, though it should be noted in this regard that enhancement of anti-vector responses was confined to the E3-specific response and was not observed in the case of the F2(G) epitope.

Deletion of four other genes by GalK insertion in MVA-BAC did not potentiate CD8+ T cell responses to the vector epitopes. C12L, A44R and B7R all carry MVA-specific mutations that do not occur in any other sequenced strain of Vaccinia virus. Although strain Ankara (from which MVA was derived) has not been sequenced, these mutations presumably arose during attenuation, and the genes affected may therefore already be inactive. Alternatively, or additionally, functional redundancy of families of immunomodulatory genes acting on the same pathways may negate the effect of the deletion mutants. Indeed, poxviruses appear to have taken a grapeshot-like approach to disruption of the TLR/IL-1R signalling pathway [6], though MVA's gun has been spiked by deletion of many components, including A52R, which like A46R can inhibit NF-κB activation, albeit by a distinct subset of stimuli [46].

The deletion of these genes also did not impair the CD8+ T cell responses to the vector, suggesting that, as well as being dispensable for growth in BHK cell culture, they are not required for circumvention of host defence mechanisms operating in vivo (though since MVA is non-replicating in mammals, this conclusion does not extend to poxviral pathogenesis). If multiple MVA genes were to be deleted, in order to overcome putative functional redundancy, one might assume that at some point the virus would be so crippled as to be only very poorly immunogenic, for example by an antiviral response occurring early during the infection cycle, which normally proceeds to immature virion formation even in non-permissive cells. On the other hand, the major evolutionary driving force for acquisition of immunomodulators by poxviruses is more likely productive replication of a virulent virus, rather than evasion of adaptive responses (i.e. antibody or cytotoxic T cell responses). The MVA-BAC recombineering system will hopefully be versatile and rapid enough to allow these subtleties to be addressed, especially as it is likely to speed up the production of 'revertant' viruses in order to confirm the genetic specificity of a novel phenotype. Furthermore, the ability to make multiple modifications by 'recycling' the GalK dual-selectable marker is likely to be of particular value.

Other applications of MVA-BAC include production of viruses carrying recombinant antigens and/or molecular adjuvants for pre-clinical evaluation and clinical trial. The use of BAC DNA and an inactivated clinical grade helper virus as the input to a manufacturing process could reduce the burden of traceability currently required by regulatory authorities, though a method for efficient removal of the BAC cassette would be needed (see Figure 1). In this setting, the issues of construct stability and helper virus contamination would doubtless require more rigorous investigation than described here. Insertion of antigen expression cassettes by recombineering additionally circumvents the requirement for a reporter gene or selectable marker and its removal by transient-dominant selection [70] when a markerless product is required. In this regard, the difficulty of insertion of the TIP antigen by GalK counterselection is rather disappointing, though there is very likely room for improvement, for example by using longer homology arms in the targeting DNA, or electroporating maximal quantities of DNA. When removing GalK without concomitant insertion of DNA (i.e., from a deletion locus), we have achieved much higher recombineering efficiencies of up to 95% (data not shown). Alternatively, recombination in RecA+ E. coli [71] is a potential means of inserting elements without the use of a selectable marker, since several rare restriction sites (required for linearisation of the BAC prior to recombination) are absent from the MVA genome. For pre-clinical work, an in vitro site specific recombination system could be developed [72].

The vaccine vector capability of Vaccinia virus expressing a recombinant antigen was demonstrated nearly 25 years ago [73], [74]. Successful vectored vaccines for diseases such as malaria, HIV-AIDS and tuberculosis will likely require combinations of antigens and delivery systems able to induce immune responses of the right character and magnitude against the recombinant antigens. Currently, MVA is one of the more promising deployable delivery systems and has potential for improvement. The MVA-BAC technology provides an opportunity to accelerate production of candidate vaccines for evaluation, and potentially a superior starting point for clinical vaccine manufacture.

Acknowledgements

We thank Arban Domi and Bernard Moss for plasmids; Richard Wade-Martins and Michele Lufino for advice and help, Jeremy Robertson for synthesis of IβT and Søren Warming for recombineering materials and advice.

Author Contributions

Conceived and designed the experiments: SG MC AS. Performed the experiments: MC JF SS AS RA. Analyzed the data: MC AS. Wrote the paper: MC. Other: Principal Investigator: AH.

References

1. Mayr A, Stickl H, Muller HK, Danner K, Singer H (1978) [The smallpox vaccination strain MVA: marker, genetic structure, experience gained with the parenteral vaccination and behavior in organisms with a debilitated defence mechanism (author's transl)]. Zentralbl Bakteriol [B] 167: 375–390.

2. Drexler I, Heller K, Wahren B, Erfle V, Sutter G (1998) Highly attenuated modified vaccinia virus Ankara replicates in baby hamster kidney cells, a potential host for virus propagation, but not in various human transformed and primary cells. J Gen Virol 79 (Pt 2): 347–352.

3. Carroll MW, Moss B (1997) Host range and cytopathogenicity of the highly attenuated MVA strain of vaccinia virus: propagation and generation of recombinant viruses in a nonhuman mammalian cell line. Virology 238: 198–211.

4. Okeke MI, Nilssen O, Traavik T (2006) Modified vaccinia virus Ankara multiplies in rat IEC-6 cells and limited production of mature virions occurs in other mammalian cell lines. J Gen Virol 87: 21–27.

5. Antoine G, Scheiflinger F, Dorner F, Falkner FG (1998) The complete genomic sequence of the modified vaccinia Ankara strain: comparison with other orthopoxviruses. Virology 244: 365–396.

6. Haga IR, Bowie AG (2005) Evasion of innate immunity by vaccinia virus. Parasitology 130 SupplS11–25.

7. Seet BT, Johnston JB, Brunetti CR, Barrett JW, Everett H, et al. (2003) Poxviruses and immune evasion. Annu Rev Immunol 21: 377–423.

8. Webb LM, Alcami A (2005) Virally encoded chemokine binding proteins. Mini Rev Med Chem 5: 833–848.

9. Wyatt LS, Carroll MW, Czerny CP, Merchlinsky M, Sisler JR, et al. (1998) Marker rescue of the host range restriction defects of modified vaccinia virus Ankara. Virology 251: 334–342.

10. Sutter G, Wyatt LS, Foley PL, Bennink JR, Moss B (1994) A recombinant vector derived from the host range-restricted and highly attenuated MVA strain of vaccinia virus stimulates protective immunity in mice to influenza virus. Vaccine 12: 1032–1040.

11. Gomez CE, Rodriguez D, Rodriguez JR, Abaitua F, Duarte C, et al. (2001) Enhanced CD8+ T cell immune response against a V3 loop multi-epitope polypeptide (TAB13) of HIV-1 Env after priming with purified fusion protein and booster with modified vaccinia virus Ankara (MVA-TAB) recombinant: a comparison of humoral and cellular immune responses with the vaccinia virus Western Reserve (WR) vector. Vaccine 20: 961–971.

12. Hirsch VM, Fuerst TR, Sutter G, Carroll MW, Yang LC, et al. (1996) Patterns of viral replication correlate with outcome in simian immunodeficiency virus (SIV)-infected macaques: effect of prior immunization with a trivalent SIV vaccine in modified vaccinia virus Ankara. J Virol 70: 3741–3752.

13. Ramirez JC, Gherardi MM, Esteban M (2000) Biology of attenuated modified vaccinia virus Ankara recombinant vector in mice: virus fate and activation of B- and T-cell immune responses in comparison with the Western Reserve strain and advantages as a vaccine. J Virol 74: 923–933.

14. Drexler I, Staib C, Sutter G (2004) Modified vaccinia virus Ankara as antigen delivery system: how can we best use its potential? Curr Opin Biotechnol 15: 506–512.

15. Earl PL, Americo JL, Wyatt LS, Eller LA, Whitbeck JC, et al. (2004) Immunogenicity of a highly attenuated MVA smallpox vaccine and protection against monkeypox. Nature 428: 182–185.

16. Mayr A (2003) Smallpox vaccination and bioterrorism with pox viruses. Comp Immunol Microbiol Infect Dis 26: 423–430.

17. Cosma A, Nagaraj R, Staib C, Diemer C, Wopfner F, et al. (2007) Evaluation of modified vaccinia virus Ankara as an alternative vaccine against smallpox in chronically HIV type 1-infected individuals undergoing HAART. AIDS Res Hum Retroviruses 23: 782–793.

18. Gilbert SC, Moorthy VS, Andrews L, Pathan AA, McConkey SJ, et al. (2006) Synergistic DNA-MVA prime-boost vaccination regimes for malaria and tuberculosis. Vaccine 24: 4554–4561.

19. Hanke T, McMichael AJ, Dorrell L (2007) Clinical experience with plasmid DNA- and modified vaccinia virus Ankara-vectored human immunodeficiency virus type 1 clade A vaccine focusing on T-cell induction. J Gen Virol 88: 1–12.

20. Kent S, De Rose R, Rollman E (2007) Drug evaluation: DNA/MVA prime-boost HIV vaccine. Curr Opin Investig Drugs 8: 159–167.

21. McShane H, Hill A (2005) Prime-boost immunisation strategies for tuberculosis. Microbes Infect 7: 962–967.

22. Skeiky YA, Sadoff JC (2006) Advances in tuberculosis vaccine strategies. Nat Rev Microbiol 4: 469–476.

23. Hill AV (2006) Pre-erythrocytic malaria vaccines: towards greater efficacy. Nat Rev Immunol 6: 21–32.

24. Reyes-Sandoval A, Harty JT, Todryk SM (2007) Viral vector vaccines make memory T cells against malaria. Immunology 121: 158–165.

25. Keating SM, Bejon P, Berthoud T, Vuola JM, Todryk S, et al. (2005) Durable human memory T cells quantifiable by cultured enzyme-linked immunospot assays are induced by heterologous prime boost immunization and correlate with protection against malaria. J Immunol 175: 5675–5680.

26. Bejon P, Mwacharo J, Kai O, Mwangi T, Milligan P, et al. (2006) A phase 2b randomised trial of the candidate malaria vaccines FP9 ME-TRAP and MVA ME-TRAP among children in Kenya. PLoS Clin Trials 1: e29.

27. Reyes-Sandoval A, Fitzgerald JC, Grant R, Roy S, Xiang ZQ, et al. (2004) Human immunodeficiency virus type 1-specific immune responses in primates upon sequential immunization with adenoviral vaccine carriers of human and simian serotypes. J Virol 78: 7392–7399.

28. Garnett CT, Greiner JW, Tsang KY, Kudo-Saito C, Grosenbach DW, et al. (2006) TRICOM vector based cancer vaccines. Curr Pharm Des 12: 351–361.

29. Kudo-Saito C, Hodge JW, Kwak H, Kim-Schulze S, Schlom J, et al. (2006) 4-1BB ligand enhances tumor-specific immunity of poxvirus vaccines. Vaccine 24: 4975–4986.

30. Abaitua F, Rodriguez JR, Garzon A, Rodriguez D, Esteban M (2006) Improving recombinant MVA immune responses: potentiation of the immune responses to HIV-1 with MVA and DNA vectors expressing Env and the cytokines IL-12 and IFN-gamma. Virus Res 116: 11–20.

31. Perera LP, Waldmann TA, Mosca JD, Baldwin N, Berzofsky JA, et al. (2007) Development of smallpox vaccine candidates with integrated interleukin-15 that demonstrate superior immunogenicity, efficacy, and safety in mice. J Virol 81: 8774–8783.

32. Chavan R, Marfatia KA, An IC, Garber DA, Feinberg MB (2006) Expression of CCL20 and granulocyte-macrophage colony-stimulating factor, but not Flt3-L, from modified vaccinia virus ankara enhances antiviral cellular and humoral immune responses. J Virol 80: 7676–7687.

33. Symons JA, Tscharke DC, Price N, Smith GL (2002) A study of the vaccinia virus interferon-gamma receptor and its contribution to virus virulence. J Gen Virol 83: 1953–1964.

34. Alcami A, Khanna A, Paul NL, Smith GL (1999) Vaccinia virus strains Lister, USSR and Evans express soluble and cell-surface tumour necrosis factor receptors. J Gen Virol 80 (Pt 4): 949–959.

35. Alcami A, Symons JA, Smith GL (2000) The vaccinia virus soluble alpha/beta interferon (IFN) receptor binds to the cell surface and protects cells from the antiviral effects of IFN. J Virol 74: 11230–11239.

36. Alejo A, Ruiz-Arguello MB, Ho Y, Smith VP, Saraiva M, et al. (2006) A chemokine-binding domain in the tumor necrosis factor receptor from variola (smallpox) virus. Proc Natl Acad Sci U S A 103: 5995–6000.

37. Kettle S, Alcami A, Khanna A, Ehret R, Jassoy C, et al. (1997) Vaccinia virus serpin B13R (SPI-2) inhibits interleukin-1beta-converting enzyme and protects virus-infected cells from TNF- and Fas-mediated apoptosis, but does not prevent IL-1beta-induced fever. J Gen Virol 78 (Pt 3): 677–685.

38. Kotwal GJ, Isaacs SN, McKenzie R, Frank MM, Moss B (1990) Inhibition of the complement cascade by the major secretory protein of vaccinia virus. Science 250: 827–830.

39. Bowie A, Kiss-Toth E, Symons JA, Smith GL, Dower SK, et al. (2000) A46R and A52R from vaccinia virus are antagonists of host IL-1 and toll-like receptor signaling. Proc Natl Acad Sci U S A 97: 10162–10167.

40. Alcami A, Smith GL (1992) A soluble receptor for interleukin-1 beta encoded by vaccinia virus: a novel mechanism of virus modulation of the host response to infection. Cell 71: 153–167.

41. Reading PC, Smith GL (2003) Vaccinia virus interleukin-18-binding protein promotes virulence by reducing gamma interferon production and natural killer and T-cell activity. J Virol 77: 9960–9968.

42. Smith VP, Bryant NA, Alcami A (2000) Ectromelia, vaccinia and cowpox viruses encode secreted interleukin-18-binding proteins. J Gen Virol 81: 1223–1230.

43. Clark RH, Kenyon JC, Bartlett NW, Tscharke DC, Smith GL (2006) Deletion of gene A41L enhances vaccinia virus immunogenicity and vaccine efficacy. J Gen Virol 87: 29–38.

44. Moore JB, Smith GL (1992) Steroid hormone synthesis by a vaccinia enzyme: a new type of virus virulence factor. Embo J 11: 1973–1980.

45. Reading PC, Moore JB, Smith GL (2003) Steroid hormone synthesis by vaccinia virus suppresses the inflammatory response to infection. J Exp Med 197: 1269–1278.

46. Stack J, Haga IR, Schroder M, Bartlett NW, Maloney G, et al. (2005) Vaccinia virus protein A46R targets multiple Toll-like-interleukin-1 receptor adaptors and contributes to virulence. J Exp Med 201: 1007–1018.

47. Tscharke DC, Reading PC, Smith GL (2002) Dermal infection with vaccinia virus reveals roles for virus proteins not seen using other inoculation routes. J Gen Virol 83: 1977–1986.

48. Froggatt GC, Smith GL, Beard PM (2007) Vaccinia virus gene F3L encodes an intracellular protein that affects the innate immune response. J Gen Virol 88: 1917–1921.

49. Chen RA, Jacobs N, Smith GL (2006) Vaccinia virus strain Western Reserve protein B14 is an intracellular virulence factor. J Gen Virol 87: 1451–1458.

50. Roper RL (2006) Characterization of the vaccinia virus A35R protein and its role in virulence. J Virol 80: 306–313.

51. Staib C, Kisling S, Erfle V, Sutter G (2005) Inactivation of the viral interleukin 1beta receptor improves CD8+ T-cell memory responses elicited upon immunization with modified vaccinia virus Ankara. J Gen Virol 86: 1997–2006.

52. Cooray S, Bahar MW, Abrescia NG, McVey CE, Bartlett NW, et al. (2007) Functional and structural studies of the vaccinia virus virulence factor N1 reveal a Bcl-2-like anti-apoptotic protein. J Gen Virol 88: 1656–1666.

53. Domi A, Moss B (2002) Cloning the vaccinia virus genome as a bacterial artificial chromosome in Escherichia coli and recovery of infectious virus in mammalian cells. Proc Natl Acad Sci U S A 99: 12415–12420.

54. Domi A, Moss B (2005) Engineering of a vaccinia virus bacterial artificial chromosome in Escherichia coli by bacteriophage lambda-based recombination. Nat Methods 2: 95–97.

55. Warming S, Costantino N, Court DL, Jenkins NA, Copeland NG (2005) Simple and highly efficient BAC recombineering using galK selection. Nucleic Acids Res 33: e36.

56. Staib C, Drexler I, Ohlmann M, Wintersperger S, Erfle V, et al. (2000) Transient host range selection for genetic engineering of modified vaccinia virus Ankara. Biotechniques 28: 1137–1142.1144–1136, 1148.

57. Staib C, Lowel M, Erfle V, Sutter G (2003) Improved host range selection for recombinant modified vaccinia virus Ankara. Biotechniques 34: 694–696, 698, 700.

58. Anderson RJ, Hannan CM, Gilbert SC, Laidlaw SM, Sheu EG, et al. (2004) Enhanced CD8+ T cell immune responses and protection elicited against Plasmodium berghei malaria by prime boost immunization regimens using a novel attenuated fowlpox virus. J Immunol 172: 3094–3100.

59. Merchlinsky M, Moss B (1989) Resolution of vaccinia virus DNA concatemer junctions requires late-gene expression. J Virol 63: 1595–1603.

60. Semprini S, Troup TJ, Kotelevtseva N, King K, Davis JR, et al. (2007) Cryptic loxP sites in mammalian genomes: genome-wide distribution and relevance for the efficiency of BAC/PAC recombineering techniques. Nucleic Acids Res 35: 1402–1410.

61. Tscharke DC, Karupiah G, Zhou J, Palmore T, Irvine KR, et al. (2005) Identification of poxvirus CD8+ T cell determinants to enable rational design and characterization of smallpox vaccines. J Exp Med 201: 95–104.

62. Tscharke DC, Woo WP, Sakala IG, Sidney J, Sette A, et al. (2006) Poxvirus CD8+ T-cell determinants and cross-reactivity in BALB/c mice. J Virol 80: 6318–6323.

63. Coupar BE, Oke PG, Andrew ME (2000) Insertion sites for recombinant vaccinia virus construction: effects on expression of a foreign protein. J Gen Virol 81: 431–439.

64. Cottingham MG, van Maurik A, Zago M, Newton AT, Anderson RJ, et al. (2006) Different levels of immunogenicity of two strains of Fowlpox virus as recombinant vaccine vectors eliciting T-cell responses in heterologous prime-boost vaccination strategies. Clin Vaccine Immunol 13: 747–757.

65. Jarmin SA, Manvell R, Gough RE, Laidlaw SM, Skinner MA (2006) Retention of 1.2 kbp of 'novel' genomic sequence in two European field isolates and some vaccine strains of Fowlpox virus extends open reading frame fpv241. J Gen Virol 87: 3545–3549.

66. Darrah PA, Patel DT, De Luca PM, Lindsay RW, Davey DF, et al. (2007) Multifunctional TH1 cells define a correlate of vaccine-mediated protection against Leishmania major. Nat Med 13: 843–850.

67. Romero P, Maryanski JL, Cordey AS, Corradin G, Nussenzweig RS, et al. (1990) Isolation and characterization of protective cytolytic T cells in a rodent malaria model system. Immunol Lett 25: 27–31.

68. Ma A, Koka R, Burkett P (2006) Diverse functions of IL-2, IL-15, and IL-7 in lymphoid homeostasis. Annu Rev Immunol 24: 657–679.

69. Zimmerli SC, Harari A, Cellerai C, Vallelian F, Bart PA, et al. (2005) HIV-1--specific IFN-gamma/IL-2-secreting CD8 T cells support CD4-independent proliferation of HIV-1-specific CD8 T cells. Proc Natl Acad Sci U S A 102: 7239–7244.

70. Falkner FG, Moss B (1990) Transient dominant selection of recombinant vaccinia viruses. J Virol 64: 3108–3111.

71. Kong Y, Yang T, Geller AI (1999) An efficient in vivo recombination cloning procedure for modifying and combining HSV-1 cosmids. J Virol Methods 80: 129–136.

72. Hartley JL, Temple GF, Brasch MA (2000) DNA cloning using in vitro site-specific recombination. Genome Res 10: 1788–1795.

73. Smith GL, Mackett M, Moss B (1983) Infectious vaccinia virus recombinants that express hepatitis B virus surface antigen. Nature 302: 490–495.

74. Smith GL, Murphy BR, Moss B (1983) Construction and characterization of an infectious vaccinia virus recombinant that expresses the influenza hemagglutinin gene and induces resistance to influenza virus infection in hamsters. Proc Natl Acad Sci U S A 80: 7155–7159.

Conserving, Distributing and Managing Genetically Modified Mouse Lines by Sperm Cryopreservation

G. Charles Ostermeier, Michael V. Wiles, Jane S. Farley
and Robert A. Taft

ABSTRACT

Background

Sperm from C57BL/6 mice are difficult to cryopreserve and recover. Yet, the majority of genetically modified (GM) lines are maintained on this genetic background.

Methodology/Principal Findings

Reported here is the development of an easily implemented method that consistently yields fertilization rates of 70±5% with this strain. This six-fold increase is achieved by collecting sperm from the vas deferens and epididymis into a cryoprotective medium of 18% raffinose (w/v), 3% skim milk (w/v) and 477 μM monothioglycerol. The sperm suspension is loaded into 0.25 mL French straws and cooled at 37±1°C/min before being plunged and then stored in LN2. Subsequent to storage, the sperm are warmed at 2,232±162°C/min and incubated in in vitro fertilization media for an hour prior to the addition of oocyte cumulus masses from superovulated females. Sperm from 735 GM mouse lines on 12 common genetic backgrounds including C57BL/6J, BALB/cJ, 129S1/SvImJ, FVB/NJ and NOD/ShiLtJ were cryopreserved and recovered. C57BL/6J and BALB/cByJ fertilization rates, using frozen sperm, were slightly reduced compared to rates involving fresh sperm; fertilization rates using fresh or frozen sperm were equivalent in all other lines. Developmental capacity of embryos produced using cryopreserved sperm was equivalent, or superior to, cryopreserved IVF-derived embryos.

Conclusions/Significance

Combined, these results demonstrate the broad applicability of our approach as an economical and efficient option for archiving and distributing mice.

Introduction

Embryo cryopreservation is an effective strategy for managing mouse lines. Its adoption has been limited by the cost, time and the number of animals required. This is especially true for those lines where embryo yields are low, e.g. BALB/c. Cryopreserving sperm is an attractive alternative. However, its widespread use has been limited by the challenge of efficiently recovering cryopreserved sperm from some commonly used inbred strains[1]. In our experience (Table 1) and in that of others[2]–[5], the impaired fertility associated with cryopreserved mouse sperm is dependent on genetic background, with sperm from the C57BL/6 backgrounds being particularly sensitive. Yet, this strain is one of the most commonly used for creating and maintaining genetically modified (GM) lines. More than 75% of the 670 mouse lines submitted to The Jackson Laboratory's Repository from January 2004 to January 2006 were maintained on a predominantly C57BL/6J background. Further, The National Institutes of Health are using C57BL/6 embryonic stem (ES) cells to create a resource containing null mutations in every gene in the mouse genome[6]. Thus, it is critical that an effective and efficient method of cryopreserving and recovering C57BL/6 sperm be developed.

Since mouse sperm survive cryopreservation with reasonable success[2], the key to an effective sperm cryopreservation and recovery scheme is maximizing post-thaw fertilization capacity. In vivo, sperm develop the capacity to fertilize oocytes during transit through the female reproductive tract. As reviewed by Visconti et al[7], sperm fertilization ability is associated with plasma membrane reorganization and increases in intracellular calcium levels and in Reactive Oxygen Species[8] (ROS). Because cryopreservation modifies aspects of sperm function associated with fertilization capacity[9], perhaps these processes can be modulated to increase the fertility of cryopreserved mouse sperm. Thus, the objective of this work was to develop economical processes to cryopreserve C57BL/6J sperm that retain or enhance fertilizing capacity.

Table 1. Dependency of impaired fertility on the genetic background of cryopreserved mouse sperm.

Genetic background of sperm donor line	# of GM lines	# females	# IVF oocytes	% of oocytes to 2-cell±StdErr
129S1/SvImJ	13	355	8,760	3±1
C57BL/6J	311	7,173	207,318	6±0
BALB/cByJ	5	71	1,261	8±3
BALB/cJ	12	159	1,937	15±3
B6129SF1/J	67	1,057	29,071	17±2
NOD/ShiLtJ	6	153	2,976	34±11
FVB/NJ	25	681	8,759	25±4
DBA/2J	9	96	2,518	70±7
Overall	448	9,743	262,600	23±8

These data (**Table S4**) were generated in The Jackson Laboratory's IVF and Cryopreservation laboratory from 6/4/01 to 1/24/07 using a method modified from Sztein et al[4] (**Figure S2**). IVF oocytes were obtained from the # of superovulated inbred females indicated. The standard error (StdErr) was calculated using the # of GM lines for each background.
doi:10.1371/journal.pone.0002792.t001

Because variable cooling and warming rates have been observed with some methods[10], our effort began by defining reproducible processes for cryopreserving and thawing mouse sperm. Our methods were then refined to enhance the ability of cryopreserved C57BL/6J sperm to fertilize C57BL/6J oocytes. The overall effectiveness of our approach as a tool for the routine management of GM mouse lines, was demonstrated by cryopreserving and recovering 735 GM lines maintained on twelve genetic backgrounds, including 527 GM lines maintained on the C57BL/6J background.

Results and Discussion

Definition of Cooling and Warming Rates

More C57BL/6J sperm retained intact membranes with moderate cooling (37±1°C/min) than with rapid cooling (94±2°C/min; Table 2). This is in accord with previous work[10], [11]. Importantly, using the device and process reported here, sperm were consistently cooled at the same rate. The warming rate also affects sperm survival[11], and encouraging results have been reported when thawing at 54°C[12]. We compared thawing cryopreserved C57BL/6J sperm at 37°C and 54°C. No differences in sample warming rate or sperm survival were observed (Table 2). However, a greater proportion of oocytes developed into 2-cell embryos when fertilized in vitro with sperm thawed at 37°C (Table 2). Even though the straws were exposed to the 54°C water bath for only 5 sec, the outer layers of the straw may have warmed to 54°C before the contents in the center had time to equilibrate. Those cells exposed to 54°C and to temperatures above 37°C likely incurred varying degrees of thermal damage, reducing their ability to fertilize. This mechanism is supported by the assertion of Jiang et al[13] that thawing sperm at 37°C reduces the risk of thermal damage.

Table 2. Suitable cooling and warming rate determination.

Treatment	Cooling rate °C/min (n)	% intact membranes (n)*	
Styrofoam box	37 ± 1^a (3)	26 ± 2^a (3)	
Stainless steel Dewar	94 ± 2^b (3)	15 ± 4^b (3)	
Water bath temp °C	**Warming rate °C/min (n)**	**% intact membranes (n)***	**% of oocytes to 2-cell (n; # oocytes)****
37	$2,232\pm162^a$ (3)	28 ± 7^a (3)	35 ± 4^a (3; 723)
54	$2,043\pm212^a$ (3)	19 ± 5^a (3)	11 ± 6^b (3; 879)

Cooling treatments are detailed in **Figure S1**. Data are means and standard errors from (**n**) replicates. Values with different superscripts are different ($p<0.05$).
*At least 300 sperm were evaluated for each replicate.
**A replicate is the average percent of CByB6F1/J oocytes developing into 2-cell embryos within four IVF drops. Each replicate represents sperm from a different set of two C57BL/6J males.
doi:10.1371/journal.pone.0002792.t002

Promoting/Maintaining Fertilization Capacity

Improved IVF success has been reported with cryopreserved C57BL/6 sperm[12], [14]. One characteristic shared by these reports is a processing step that delayed the time between thawing and combining eggs and sperm. To determine if pre-incubating cryopreserved C57BL/6J sperm enhances fertilizing ability, fresh and cryopreserved sperm samples were incubated for 0, 30, 60, or 80 min in Mouse Vitro Fert medium (MVF; Cook Medical; Brisbane, Australia). Freshly collected C57BL/6J sperm did not benefit from incubation. However, cryopreserved sperm exhibited considerable enhancement, with maximum fertility of cryopreserved C57BL/6J sperm being achieved after one hour of incubation (Figure 1a). Thus, after thawing, cryopreserved C57BL/6J sperm acquire fertilization competence over time. Similar findings have been reported with sperm incubated in media containing methyl-beta-cyclodextrin[15]. Together, these studies indicate that cryopreserved C57BL/6 sperm must undergo maturational events prior to the addition of oocytes. It is postulated that sperm must commence maturation prior to the addition of oocytes because of the spontaneous cortical granule release, which results in zona hardening and fertilization capacity reduction[16], [17]. In fact, we have observed strain specific differences in the time required for zona pellucida dissolution (personal unpublished data), which may be due to differences in the level of spontaneous cortical granule release or may simply reflect differences in the zona pellucidae from different strains. Nonetheless, if the sperm maturational events are not given time to occur prior to zona hardening, a reduced number of 2-cell embryos or a complete fertilization block would be observed. The need to incubate cryopreserved C57BL/6 sperm is in contrast to previous studies, in which cooling and cryopreservation reduced the time necessary for sperm to obtain fertilization competency[9], [18]. Conceivably, inherent genetic components regulate the differences observed in this phenotype.

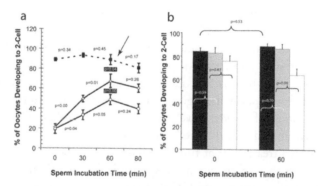

Figure 1. Influence of sperm pre-incubation on in vitro fertilization success. (a). Variability among males was limited by creating 3 pools of sperm, each containing sperm from 2 C57BL/6J males. Sperm treatments included cryopreservation in 3% skim milk, 18% raffinose (o); 3% skim milk, 18% raffinose supplemented with 477 µM monothioglycerol (◊) and freshly collected in Cook's Mouse Vitro Fert (MVF; ■). Sperm were cryopreserved and thawed using previously optimized cooling and warming rates. Three IVFs per treatment, each using cumulus oocyte masses from 2 superovulated C57BL/6J females (mean±standard deviation; 66±20 oocytes/IVF), were performed following 0, 30, 60, or 80 min of sperm incubation in MVF. Treatment effects were determined using ANOVA and preplanned contrasts between consecutive time points. The arrow and black boxed p-values indicate the reduced ability of cryopreserved C57BL/6J sperm to fertilize oocytes after 60 min of incubation when compared to freshly collected sperm using Dunnett's Method[32]. (b). Sperm were collected from 6 C57BL/6J males and 3 pools created, each containing sperm from 2 males. The pools were subdivided into 3 treatments: i) Control – no treatment (black); ii) Diluted live sperm – sperm were diluted 1:4 in MVF to mimic the concentration of live sperm equivalent to that found in cryopreserved samples (gray); and iii) Diluted live+dead sperm – live sperm were diluted 1:4 with killed sperm (flash frozen in liquid nitrogen) to simulate the number and ratio of live and dead sperm observed in the cryopreserved samples (white). The various treatments received either no incubation or 60 min of incubation in MVF before adding cumulus oocyte masses from two superovulated C57BL/6J oocytes (mean±standard deviation; 69±23 oocytes/IVF). In vitro fertilization was done at least six times for each treatment, twice per pool of males. Comparisons indicated were evaluated using preplanned contrast.

Approximately 75% of C57BL/6J sperm do not survive cryopreservation (Table 2). Consequently, it may be necessary to increase sperm concentration to ensure that enough viable sperm are present to maximize fertility. However, this could also increase the concentration of dead sperm, which may be detrimental[19]. This hypothesis is supported by reports demonstrating improved fertilization following the separation of viable from non-viable sperm prior to IVF[12], [15]. To determine whether the requirement for a pre-incubation was linked to limited concentrations of viable sperm or increased concentrations of dead sperm, freshly collected sperm were a) diluted to limit the number of live sperm or b) were mixed with dead sperm to simulate the proportion and concentration of dead sperm present following cryopreservation and thawing. Sperm in the various treatments were incubated for 0 or 60 min before C57BL/6J oocytes were added. The proportion of oocytes developing into 2-cell embryos was determined after overnight incubation.

The percentage of oocytes developing into 2-cell embryos did not differ between 0 and 60 min for Control (Figure 1b), confirming our previous observation that freshly collected C57BL/6J sperm do not require incubation prior to being combined with oocytes. Further, no difference in the proportion of oocytes developing into 2-cell embryos was detected between Control and Diluted live sperm at time 0 or 60 min (Figure 1b). Thus, a broad range of viable sperm concentrations yield similar IVF results and reduced sperm concentrations do not lead to a requirement for sperm pre-incubation. Lastly, we determined if dead sperm or substances released from sperm damaged by cryopreservation interfere with fertilization. The percentage of oocytes developing into 2-cell embryos using Diluted live sperm was compared with Diluted live+dead sperm. No difference was detected between the two treatments at time 0 (Figure 1b), indicating that dead sperm and the substances they release are not responsible for the initial reduction in fertilization capacity of cryopreserved C57BL/6J sperm. In contrast, after 60 min of sperm incubation, the Diluted live+dead sperm fertilized ~26% fewer oocytes than the Diluted live sperm (Figure 1b). This demonstrates that dead sperm or substances they liberate can be deleterious to the fertilizing capacity of viable sperm following prolonged exposure. Based on these results, it can be concluded that the benefit of a pre-incubation step following thawing is separable from the effects of limiting viable sperm concentrations or increasing the concentration of dead sperm.

While the generation of ROS is necessary for fertilization[20], overproduction of ROS during cryopreservation[21] could alter this process or damage sperm, impairing their ability to fertilize or to support subsequent embryo development. To determine if reducing ROS generation can increase the fertilizing capacity of C57BL/6J sperm, cryopreservation was carried out in the presence of varying concentrations of the reducing agent alpha-monothioglycerol (MTG). This re-agent was selected because previous work has shown it is necessary for the culture of ES-cells in sensitive/stressful serum-free systems[22]. The presence of at least 318 µM MTG improved the fertilization proficiency of C57BL/6J sperm (Figure 2). This improvement was not associated with an increase in the proportion of motile sperm (p = 0.12), providing further evidence that it is essential to evaluate endpoints other than sperm survival and motility when optimizing cryopreservation methods.

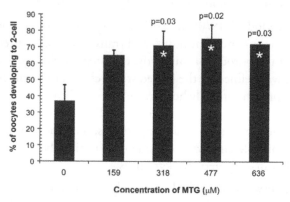

Figure 2. Dose response of monothioglycerol addition to mouse sperm cryopreservation media. Sperm were collected from 6 C57BL/6J males. To reduce male-to-male variability, 3 pools of sperm were created by combining 2 males each. The samples were split 5 ways and the sperm were cryopreserved in the concentrations of monothioglycerol (MTG) illustrated on the x-axis. The cryopreserved sperm were thawed, placed into MVF and incubated for 60 min prior to the addition of cumulus oocyte masses from two superovulated C57BL/6J females (mean±standard deviation; 50±22 oocytes/IVF). Differences among the treatments were determined using ANOVA and preplanned comparisons to the control of no MTG using Dunnett's method[32].

Excess ROS generation may be linked to the requirement for a sperm pre-incubation step, as observed in Figure 1a. To test this, C57BL/6J sperm were cryopreserved in the presence of 477 µM MTG and incubated in MVF media for 30, 60, 90 or 120 minutes prior to adding C57BL/6J oocytes. As shown in Figure 1a, the presence of MTG did not change the incubation time required to reach maximum fertility. However, the percentage of 2-cell embryos was greater for sperm frozen in the presence of MTG. Collectively, these observations indicate that damage by ROS may be partly responsible for the reduction in fertility of cryopreserved mouse sperm. This effect could be due to the oxidation of lipids[23] and/or proteins[24] that participate in the fusion and subsequent penetration of the oocyte by sperm. Reactive oxygen species can also oxidize DNA[25], poten-tially leading to a reduction in embryo development. Nonetheless, the observa-tion that cryopreserved C57BL/6J sperm require post-thaw incubation to reach maximum fertility does not appear linked to ROS generation.

Application of New Sperm Cryopreservation and Recovery Methods

The methods reported here yield fertilization rates of 70±5% with C57BL/6J oo-cytes and cryopreserved C57BL/6J sperm. This is a significant improvement over many previously published approaches [2]–[5], but these data, and those reported by others, may not reflect performance "in the field". Indeed, the demand for im-proved sperm cryopreservation methods comes from the necessity to cryopreserve GM lines, not inbred strains. Thus, before reaching conclusions on applicability, we believe it is crucial to test methods as they will be used in practice.

Over a 15-month period (7/19/06 to 10/8/07), sperm were cryopreserved from 994 GM lines (predominantly knockouts and transgenics) submitted to The Jackson Laboratory. For the purpose of this study, the predominant genetic back-ground of the GM line was defined by the inbred strain chosen to be the source of oocytes for IVF. This approach generally holds true, but more importantly, reflects the situation encountered by repositories and core facilities responsible for cryo-preservation. To allow for robust statistical comparisons, only those backgrounds represented by five or more GM lines were considered, limiting the data set to 735 GM lines and 12 predominant genetic backgrounds.

Sperm were cryopreserved as described previously and stored in liquid nitro-gen for at least 24 hours. In vitro fertilization was performed using 60 minutes of sperm pre-incubation and oocytes from superovulated inbred females corre-sponding to the predominant genetic background of each of the various GM lines. For comparison, IVF data was obtained using freshly collected sperm from 847 GM lines being maintained on the same genetic backgrounds. No difference in fertility was detected between freshly collected or cryopreserved sperm for 10 of the 12 genetic backgrounds (129S1/SvImJ; C57BLKS/J; 129X1/SvJ; BALB/cJ; B6129SF1/J; NOD/ShiLtJ; FVB/NJ; C3H/HeJ; C3HeB/FeJ; DBA/2J; Figure 3a). Cryopreserved sperm from two of the genetic backgrounds (C57BL/6J and BALB/cByJ) showed slight reductions in fertility. However, when the proportion of 2-cell embryos generated on these two backgrounds using our new method and a method modified from Sztein et al[4] were compared, six- and five-fold increases were observed, respectively (Figure 3a vs Table 1). When frequency dis-tributions are compared, a similar proportion of lines perform poorly (defined as fertilization rate of five percent or less) whether fresh or frozen sperm are used, demonstrating the reliability of the method. The difference observed between the fertilization capacity of the GM C57BL/6 lines (Figure 3a, Table 1) and inbred C57BL/6J mice (Figure 1a, Figure 2) is likely due to genetic modifications that may have affected fertility and the presence of alleles from other strains in lines that were not fully congenic. Relying on data from inbred strains alone would have overestimated the capability of the method in practice.

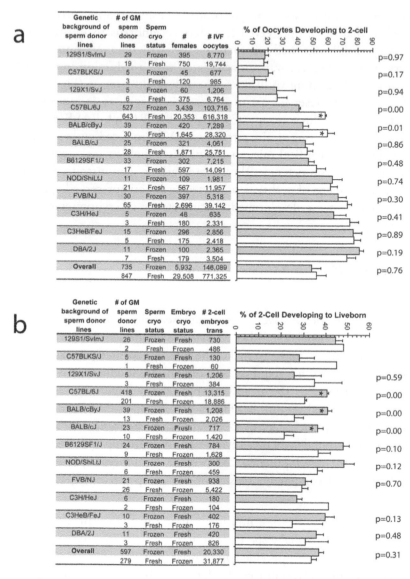

Figure 3. High throughput application of cryopreserved mouse sperm. (a). Genetically modified (GM) male mice were grouped based on their predominant background (first column), and the number of unique GM lines within each of the backgrounds is shown (second column). In vitro fertilization was preformed employing either our new sperm cryopreservation and recovery methods (Frozen) or freshly collected sperm (Fresh). The number of females and oocytes utilized are shown in the fourth and fifth columns. Differences between the Frozen and Fresh sperm treatments were assessed using ANOVA and preplanned contrasts. Asterisks within a bar indicate a significant difference between the treatments within a background. (b). A subset of the previously created embryos using cryopreserved sperm was transferred directly to pseudopregnant recipients. Those embryos created using freshly collected sperm were cryopreserved prior to transfer, as shown in column four. Differences between the two treatments were assessed using ANOVA and preplanned contrasts. Asterisks within a bar indicate a significant difference between the treatments within a background. No p-value is reported in those instances where the sample size was too small for statistical analysis.

The developmental capacity of embryos produced using cryopreserved sperm was compared to that of cryopreserved IVF-derived embryos to illustrate the relative efficiency of different methods available for archiving GM mouse lines. The use of cryopreserved sperm was not associated with a reduction in the percentage of embryos developing to liveborn for any of the groups (Figure 3b). However, the use of cryopreserved sperm was associated with an increase in the production of liveborn in three groups (C57BL/6J, BALB/cByJ, BALB/cJ). Perhaps genetic differences among strains predispose IVF produced embryos from C57BL/6J, BALB/cByJ, BALB/cJ mice to cryopreservation damage, resulting in lowered developmental competency. These observations demonstrate that the developmental competence of embryos produced using cryopreserved sperm is equal to or greater than the developmental competence of cryopreserved embryos produced by IVF.

Implications of Efficient Mouse Sperm Cryopreservation and Recovery

We report here a simple, inexpensive sperm cryopreservation and recovery method that is easier to implement than other reported methods[12], [14], [15]. The cryopreservation medium is simple to make, the freezing apparatus is easily constructed, sperm separation is not required, and only minor modifications to standard IVF protocols are needed. The breadth of applicability (Figure 3a, Figure 3b) and efficiency demonstrated (Figure 3a, Figure 3b) illustrate the general utility and robustness of the approach. Comparable rates of fertilization were observed for most genetic backgrounds whether cryopreserved or freshly collected sperm was used. Likewise, the developmental competence of embryos derived using cryopreserved sperm was comparable or superior to cryopreserved, IVF-derived embryos.

Unless it is necessary to preserve multiple mutations or the entire genome, the efficiencies reported here make sperm cryopreservation preferable to embryo cryopreservation in many cases, provided appropriate females will be available in the future. Sperm cryopreservation can also be preferable to maintaining small colonies. Often, once strains are no longer under active investigation, they are reduced to only a few breeding pairs, placing them at risk from breeding problems, genetic drift, and genetic contamination. When needed, these colonies are scaled up; a process that often takes months. Sperm cryopreservation offers an economical alternative by eliminating ongoing breeding costs and reducing the time to produce experimental cohorts, as the number of oocyte donors for IVF or the number of females used for artificial insemination can easily be scaled up to produce the desired number of animals.

Widespread adoption of sperm cryopreservation will hopefully make the pro-active cryopreservation of strains routine. This will benefit facilities and investigators by reducing operating costs, providing protection against disease outbreaks or disasters, and improving collaboration by reducing barriers to distribution. A system in which investigators cryopreserve lines and deposit the cryopreserved material within one of the many publicly funded resource centers may now be possible. This will reduce the operating costs of resource centers, provide secure off-site backup of lines, and relieve investigators and core facilities of the burden of maintaining and distributing lines. The development of an effective, efficient sperm cryopreservation method marks a new paradigm in repository operations, facilitating the global need to archive and distribute mouse models.

Methods

Animals

Mice were maintained at The Jackson Laboratory (Bar Harbor, ME, USA) in accordance with The Jackson Laboratory institutional protocols and the Guide for the Care and Use of Laboratory Animals [26]. The animals were housed in a minimum barrier facility with a light cycle of 14 hrs on and 10 hrs off (on at 05:00 AM).

Sperm Collection and Cryopreservation

Briefly, 2 mL of CryoProtective Medium [CPM - 18% w/v raffinose (Sigma Aldrich; cat # R7630); 3% w/v skim milk (BD Diagnostics; cat # 232100); 477 μM monothioglycerol (Sigma cat # M6145); water (Invitrogen, cat # 15230-238)] were used to collect the sperm from the epididymides and vas deferentia of two 3–5 month-old male mice. Sperm were allowed to disperse from the tissue for 10 min and then 10 μL of sperm+CPM and ballast were loaded into 0.25 mL French straws (IMV; Maple Grove, MN; cat# AAA201). Straws were sealed with an impulse heat sealer (American International Electric; Whittier, CA; model AIE-305HD) and five straws per cassette (Zanders Medical Supplies; Vero Beach, FL; cat #16980/0601) were placed onto a raft (polystyrene 15.5 cm wide×20 cm long×2.5 cm deep) floating in LN2 for ≥10 min. This resulted in the sperm being cooled from −10 to −60°C at a rate of 37±1°C/min before being plunged and then stored in LN2.

Cooling rates were determined by placing the wire lead of an Ertco TC4000 thermocouple (Dubuque, IA) into the 10 μL volume of sperm+CPM within a straw that was inside a cassette with 5 loaded straws. Data points were obtained every two seconds, and those obtained from a temperature of –10°C to a temperature of –60°C were analyzed by linear regression (JMP 6; SAS Institute, Cary, NC). The calculated slope defined the cooling rate.

Warming rates were determined by moving the straw equipped with the thermocouple directly from the cassette within LN2 into a water bath. Data points were obtained every two seconds, and those obtained from a temperature of –175°C to –2°C were analyzed with a linear regression to provide a warming rate.

In Vitro Fertilization

In vitro fertilizations were performed using a modified version of that described previously[27]. In vitro fertilization culture medium, Mouse Vitro Fert (Cook Medical; Brisbane, Australia), was used for sperm incubation, IVF and zygote culture. The components of MVF are the same as those listed for modified human tubal fluid[28], except for the substitution of gentamicin for penicillin and streptomycin (personal communication; Cook Medical; Brisbane, Australia). The IVF dishes contained one 500 µL fertilization drop.

Sperm samples were thawed in a 37°C water bath for ~30 sec. The CPM+sperm was pushed out of the straw into the IVF drop and incubated for 1 hr. The number of sperm within an IVF drop (mean±standard deviation; 6.1±2.5×105) varied depending on the sperm count of the males. Sperm count variation was controlled among treatments by applying treatments to be compared to a single pool of sperm. This resulted in the same sperm concentration being utilized across treatments. Superovulated 17- to 27-day-old female mice were used as oocyte donors. After 4 hrs of co-incubation, the presumptive zygotes were washed, and only those appearing normal were cultured overnight in a 150 µL MVF drop. Approximately 18 hrs after washing, the proportion of oocytes fertilized was calculated by dividing the number of 2-cell embryos by the sum of 2-cell embryos and normally appearing presumptive 1-cell oocytes. Polyspermy and parthenogenesis are negligible in mouse IVF systems[15] (personal unpublished data) and assumed consistent across treatments.

Sperm Membrane Assessment

Sperm were stained according to the Live/Dead Spermatozoa Viability Kit (Molecular Probes; Eugene, OR). Red (compromised membrane) and green (intact membrane) sperm were visualized and for each analysis three different individuals determined the number of green sperm in a total of at least 100 cells.

Sperm Motility Assay

Sperm were diluted at least 1:10 in MVF. The samples were then drawn into capillary tubes (Microslide 1099, VitroCom Inc; Mt Lakes, NJ) or loaded into 80 micron 2X-CEL chambers (Hamilton Thorne; Beverly, MA). Three fields per sample were manually selected and analyzed using the Hamilton Thorne IVOS computerized semen analyzer (Beverly, MA).

Embryo Cryopreservation

Two-cell embryos were cryopreserved in 1.5 M propylene glycol as described by Glenister and Rall[29].

Embryo Transfer

Pseudopregnant CByB6F1/J mice of 9 to 13 weeks of age were used as embryo transfer recipients. Four to 15 embryos were transferred into one oviduct of each female as described by Nagy et al.[30].

Statistical Analysis

Percentages were arcsine transformed[31], and treatment differences were detected by analysis of variance (ANOVA) and preplanned comparisons in JMP 6 (SAS Institute, Cary, NC). For presentation, means and standard errors of percentages are shown.

Acknowledgements

The authors are grateful for the discussion and insight provided by Shannon Byers, Sian Clements, Tom Gridley, Mary Ann Handel, John Kulik and Stanley Leibo. In addition, they appreciate the assistance provided by the Reproductive Sciences Service at The Jackson Laboratory. Their skill in mouse husbandry, as well as performing mouse surgeries, in vitro fertilizations, and sperm recoveries are second to none.

Author Contributions

Conceived and designed the experiments: GCO MVW JSF RAT. Performed the experiments: GCO JSF. Analyzed the data: GCO MVW JSF RAT. Contributed reagents/materials/analysis tools: MVW RAT. Wrote the paper: GCO MVW RAT.

References

1. Critser JK, Mobraaten LE (2000) Cryopreservation of murine spermatozoa. ILAR J 41: 197–206.

2. Nakagata N, Takeshima T (1993) Cryopreservation of mouse spermatozoa from inbred and F1 hybrid strains. Jikken dobutsu 42: 317–320.

3. Songsasen N, Leibo SP (1997) Cryopreservation of mouse spermatozoa. II. Relationship between survival after cryopreservation and osmotic tolerance of spermatozoa from three strains of mice. Cryobiology 35: 255–269.

4. Sztein JM, Farley JS, Mobraaten LE (2000) In vitro fertilization with cryopreserved inbred mouse sperm. Biol Reprod 63: 1774–1780.

5. Yildiz C, Ottaviani P, Law N, Ayearst R, Liu L, et al. (2007) Effects of cryopreservation on sperm quality, nuclear DNA integrity, in vitro fertilization, and in vitro embryo development in the mouse. Reproduction 133: 585–595.

6. International Mouse Knockout Consortium T (2007) A Mouse for All Reasons. Cell 128: 9–13.

7. Visconti PE, Westbrook VA, Chertihin O, Demarco I, Sleight S, et al. (2002) Novel signaling pathways involved in sperm acquisition of fertilizing capacity. J Reprod Immunol 53: 133–150.

8. Ecroyd HW, Jones RC, Aitken RJ (2003) Endogenous redox activity in mouse spermatozoa and its role in regulating the tyrosine phosphorylation events associated with sperm capacitation. Biol Reprod 69: 347–354.

9. Bailey JL, Bilodeau JF, Cormier N (2000) Semen cryopreservation in domestic animals: a damaging and capacitating phenomenon. J Androl 21: 1–7.

10. Stacy R, Eroglu A, Fowler A, Biggers J, Toner M (2006) Thermal characterization of Nakagata's mouse sperm freezing protocol. Cryobiology 52: 99–107.

11. Koshimoto C, Mazur P (2002) Effects of cooling and warming rate to and from –70 degrees C, and effect of further cooling from –70 to –196 degrees C on the motility of mouse spermatozoa. Biol Reprod 66: 1477–1484.

12. Bath ML (2003) Simple and efficient in vitro fertilization with cryopreserved C57BL/6J mouse sperm. Biol Reprod 68: 19–23.

13. Jiang MX, Zhu Y, Zhu ZY, Sun QY, Chen DY (2005) Effects of cooling, cryopreservation and heating on sperm proteins, nuclear DNA, and fertilization capability in mouse. Mol Reprod Dev 72: 129–134.

14. Kaneko T, Yamamura A, Ide Y, Ogi M, Yanagita T, et al. (2006) Long-term cryopreservation of mouse sperm. Theriogenology 66: 1098–1101.

15. Takeo T, Hoshii T, Kondo Y, Toyodome H, Arima H, et al. (2008) Methyl-Beta-cyclodextrin improves fertilizing ability of C57BL/6 mouse sperm after freezing and thawing by facilitating cholesterol efflux from the cells. Biol Reprod 78: 546–551.

16. Gianfortoni JG, Gulyas BJ (1985) The effects of short-term incubation (aging) of mouse oocytes on in vitro fertilization, zona solubility, and embryonic development. Gamete Res 11: 59–68.

17. Wolf DP, Hamada M (1976) Age dependent losses in the penetrability of mouse eggs. J Reprod Fertil 48: 213–214.

18. Fuller SJ, Whittingham DG (1997) Capacitation-like changes occur in mouse spermatozoa cooled to low temperatures. Mol Reprod Dev 46: 318–324.

19. Szczygiel MA, Kusakabe H, Yanagimachi R, Whittingham DG (2002) Separation of motile populations of spermatozoa prior to freezing is beneficial for subsequent fertilization in vitro: a study with various mouse strains. Biol Reprod 67: 287–292.

20. De Lamirande E, Gagnon C (1995) Capacitation-associated production of superoxide anion by human spermatozoa. Free Radical Bio Med 18: 487–495.

21. Chatterjee S, Gagnon C (2001) Production of reactive oxygen species by spermatozoa undergoing cooling, freezing, and thawing. Mol Reprod Dev 59: 451–458.

22. Wiles MV, Johansson BM (1999) Embryonic stem cell development in a chemically defined medium. Exp Cell Res 247: 241–248.

23. Alvarez JG, Storey BT (1984) Lipid peroxidation and the reactions of superoxide and hydrogen peroxide in mouse spermatozoa. Biol Reprod 30: 833–841.

24. Mammoto A, Masumoto N, Tahara M, Ikebuchi Y, Ohmichi M, et al. (1996) Reactive oxygen species block sperm-egg fusion via oxidation of sperm sulfhydryl proteins in mice. Biol Reprod 55: 1063–1068.

25. Loft S, Kold-Jensen T, Hjollund NH, Giwercman A, Gyllemborg J, et al. (2003) Oxidative DNA damage in human sperm influences time to pregnancy. Hum Reprod 18: 1265–1272.

26. Clark JD, Baldwin RL, Bayne KA, Brown MJ, Gebhart GF, et al. (1996) Guide for the Care and Use of Laboratory Animals; In: Grossblatt N, editor. Washington, D.C.: National Academy Press.

27. Byers SL, Payson SJ, Taft RA (2006) Performance of ten inbred mouse strains following assisted reproductive technologies (ARTs). Theriogenology 65: 1716–1726.

28. Kito S, Ohta Y (2005) Medium effects on capacitation and sperm penetration through the zona pellucida in inbred BALB/c spermatozoa. Zygote 13: 145–153.

29. Glenister PH, Rall WF (1999) Cryopreservation and rederivation of embryos and gametes. In: Jackson IJ, Abbott CA, editors. Mouse genetics and transgenics: a practical approach 2nd ed:. Oxford University Press.

30. Nagy A, Gertsenstein M, Vintersten K, Behringer R (2003) Manipulating the mouse embryo: a laboratory manual (3rd ed.):. Cold Spring Harbor Laboratory Press. pp. 175, 263–177, 571.

31. Amann RP (2005) Weaknesses in reports of "fertility" for horses and other species. Theriogenology 63: 698–715.

32. Dunnett CW (1955) A multiple comparison procedure for comparing several treatments with a control. J Am Stat Assoc 50: 1096–1121.

c-MycERTAM Transgene Silencing in a Genetically Modified Human Neural Stem Cell Line Implanted into MCAo Rodent Brain

Lara Stevanato, Randolph L. Corteling, Paul Stroemer,
Andrew Hope, Julie Heward, Erik A. Miljan and John D. Sinden

ABSTRACT

Background

The human neural stem cell line CTX0E03 was developed for the cell based treatment of chronic stroke disability. Derived from fetal cortical brain tissue, CTX0E03 is a clonal cell line that contains a single copy of the c-mycER^{TAM} transgene delivered by retroviral infection. Under the conditional *regulation by 4-hydroxytamoxifen (4-OHT), c-mycER^{TAM} enabled large-scale stable banking of the CTX0E03 cells. In this study, we investigated the fate of this transgene following growth arrest (EGF, bFGF and 4-OHT withdrawal) in vitro and following intracerebral implantation into a mid-cerebral artery occluded (MCAo) rat brain. In vitro, 4-weeks after removing growth factors and 4-OHT from the culture medium, c-mycER^{TAM} transgene transcription is reduced by ~75%. Furthermore, immunocytochemistry and western blotting demonstrated a concurrent decrease in the c-MycER^{TAM} protein. To examine the transcription of the transgene in vivo, CTX0E03 cells (450,000) were implanted 4-weeks post MCAo lesion and analysed for human cell survival and c-mycER^{TAM} transcription by qPCR and qRT-PCR, respectively.*

Results

The results show that CTX0E03 cells were present in all grafted animal brains ranging from 6.3% to 39.8% of the total cells injected. Prior to implantation, the CTX0E03 cell suspension contained 215.7 (SEM = 13.2) copies of the c-mycER^{TAM} transcript per cell. After implantation the c-mycER^{TAM} transcript copy number per CTX0E03 cell had reduced to 6.9 (SEM = 3.4) at 1-week and 7.7 (SEM = 2.5) at 4-weeks. Bisulfite genomic DNA sequencing of the in vivo samples confirmed c-mycER^{TAM} silencing occurred through methylation of the transgene promoter sequence.

Conclusion

In conclusion the results confirm that CTX0E03 cells downregulated c-mycER^{TAM} transgene expression both in vitro following EGF, bFGF and 4-OHT withdrawal and in vivo following implantation in MCAo rat brain. The silencing of the c-mycER^{TAM} transgene in vivo provides an additional safety feature of CTX0E03 cells for potential clinical application.

Background

Stem cell therapy is a facet of regenerative medicine that aims to ameliorate the damage caused to the brain by the grafting of healthy "reparative" cells. Pioneering studies implanting mouse neural stem cells (NSCs) into the brains of stroke animals have demonstrated significant recovery in motor and cognitive tests [1-5]. These findings provide a rational approach to the development of a cell based therapy for ischemic stroke. A substantial and consistent supply of allogeneic NSCs is required in order to treat a large patient population. Unfortunately, human NSCs are somatic stem cells and susceptible to genetic and phenotypic changes and loss of biological activity following extensive tissue culture expansion [6-8].

Immortalization of NSCs overcomes their restricted expansion potential. The myc proto-oncogene has proven a successful tool for immortalization of NSCs [9-16]. In these research applications, both v- and c-myc promote consistent and enhanced tissue culture expansion of human NSCs while maintaining a stable karyotype and retaining biological activity. The c-mycERTAM technology has proven to be as successful in generating immortalized neural stem cell lines as c-myc alone [17-20]. The c-mycERTAM transgene is better suited as an immortalizing agent for clinical applications because c-Myc protein function is conditional on the presence of the tamoxifen metabolite, 4-hydroxytamoxifen (4-OHT). The translated c-MycERTAM recombinant protein is a conjugation of human c-Myc and modified mouse estradiol receptor (ER) and is present as an inactive monomer in the cytosol of the cell. When added to the tissue culture medium, 4-OHT specifically binds to the ER moiety causing the c-MycERTAM protein to dimerize and subsequently translocate to the cell nucleus where c-Myc is active as a transcription factor [21].

CTX0E03 is a human NSC line that was derived using c-mycERTAM technology and showed recovery in sensorimotor deficits following grafting in MCAo rodent brain [19,22]. One copy of the c-mycERTAM transgene is integrated into the CTX0E03 cell genome (chromosome 13) under the cytomegalovirus (CMV) immediate early promoter [19]. We have shown that the c-mycERTAM transgene is expressed in proliferating CTX0E03 cell cultures in vitro; however, the expression of this transgene following the grafting of CTX0E03 cells in vivo was not characterized. Although neoplasm formation has never been observed with CTX0E03 cells in multiple pre-clinical studies, this information would nonetheless be important in assessing the inherent risk of using a genetically modified cell line for clinical applications.

Analysis of transgene expression following implantation of genetically modified cells is challenging because of the relatively small proportion of implanted cells compared to the host. Using a fluorescent reporter transgene, the CMV promoter has been shown to undergo silencing in grafted hematopoietic cells in vivo [23]. In another example, Lee et al. demonstrated by immunohistochemistry (IHC) that v-myc was silenced following intracerebral implantation of immortalized HBF1.3 cells [24]. Unfortunately, these methodologies were not appropriate for our application. The c-mycERTAM transgene in CTX0E03 cells does not contain a fluorescent reporter that could be monitored. With respect to IHC, we were concerned it would be inconclusive due to the unknown sensitivity and potential cross-reactivity of the antibodies. Furthermore, IHC was ruled out since the analysis would be limited to a finite number of sections and would not encompass all cells present in the graft.

In this study, we developed a novel and sensitive method that is able to analyze absolute gene expression within human genetically modified cells grafted into rodent brain. This method used the human specific Alu genomic sequence to detect and quantify CTX0E03 cells grafted within the rodent brain background [25,26]. Absolute quantification of c-mycERTAM transcription made it possible to calculate the number of c-mycERTAM transcript copies per CTX0E03 cell. Using this methodology, c-mycERTAM transgene silencing was shown following grafting of CTX0E03 cells into MCAo rat brain. Although not intentionally incorporated into the design technology of c-mycERTAM, silencing of the CMV transgene promoter in vivo demonstrates an additional safety feature of this technology and of the CTX0E03 cells.

Results

Alu and c-mycERTAM Assay Design

The analysis of the c-mycER^{TAM} transcript expression per CTX0E03 cell, both in vivo and in vitro, was dependent upon the ability to sequentially purify genomic DNA (gDNA) and RNA from the same biological sample using the All DNA/RNA Kit (Qiagen). The gDNA preparation yielded one single band by agarose gel electrophoresis with no evidence of degradation or contaminating RNA (Figure 1A). The RNA preparation yielded the expected 28S and 18S bands, again with no degradation or DNA contamination visible (Figure 1B). Standard curves were then generated to determine: 1) the number of CTX0E03 cells present in a gDNA sample by Alu qPCR (Figure 1C); and 2) the number of c-mycER^{TAM} transcripts present in a retro-transcribed RNA reaction sample (Figure 1D). The Alu sequence assay could reproducibly detect human CTX0E03 gDNA with as few as one CTX0E03 cell per reaction; whereas, the c-mycER^{TAM} assay could detect as few as 10 copies per reaction.

Validation of Alu and c-mycERTAM Assays

The Alu sequence and c-mycER^{TAM} transcription assays were developed in the presence of rat brain gDNA or cDNA, as described in the Methods. These experiments involved mixing purified CTX0E03 DNA/cDNA with purified rat brain DNA/cDNA. However, in order to verify that CTX0E03 cells grafted in rat brain could be processed and analysed effectively together a positive control experiment was carried out. This experiment consisted of harvesting and processing the rat brain tissue immediately after a 6 μl injection of CTX0E03 cells was delivered. It was determined by counting using a haemocytometer that the cell suspension prior to delivery contained 46,600 cells/µl – i.e. total of 279,600 cells were present in 6 µl loaded into the syringe. By Alu sequence analysis, the injected rat brain tissue sections contained an average of 241,116 (SEM = 15,487) CTX0E03 cells (Figure 1F). The measured number of CTX0E03 cells is within the range expected. Analysis of the total RNA isolated from these samples showed CTX0E03 cells to express on average 236.9 (SEM = 7.8) c-mycERTAM transcript copies per cell (Figure 1G). Control CTX0E03 cells measured in culture were shown to have a similar value of 242.3 (SEM = 47.8) c-mycERTAM transcript copies per cell.

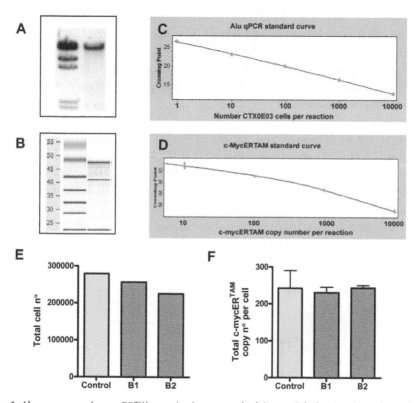

Figure 1. Alu sequence and c-mycERTAM assay development and validation. Gels showing the quality and purity of gDNA by agarose gel electrophoresis (A) and RNA by virtual gel produced by Agilent 2100 Bioanalyzer (B, RNA Integrity Number >9.4 as analyzed by Agilent RNA 600 nano kit [32]) isolated from the same sample. Standard curves used to determine: cell number by Alu sequence qPCR (C, Error 0.0300, efficiency 1.993; 3 replicates); absolute c-mycERTAM copy number by c-mycERTAM qRT-PCR (D, Error 0.0837 and efficiency 2.131; 3 replicates). All standard curves were generated from CTX0E03 gDNA diluted in rat gDNA or cDNA. Crossing point refers to the number of PCR cycles required to generate a detectable fluorescent signal generated on a Roche LC480 instrument. Positive control rat brain samples (B1 and B2) were grafted with approximately 300,000 CTX0E03 cells each and harvested immediately (E, F). Data shown is the total number of CTX0E03 cells in each tissue section as determined by Alu, where control is the number of viable cells in the cell suspension prior to injection as determined by counting using a haemocytometer (E); and total c-mycERTAM transcript copy number calculated per CTX0E03 cell detected in brain samples, where control is the number of copies per cell detected in vitro culture (F).

CTX0E03 Cell Detection and Quantification in Vivo

Four weeks after the MCAo, a total of 450,000 CTX0E03 cells were implanted bilaterally into the putamen. The rats were sacrificed 1-week or 4-weeks post-implantation, their brains were removed and sections encompassing the injection tracts were dissected and snap frozen in liquid nitrogen. The Alu and c-mycER[TAM] assays were then carried out on the gDNA and cDNA isolated from each tissue section, respectively. CTX0E03 cells were detected in all implanted animals at both time points, with a range of 28,151 to 179,184 CTX0E03 cells detected per brain. The total number of CTX0E03 cells detected, per rat brain, is shown at 1-week (Figure 2A) and 4-weeks (Figure 2B) post-implantation. An aliquot of the CTX0E03 cell suspension prepared for implantation was retained and used to determine the pre-implantation c-mycER[TAM] transcript copy number. The average c-mycER[TAM] transcription level for the three CTX0E03 cell suspensions prepared in this study was found to be 215.7 (SEM = 13.2) copies per cell (Figure 2C). This is compared to the population of CTX0E03 cells detected in vivo, where the average number of c-mycER[TAM] transcript copies per CTX0E03 cell was calculated to be 6.9 (SEM = 3.4) at 1-week and 7.7 (SEM = 2.5) at 4-weeks (Figure 2C). The number of CTX0E03 cells, the number c-mycERTAM transcripts calculated per tissue section and the resultant number of c-mycERTAM transcript copies per CTX0E03 cell, are shown in Table 1. c-mycER[TAM] transcripts were detected within approximately 50% of the tissue sections containing CTX0E03 cells at both time points. It is important to note, in no instance was a c-mycER[TAM] transcript signal detected where CTX0E03 cells were absent.

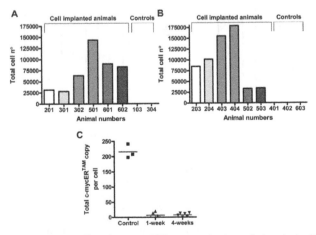

Figure 2. Detection of CTX0E03 cells and c-mycER[TAM] transcription in grafted rat brain. Total cells found at 1-week (A) and 4-weeks (B) post-implantation by Alu qPCR. Animal numbers 103, 304, 401, 402 and 603 were vehicle injected only control brain samples. Absolute quantification of c-mycER[TAM] transcript level per CTX0E03 cell, in cell suspensions prior to implantation (control) and in vivo (C).

Table 1. c-mycER^{TAM} transcript copy number and number of CTX0E03 cells calculated in the grafted rat brain sections at 1- and 4-weeks post-implantation.

Week 1					Week 4				
Rat# region	Total c-mycER^{TAM} copy #	Total cell # Alu assay	Total c-mycER^{TAM} copy #/Total cell # Alu assay	c-mycER^{TAM} copy # per cell per animal	Rat# region	Total c-mycER^{TAM} copy #	Total cell # Alu assay	Total c-mycER^{TAM} copy #/Total cell # Alu assay	c-mycER^{TAM} copy # per cell per animal
201 F1	N.D.¹	N.D.²	N/A		203 F1	N.D.	N.D.	N/A	
201 F2	N.D.	1066.9	0.0		203 F2	N.D.	N.D.	N/A	
201 FN1	18671.8	18821.5	1.0		203 FN1	N.D.	1776.9	0.0	
201 FN2	N.D.	11695.4	0.0	0.8	203 FN2	N.D.	15046.2	0.0	
301 F1	N.D.	530.8	0.0		203 C	521495.6	68123.1	7.7	6.1
301 F2	N.D.	356.8	0.0		204 F1	902695.4	45261.5	20.0	
301 FN1	N.D.	16269.2	0.0		204 F2	N.D.	5563.1	0.0	
301 FN2	N.D.	10993.9	0.0	0.0	204 FN1	N.D.	4850.8	0.0	
302 F1	N.D.	1075.4	0.0		204 FN2	418067.7	45661.5	9.2	13.0
302 F2	N.D.	837.7	0.0		403 F1	N.D.	N.D.	N/A	
302 FN1	909865.9	42992.3	21.1		403 F2	264100.0	5501.5	48.0	
302 FN2	N.D.	18664.6	0.0	14.3	403 FN1	388332.3	85000.0	4.6	
501 F1	N.D.	N.D.	N/A		403 FN2	224224.0	17169.2	13.1	
501 F2	550.1	17846.2	0.0		403 C	1324893.5	47584.6	27.8	14.2
501 C1	2858994.9	123000.0	23.2		404 F1	N.D.	2738.5	0.0	
501 C2	N.D.	2701.5	0.0	19.9	404 F2	31223.6	11076.9	2.8	
601 F1	N.D.	N.D.	N/A		404 FN1	N.D.	146750.0	0.0	
601 F2	N.D.	964.2	0.0		404 FN2	218605.5	15061.5	14.5	
601 FN1	10151.6	70061.5	0.9		404 F ANT	N.D.	3556.9	0.0	0.4
601 FN2	109810.0	18692.3	5.9	1.3	502 F1	N.D.	241.4	0.0	
602 F1	380824.6	41123.1	9.3		502 F2	N.D.	1046.2	0.0	
602 F2	N.D.	24076.9	0.0		502 FN1	150198.4	12776.9	11.7	
602 C	40085.1	11253.9	3.6		502 FN2	224974.8	19261.5	11.7	11.3
602 FL	N.D.	6541.5	0.0	5.1	503 F1	N.D.	N.D.	N/A	
					503 F2	N.D.	N.D.	N/A	
					503 FN1	N.D.	1230.8	0.0	
					503 FN2	N.D.	13292.3	0.0	
					503 C	N.D.	19692.3	0.0	
		Mean	6.9				Mean	7.7	
		SEM	3.4				SEM	2.5	

¹N.D. Not Detected, signal was below the level of 10 c-mycER^{TAM} copies per reaction
²N.D. Not Detected, signal was below the level of one CTX0E03 cell per reaction

Transgene Methylation Analysis

Direct modification of the DNA by methylation is an epigenetic mechanism to downregulate or silence gene expression. Using bisulfite sequencing, the methylation status of the CMV transgene promoter was investigated in CTX0E03 cells pre- and post-implantation. The CMV promoter of the transgene contains numerous potential methylation sites within a CpG rich island (Figure 3A). Bisulfite sequencing of gDNA clones taken from control non-implanted CTX0E03 cells showed only 2% of these CpG sites were methylated (Figure 3B). However, the same analysis carried out on clones obtained from the CTX0E03 cell implanted MCAo rat brain samples showed that 86.8% (SEM = 7.1) and 75.0% (SEM = 1.4) of the CpG islands were methylated at 1-week and 4-weeks, respectively (Figure 3B). There was no statistical difference between 1- and 4-weeks, p = 0.14, indicating that the methylation had occurred within the first week and was constant to the 4-week time point. Collectively, only four gDNA transgene promoter clones in vivo were found to be devoid of any methylated sites from a total of 99 analysed; in other words, ~96% of the transgene promoter sequences analysed in vivo where found to contain at least one methylated CpG island.

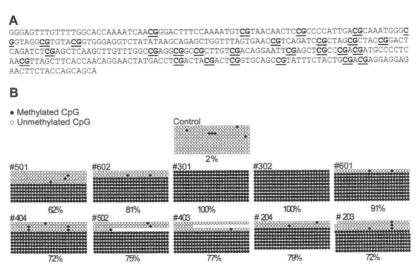

Figure 3. In vivo analysis of CpG methylation within the c-mycERTAM promoter region in CTX0E03 cells. CpG dinucleotide-containing regions (underlined) examined in CMV promoter of the c-mycER[TAM] transgene (A). Methylation sequence analysis of the CMV transgene promoter region of in vivo and in vitro samples (B). Ten animal grafted brains were analyzed, five at 1-week (top set) and five at 4-weeks (bottom set), in addition, a non-implanted CTX0E03 cell sample (control). Ten clones containing the sequence depicted in (A) from each grafted rat brain were analyzed. Each row shown represents a clone. Filled circles = methylated CpG island; open circles = unmethylated CpG island. Percentages of global methylation are reported at the bottom of each sample.

In Vitro Characterization

Additional experiments were carried out in vitro to determine if a reduction in transgene transcription leads to a concomitant reduction in translated protein.

Analyses were carried out on control CTX0E03 cells and those cultured for 1- and 4-weeks in the absence of growth factors and 4-OHT. In this non growth permissive condition, transgene transcription was found to decrease by 59.6% at 1-week and 73.4% at 4-weeks relative to control by qRT-PCR (Figure 4A). Analysis of c-mycER^{TAM} protein was performed by colocalization of c-Myc and ERα antigen staining by immunocytochemistry (ICC; Figure 4B, C, D, E). The c-Myc and ERα antigens were always found to colocalize in the nucleus of a proportion of CTX0E03 cells, indicating that the antibodies were detecting the c-mycER^{TAM} target (Figure 4E). Immunoreactive cells were counted and expressed as a proportion of total cells. These data showed control CTX0E03 cell cultures were approximately 52.0% positive for c-mycER^{TAM}; whereas, following EGF, bFGF and 4-OHT withdrawal the number of CTX0E03 cells positive for c-mycER^{TAM} were 42.5% at 1-week and 12.8% at 4-weeks (Figure 4B). Western blot analysis using the c-Myc and ERα antibodies demonstrated that total c-mycER^{TAM} protein levels followed more closely the progressive drop observed by gene expression and were reduced by 53.8% at 1-week and 73.0% at 4-weeks following EGF, bFGF, and 4-OHT withdrawal (Figure 4F, G).

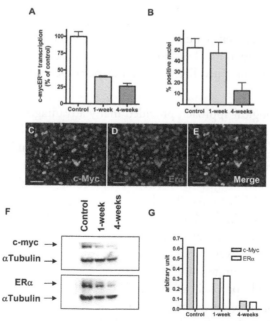

Figure 4. In vitro characterisation of c-mycER^{TAM} transcript and protein expression. CTX0E03 cells were cultured in growth medium (control) and in non-growth promoting medium (in the absence of growth factors and 4-OHT) for 1- and 4-weeks. Evaluation of c-mycER^{TAM} gene transcript and protein levels were performed by qRT-PCR (A), ICC (B to E) and western blot (F). CTX0E03 cell ICC images shown in panels C to E are representative images of the control c-Myc (green), ERα (red) and overlay (Merge). Scale bars represent 50 μm. The western blots in panel F were quantified using densitometry normalised by α-tubulin (G).

Discussion

The c-mycER[TAM] immortalization technology has been successfully utilized to derive a clonal human neural stem cell line from fetal cortical brain. This cell line, referred to as CTX0E03, possess biological attributes which are desirable for clinical applications for stroke therapy. Namely, this cell line can be continuously expanded in vitro, and following implantation in MCAo rodent brain, results in sensorimotor recovery in the absence of cell treatment-related histopathology [19,22]. The in vitro effects of c-mycER[TAM] in the CTX0E03 cell line have been well characterized. Here we investigated the expression of this transgene following implantation of CTX0E03 cells into MCAo rat brain.

In this paper we report a novel method to examine absolute gene expression within human cells implanted into rodent brain. In this example, we analyzed the number of c-mycERTAM transcript copies per CTX0E03 cell both in vitro and in vivo. The calculated c-mycERTAM copies per CTX0E03 cell prepared directly from fresh culture (242.3; SEM = 47.8), cell suspension prior to implantation (215.7; SEM = 13.2) or immediately following implantation (236.9; SEM = 7.8) had comparable values. This validated that the assays employed were CTX0E03 cell specific and the rodent brain background did not cause any interference. Using these established assays, c-mycERTAM transgene transcription was analyzed in the surviving CTX0E03 cell population at 1- and 4-weeks post-implantation in MCAo rodent brain. At both time points there was a substantial reduction in the calculated number of c-mycERTAM transcript copies per CTX0E03 cell detected.

This molecular biological approach provided many advantages over immunohistochemistry and in situ hybridization techniques. The results obtained by RT-PCR and PCR were quantitative. Immunohistochemistry or in situ hybridization techniques do not permit the quantitative assessment of c-mycERTAM levels within a given cell. In addition, the sensitivity of molecular biological techniques were defined, as a single CTX0E03 cell and ten copies of c-mycERTAM per reaction, respectively; whereas, this level of sensitivity would have been difficult, if not impossible, to achieve by immunohistochemistry or in situ hybridization techniques. Our in vivo samples were processed rapidly, immediately after harvesting, thus capturing the most accurate data. Furthermore, from a practical perspective this molecular biological approach analyzed all CTX0E03 cells present in the rodent brain, whereas, histological preparation methods would require a large number of sections to analyze all cells present. Although the average transgene expression significantly decreased in vivo, we cannot exclude the possibility that the expression observed may have arisen from a limited number of CTX0E03 cells. However, the silencing of c-mycERTAM expression observed in vivo is consistent with the absence of tumor formation derived from CTX0E03 cells when implanted in vivo [19,22].

Transgene silencing can occur at a transcriptional or post-transcriptional level. Long-term transcriptional silencing of retrovirus-delivered genes has been shown largely to result from methylation of the cytosine nucleotides within the promoter regions [27]. The factors responsible for DNA methylation is reviewed in detail elsewhere [28]. As the c-mycERTAM transgene was delivered into the CTX0E03 cell line using a retrovirus, it was reasonable to speculate that transgene silencing had occurred through methylation of CpG islands within the transgene promoter region. Bisulfite DNA sequencing was used to determine the methylation status of the CMV promoter region of the c-mycERTAM transgene within CTX0E03 cells. In culture, only 6 of the total CpG sites screened were found to be methylated. In the samples prepared from CTX0E03 grafted MCAo rodent brain 62 to 100% of the sites were methylated. These data unequivocally show that c-mycERTAM expression is silenced at a transcriptional level in vivo as a result of DNA methylation within the CMV transgene promoter region.

The c-mycERTAM technology acts at a protein rather than a transgene expression level. To mitigate the risk of the technology following intracerebral implantation of CTX0E03 cells it was important to investigate if c-mycERTAM transcription level was a predictive measure of the recombinant protein level. Western blot analysis of the in vivo samples proved impossible because the c-mycERTAM protein would be too dilute to detect in whole brain extracts. In addition, IHC analysis of the in vivo samples was impractical because of the substantial number of sections required to perform a global analysis of c-mycERTAM protein expression. To overcome these challenges we developed and in vitro model of c-mycERTAM silencing in CTX0E03 cells. The kinetics of transgene silencing in vitro was slower than that observed in vivo, which provided the ideal cell culture model to directly compare the progressive decline in c-mycERTAM transcription with the translated protein level. Although we observed a decrease in the number of c-mycERTAM positive cells by ICC, the relative percentage of positive cells did not closely follow relative transgene expression levels. These ICC findings further highlight the drawback of histological approaches because the cells were scored positive independent of the quantity of c-mycERTAM protein present within the cell. On the other hand, when total c-mycERTAM protein levels were measured by western blot they were found to decrease in parallel with transgene transcript level over the 1- to 4-week period. This data indicates that the c-mycERTAM transcript and protein levels are closely linked in CTX0E03 cells.

Conclusion

In conclusion, we have demonstrated that c-mycER[TAM] transgene expression levels in CTX0E03 cells were significantly reduced in vitro, upon growth factor and 4-OHT withdrawal, and in vivo following implantation. We demonstrated that in vivo transgene silencing occurred through DNA methylation. In addition, the in vitro results showed that c-mycER[TAM] transcription level was representative of the c-mycER[TAM] protein level. The data demonstrate that uncontrolled in vivo cell growth arising from c-mycER[TAM] technology is unlikely based on both the silencing of the c-mycER[TAM] gene expression reported here and the specific requirement for the 4-OHT ligand to activate the c-mycER[TAM] protein. Furthermore, these data demonstrate that the retroviral insertion of stable transgene constructs into human neural stem cells with the CMV promoter element is not an inherently unsafe procedure for the production of cellular therapies.

Methods

Ethics Statement

Fetal tissue used to derive CTX0E03 cells was obtained in accordance with nationally (UK and/or USA) approved ethical and legal guidelines. All animal work was performed in accordance with the UK Animals (Scientific Procedures) Act (1986) and approved by the local animal Ethical Review Committee.

Derivation of CTX0E03

The CTX0E03 clonal cell line was derived from human fetal brain cortical tissue of 12 weeks' gestation, as described previously [19]. Briefly, a mixed culture was initiated from the dissociated cortex tissue and was infected with an amphotropic replication-incompetent retroviral vector (pLNCX-2; Clontech) encoding the c-mycER[TAM] transgene. The CTX0E03 clonal cell line was isolated using a ring cloning method. The CTX0E03 cells were routinely cultured as a monolayer on mouse laminin (Trevigen, AMS Biotechnology, UK) freshly coated flasks in serum free medium (RMM) containing the growth factor supplements 10 ng/ml bFGF (Peprotech), 20 ng/ml EGF (Peprotech), and 100 nM 4-OHT (Sigma). All experiments were carried out using CTX0E03 cells between passage number 33 to 37, which corresponds to a population doubling level (PDL) ranging between 65 to 75.

MCAo and Cell Grafting

Male Sprague-Dawley rats (Charles River) were group housed with food and water ad libitum. Briefly, animals were subjected to 60 minutes of unilateral MCAo initiated under halothane anesthesia [29]. At 50 minutes into the occlusion, animals were evaluated for behavioral dysfunction (forelimb flexion and contralateral circling behavior). Animals that did not demonstrate dysfunction were removed from the study.

CTX0E03 cells were harvested and suspended at a concentration of ~50,000 cells/µl in HBSS-NAC (HBSS without calcium, magnesium and containing 0.5 mM N-acetyl cysteine; Sigma) and implanted 4-weeks ± 3 days after the occlusion. Each grafted animal received two 4.5 µl bolus injections (approximately 225,000 cells/site with control lesioned animals receiving vehicle only) over 68 sec with 4 min rest before withdrawal. Coordinates from bregma were as follows: Site 1) AP 0.0 mm, L -3.6 mm, V -5.5 mm (from skull surface at bregma); and Site 2) AP 0.0 mm, L +3.6 mm, V -5.5 mm (from skull surface at bregma). Animal brains were collected at 1-week (6 grafted + 2 non-grafted controls) or 4-weeks (6 grafted + 3 non-grafted controls) post-implantation. All animals were administered Cyclosporin A (20 mg/kg, Sandoz Pharmaceuticals) in cremaphore L (Sigma) a day prior to grafting, the day of grafting and then three times weekly for the remaining duration of the study. In addition, Methylprednisolone (Pharmacia Upjohn) was administered at 20 mg/kg for the first 14 days, 10 mg/kg for days 15–17 and 5 mg/kg for days 18–19 post-grafting.

Alu Sequence Assay to Detect and Quantify CTX0E03 Cells

The qPCR Alu assay was carried out using the LightCycler 480 and the Probes Master mix (Roche). Briefly, an average of 250 ng of genomic DNA was amplified using specific primers (Forward-TGAGGCAGGCGAATCGCTTGAA, Reverse-GACGGAGTTTCGCTCTTGTTG) and a FAM-labeled fluorogenic probe (CGCGATCTCGGCTCACTGCAACCTCCATCG) (PrimerDesign) against a conserved region of the human Alu-Sq sequence (Accession number U14573). The PCR was preformed under the following conditions: 95°C for 30 seconds, followed by 45 cycles of 95°C for 15 s, 58°C for 30 s, 72°C for 15 s. Quantification of the human CTX0E03 DNA in rat tissue was based on a standard curve using serial dilutions that ranged from 1 to 10000 CTX0E03 cells, calculated on the assumption that one human cell contains 6.6 pg of gDNA [30]. The standard curve was performed in the presence of 250 ng of rat gDNA.

CTX0E03 c-mycER^TAM Transgene Transcription Assay

Primers were designed to specifically detect the relative c-mycER^TAM transgene transcription level in CTX0E03 cells using the Roche Diagnostics Lightcycler LC480 system. Using SYBR Green Master mix (Roche) and primers (Forward-AAAGGCCCCCAAGGTAGTTA, Reverse- AAGGACAAGGCAGGGCT-ATT), which amplified the c-myc and estrogen receptor (ER) junction, qPCR was performed under the following conditions: 95°C for 5 s, followed by 45 cycles at 95°C for 10 s, 60°C for 20 s, 72°C for 20 s, 82°C for 5 s, measuring the fluorescence. A melt curve analysis was also performed after the assay to check the specificity of the reaction using the following conditions: 95°C for 5 s, 65°C for 60 s (1°C increments) followed by 97°C continuous. In order to carry out absolute quantification of c-mycER^TAM transcription, a known concentration standard was required to plot a standard curve. Based on the assumption that CTX0E03 cell contains one genomic copy of the c-mycER^TAM transgene [19] serial dilutions ranging from 10 to 10,000 copies of c-mycER^TAM were used per reaction to produce the standard curve. The standard curve was performed in the presence of rat cDNA.

Calculation of c-mycER^TAM Transcription Level in Vivo

At 1- or 4-weeks post-implantation, the rat brains were removed, cut into 50–250 mg sections and snap frozen in liquid nitrogen. The gDNA and total RNA was then sequentially isolated from each tissue piece using the AllPrep DNA/RNA mini kit (Qiagen). Four tissue sections were processed per rat brain (i.e. two tissue sections encompassing each injection tract). In brief, the tissue samples were homogenised in the supplied lysis buffer according to sample weight. The lysed sample was processed first through an AllPrep DNA spin column, to isolate the gDNA, and subsequently through an RNeasy spin column to selectively isolate RNA. The gDNA and total RNA samples (1 μl/reaction) were then analyzed by qPCR for Alu signal and qRT-PCR for c-mycERTAM, respectively. Based on the Alu signal, the number of CTX0E03 cells present per reaction was calculated from the standard curve. Similarly, the number of c-mycER^TAM transcripts present per reaction was interpolated from the c-mycER^TAM copy number standard curve. A calculation was then carried out to correct for the amount analyzed and the total volume of tissue sample lysate (for example, 1 μl from a total of 650 μl). The corrected figures were then expressed as the number of c-mycER^TAM transcripts present over the number of CTX0E03 cells present in each tissue section. Standard error of the mean was calculated for all tissue sections found to contain cells (Microsoft Excel version 2003).

Transgene Methylation Analysis by Bisulfite Sequencing

Bisulfite treatment of DNA followed by sequencing was used to determine the endogenous transgene promoter pattern of methylation in CTX0E03 cells. Methylation is implicated in repression of genetic activity and involves the addition of a methyl group to the carbon-5 position of cytosine residues of the dinucleotide CpG. Bisulfite sequencing is a technique used to identify specific gDNA methylation sites based on the fact that treatment of DNA with bisulfite converts cytosine residues to thymidine, but leaves 5-methylcytosine residues unaffected. Bisulfite treatment thus introduces a specific change in the DNA sequence which is dependent on the methylation status of individual cytosine residues. A bisulfite reaction was performed using Epitect Bisulfite (Qiagen). Up to 2 µg of genomic DNA was used for conversion with the bisulfite reagent. Bisulfite-converted DNA was used as a template for PCR amplification. PCR primers were designed using the web program METHPRIMER ([31]; http://www.urogene.org/methprimer/index1.html webcite). Primers used were: Forward – GGGAGTTTGTTTTGGTAT-TAAAATTAA and Reverse – ACTACTACTAATAAAAATTCTCCTCCTC. PCR conditions were 94°C for 5 minutes, 30 cycles of 94°C for 30 seconds, 60°C for 1 min, and 72°C for 30 seconds followed by 10 min at 72°C. PCR fragments were cloned into TA-cloning vector using TOPO TA Cloning Kit for Sequencing (Invitrogen). Ten individual clones were sequenced by GATC Biotech (Germany) using the T3 universal PCR primers. The resulting sequences were analyzed using blast software http://blast.ncbi.nlm.nih.gov/Blast.cgi. Sequence analysis that identified an unchanged cytosine base (i.e. protected from the bisulfite conversion reaction) was considered to be methylated, but a cytosine conversion to thymidine in the sequence indicated no methylation.

In Vitro Characterisation: Western Blot and Immunocytochemical Analysis

In vitro, CTX0E03 cell samples were prepared by maintaining the cells in non-growth permitting medium alone, without the addition of EGF, bFGF, and 4-OHT, for 1- and 4-weeks. Western blot analysis for c-mycER[TAM] protein in CTX0E03 cells was carried out using the whole-cell lysate. To prepare cell lysate, CTX0E03 cell monolayers in a T75 flask were rinsed with cold PBS (4°C), lysed with 750 µl 1× SDS Sample Buffer and dithiothreitol (DTT) Reducing Agent (PAGEgel Inc, AMS Biotechnology, UK), and collected in a 1.5 ml centrifuge tube. Samples were triturated multiple times to lyse and shear the sample, heated at 100°C for 2 min and centrifuged. Cell lysates were loaded onto a 4%–20% Tris-Tricine gel (PAGEgel Inc, AMS Biotechnology, UK), 10 µl per well. After electrophoresis, the protein was transferred onto a nitrocellulose membrane with

0.2 μm pore size by electroelution. Immunodetection was performed with a rabbit anti-ERα polyclonal or a mouse anti-c-Myc monoclonal antibody (Santa Cruz Biotechnology Inc, both used at 1:100). Transgene product was detected using horseradish peroxidase-conjugated anti rabbit-IgG (Cell Signaling Technology, 1:2000) or anti mouse-IgG (Pierce Biotechnology Inc., 1:800) secondary antibodies, as appropriate. The nitrocellulose membrane was then processed using chemiluminescence detection reagents (Thermo scientific). The blots were then stripped and reprobed using anti-α-tubulin (Sigma, 1:1000) to act as an internal loading level standard. Western blot images were captured using BioRad FluorS Imaging Unit and densitometry carried out using Scion Image software (Scion Corporation, version Alpha 4.0.3.2, USA).

Immunocytochemistry was carried out on CTX0E03 cells fixed in a 96-well plate (BD) using 4% PFA/4% sucrose in PBS for 15 min after 3 days growth (control) and following 1- and 4-weeks in medium containing no growth factors (EGF, bFGF) or 4-OHT. Anti-c-myc (1:100) and anti-ERα (1:500) were incubated in 1xPBS containing 1% normal goat serum (Vector laboratories) overnight at room temperature. After rinsing, the CTX0E03 cells were incubated for 1.5 h at room temperature using anti-mouse conjugated Alexa Fluor 488 (Invitrogen, 1:2000) for c-myc and anti-rabbit conjugated Alexa Fluor 568 (Invitrogen, 1:2500) for ERα. CTX0E03 cells were then counterstained by incubating with Hoechst 33342 (Sigma, 1 μM) for 2 min. Quantification of c-myc and ERα was carried out by manual counting of three representative fields using an Olympus IX70 fluorescent microscope.

Authors' Contributions

EM, LS, AH and JS conceived the experiments. LS, EM, RC, JH and PS designed and performed the experiments and analyzed the data. All authors contributed to the preparation of the written manuscript.

Acknowledgements

We thank Prof. Jack Price (King's College London, United Kingdom) for his general support and helpful discussions. In addition, our gratitude extends to Dr. Aaron Jeffries (King's College London, United Kingdom) for advice and guidance on DNA methylation analysis; and to PrimerDesign for their help in developing the Alu primers. We also extend our thanks to Susan Hines (Imperial College London, United Kingdom) and Dr. Marc Oliver Baradez (Laboratory of the Government Chemist, LGC, United Kingdom) for their kind assistance in the preparation of the cells used in this study. This study was supported and designed by ReNeuron (RENE.L) who where involved in study design, data collection, analysis, and interpretation, as well as manuscript preparation and decision to submit for publication.

References

1. Wong AM, Hodges H, Horsburgh K: Neural stem cell grafts reduce the extent of neuronal damage in a mouse model of global ischaemia. Brain Res 2005, 1063(2):140–150.

2. Hodges H, Nelson A, Virley D, Kershaw TR, Sinden JD: Cognitive deficits induced by global cerebral ischaemia: prospects for transplant therapy. Pharmacol Biochem Behav 1997, 56(4):763–780.

3. Sinden JD, Rashid-Doubell F, Kershaw TR, Nelson A, Chadwick A, Jat PS, Noble MD, Hodges H, Gray JA: Recovery of spatial learning by grafts of a conditionally immortalized hippocampal neuroepithelial cell line into the ischaemia-lesioned hippocampus. Neuroscience 1997, 81(3):599–608.

4. Modo M, Stroemer RP, Tang E, Patel S, Hodges H: Effects of implantation site of stem cell grafts on behavioral recovery from stroke damage. Stroke 2002, 33(9):2270–2278.

5. Veizovic T, Beech JS, Stroemer RP, Watson WP, Hodges H: Resolution of stroke deficits following contralateral grafts of conditionally immortal neuroepithelial stem cells. Stroke 2001, 32(4):1012–1019.

6. De Filippis L, Ferrari D, Rota Nodari L, Amati B, Snyder E, Vescovi AL: Immortalization of human neural stem cells with the c-myc mutant T58A. PLoS ONE 2008, 3(10):e3310.

7. Bai Y, Hu Q, Li X, Wang Y, Lin C, Shen L, Li L: Telomerase immortalization of human neural progenitor cells. Neuroreport 2004, 15(2):245–249.

8. Villa A, Navarro-Galve B, Bueno C, Franco S, Blasco MA, Martinez-Serrano A: Long-term molecular and cellular stability of human neural stem cell lines. Exp Cell Res 2004, 294(2):559–570.

9. Donato R, Miljan EA, Hines SJ, Aouabdi S, Pollock K, Patel S, Edwards FA, Sinden JD: Differential development of neuronal physiological responsiveness in two human neural stem cell lines. BMC Neurosci 2007, 8:36.

10. De Filippis L, Lamorte G, Snyder EY, Malgaroli A, Vescovi AL: A novel, immortal, and multipotent human neural stem cell line generating functional neurons and oligodendrocytes. Stem Cells 2007, 25(9):2312–2321.

11. Cho T, Bae JH, Choi HB, Kim SS, McLarnon JG, Suh-Kim H, Kim SU, Min CK: Human neural stem cells: electrophysiological properties of voltage-gated ion channels. Neuroreport 2002, 13(11):1447–1452.

12. Ryder EF, Snyder EY, Cepko CL: Establishment and characterization of multipotent neural cell lines using retrovirus vector-mediated oncogene transfer. J Neurobiol 1990, 21(2):356–375.

13. Snyder EY, Deitcher DL, Walsh C, Arnold-Aldea S, Hartwieg EA, Cepko CL: Multipotent neural cell lines can engraft and participate in development of mouse cerebellum. Cell 1992, 68(1):33–51.

14. Rao MS, Anderson DJ: Immortalization and controlled in vitro differentiation of murine multipotent neural crest stem cells. J Neurobiol 1997, 32(7):722–746.

15. Jung M, Kramer EM, Muller T, Antonicek H, Trotter J: Novel pluripotential neural progenitor lines exhibiting rapid controlled differentiation to neurotransmitter receptor-expressing neurons and glia. Eur J Neurosci 1998, 10(10):3246–3256.

16. Villa A, Snyder EY, Vescovi A, Martinez-Serrano A: Establishment and properties of a growth factor-dependent, perpetual neural stem cell line from the human CNS. Exp Neurol 2000, 161(1):67–84.

17. Nakafuku M, Nakamura S: Establishment and characterization of a multipotential neural cell line that can conditionally generate neurons, astrocytes, and oligodendrocytes in vitro. J Neurosci Res 1995, 41(2):153–168.

18. Miljan EA, Hines SJ, Pande P, Corteling RL, Hicks C, Zbarsky V, Umachandran M, Sowinski P, Richardson S, Tang E, et al.: Implantation of c-mycER-TAM immortalized human mesencephalic derived clonal cell lines ameliorates behaviour dysfunction in a rat model of Parkinson disease. Stem Cells Dev 2008, 18(2):307–319.

19. Pollock K, Stroemer P, Patel S, Stevanato L, Hope A, Miljan E, Dong Z, Hodges H, Price J, Sinden JD: A conditionally immortal clonal stem cell line from human cortical neuroepithelium for the treatment of ischemic stroke. Exp Neurol 2006, 199(1):143–155.

20. Maurer J, Fuchs S, Jager R, Kurz B, Sommer L, Schorle H: Establishment and controlled differentiation of neural crest stem cell lines using conditional transgenesis. Differentiation 2007, 75(7):580–591.

21. Littlewood TD, Hancock DC, Danielian PS, Parker MG, Evan GI: A modified oestrogen receptor ligand-binding domain as an improved switch for the regulation of heterologous proteins. Nucleic Acids Res 1995, 23(10):1686–1690.

22. Stroemer P, Patel S, Hope A, Oliveira C, Pollock K, Sinden J: The Neural Stem Cell Line CTX0E03 Promotes Behavioral Recovery and Endogenous Neurogenesis after Experimental Stroke in a Dose-Dependent Fashion. Neurorehab Neural Repair, in press.

23. Zhang F, Thornhill SI, Howe SJ, Ulaganathan M, Schambach A, Sinclair J, Kinnon C, Gaspar HB, Antoniou M, Thrasher AJ: Lentiviral vectors containing an enhancer-less ubiquitously acting chromatin opening element (UCOE) provide highly reproducible and stable transgene expression in hematopoietic cells. Blood 2007, 110(5):1448–1457.

24. Lee HJ, Kim KS, Kim EJ, Choi HB, Lee KH, Park IH, Ko Y, Jeong SW, Kim SU: Brain transplantation of immortalized human neural stem cells promotes functional recovery in mouse intracerebral hemorrhage stroke model. Stem Cells 2007, 25(5):1204–1212.

25. Nicklas JA, Buel E: Development of an Alu-based, real-time PCR method for quantitation of human DNA in forensic samples. J Forensic Sci 2003, 48(5):936–944.

26. Schneider T, Osl F, Friess T, Stockinger H, Scheuer WV: Quantification of human Alu sequences by real-time PCR – an improved method to measure therapeutic efficacy of anti-metastatic drugs in human xenotransplants. Clin Exp Metastasis 2002, 19(7):571–582.

27. Lavie L, Kitova M, Maldener E, Meese E, Mayer J: CpG methylation directly regulates transcriptional activity of the human endogenous retrovirus family HERV-K(HML-2). J Virol 2005, 79(2):876–883.

28. Strathdee G, Brown R: Aberrant DNA methylation in cancer: potential clinical interventions. Expert Rev Mol Med 2002, 4(4):1–17.

29. Longa EZ, Weinstein PR, Carlson S, Cummins R: Reversible middle cerebral artery occlusion without craniectomy in rats. Stroke 1989, 20(1):84–91.

30. Applied_Biosystems: Creating standard curves with genomic DNA or plasmid DNA templates for use in quantitative PCR. [http://www.appliedbiosystems.com/support/tutorials/pdf/quant_pcr.pdf]

31. Li LC, Dahiya R: MethPrimer: designing primers for methylation PCRs. Bioinformatics 2002, 18(11): 1427–1431.

32. Schroeder A, Mueller O, Stocker S, Salowsky R, Leiber M, Gassmann M, Lightfoot S, Menzel W, Granzow M, Ragg T: The RIN: an RNA integrity number for assigning integrity values to RNA measurements. BMC Mol Biol 2006, 7:3.

Evaluation of the Sensitization Rates and Identification of IgE-Binding Components in Wild and Genetically Modified Potatoes in Patients with Allergic Disorders

Soo-Keol Lee, Young-Min Ye, Sung-Ho Yoon, Bou-Oung Lee, Seung-Hyun Kim and Hae-Sim Park

ABSTRACT

Background

The potato is one of the most common types of genetically modified (GM) food. However, there are no published data evaluating the impact of genetic manipulations on the allergenicity of GM potatoes. To compare the

allergenicity of GM potatoes with that of wild-type potatoes using in vivo and in vitro methods in adult allergy patients sensitized to potatoes.

Methods

A total of 1886 patients with various allergic diseases and 38 healthy controls participated in the study. Skin-prick testing and IgE-ELISA were carried out with extracts prepared from wild-type and GM potatoes. An ELISA inhibition test was used to confirm the binding specificity. IgE-binding components in extracts from the two types of potato were identified by SDS-PAGE and IgE-immunoblotting. The effects of digestive enzymes and heat on the allergenicity of the extracts was evaluated by preincubating the potatoes with or without simulated gastric and intestinal fluids in the absence or presence of heat.

Results

Positive responses (ratio of the wheal size induced by the allergen to that induced by histamine (A/H) ≥ 2+) to wild-type or GM potato extracts, as demonstrated by the skin-prick test, were observed in 108 patients (5.7%). Serum-specific IgE was detected in 0–88% of subjects who tested positively. ELISA inhibition tests indicated significant inhibition when extract from each type of potato was added. IgE-immunoblot analysis demonstrated the presence of 14 IgE-binding components within the wild-type potato and 9 within the GM potato. Furthermore, a common 45-kDa binding component that yielded similar IgE-binding patterns was noted in more than 80% of the reactions using sera from patients sensitized to wild-type or GM potato. Exposure to simulated gastric fluid and heat treatment similarly inhibited IgE binding by extracts from wild-type and GM potatoes, whereas minimal changes were obtained following exposure of the extracts to simulated intestinal fluid.

Conclusion

Our results strongly suggest that genetic manipulation of potatoes does not increase their allergenic risk. The sensitization rate of adult allergy patients to both types of extract was 5.7%, and a common major allergen (45 kDa) was identified.

Background

Food-induced allergic reactions are responsible for a variety of symptoms involving the skin, gastrointestinal tract, and respiratory tract, and proceed through IgE- and non-IgE-mediated mechanisms. The major foods responsible for food-induced allergic reactions are milk, eggs, peanuts, fish, and tree nuts in children, and peanuts, tree nuts, fish, and shellfish in adults [1]. There are some reports in the literature on allergic reactions to potatoes; for example, a child developed urticaria and angioedema after eating a potato [2,3], bronchial asthma was induced in an individual while peeling a raw potato [4], and anaphylaxis developed in response to raw potatoes [5]. Furthermore, potatoes have been found to cross-react with birch pollen, fruits, and latex [6,7].

Agricultural biotechnology has tremendous implications for both agriculture and the general public. Insect-resistant corn and herbicide-tolerant soybeans are grown on 30–50% of the total acreage planted with these crops in North America [8]. Previous studies comparing the allergenicity of wild-type and genetically modified (GM) corn demonstrated that the allergic risk was not increased after genetic manipulation [9-11]. In Korea, potato, soybean, and corn are the most commonly exposed GM foods; however, to date, there are no reports on the allergenic risk of GM potatoes.

In this study, the sensitization rates of adult allergy patients in response to wild-type and GM potatoes were evaluated by skin-prick test and ELISA (enzyme linked immunosorbent assay). SDS-PAGE (sodium dodecyl sulfate-polyacrylamide gel electrophoresis) and IgE-immunoblotting were carried out to identify the major allergens present in the potato extracts. To evaluate the effects of digestive enzymes and heat on the allergenicity of the two types of potato, the extracts were preincubated with or without simulated gastric and intestinal fluids in the presence or absence of heat.

Methods

Subjects

Sensitization rates to the two potato extracts were evaluated in 1886 allergy patients and in 38 healthy non-atopic subjects. The participants, who ranged in age from 15 to 65 years, were enrolled in the study by the Department of Allergy and Rheumatology, Ajou University School of Medicine, Suwon, Korea. The GM potato, carrying the neomycin phosphotransferase II (NPTII) and phosphinothricin acetyltransferase (PAT) genes (Table 1), was provided by ChonBuk National University, Chunju, Korea. The wild-type potato was produced in Korea. From January 2004 to October 2004, 1886 patients admitted to the hospital for the treatment of various allergic diseases, including asthma, allergic rhinitis, and food and drug allergy, were skin-prick tested with common inhalant allergens and with extracts from GM and wild-type potatoes. In the skin-prick tests, 50 common inhalant allergens, 30 food allergens, and the potato extracts were applied using 26-gauge needles to the backs of the patients. The results were read 15 min later. A positive reaction was defined as a mean wheal diameter of ≥3 mm. The size of the wheal produced by each antigen or by the positive control histamine was expressed in terms of maximum diameter and vertical length at the midportion of the maximal length. Skin reactivity was expressed as the ratio of the wheal size induced by the allergen and that induced by histamine (A/H). The data were recorded in accordance with the recommendations of the Standardization Committee of the Northern Society of Allergology [12]. An A/H of 0.1–1 and an erythema diameter of <21 mm was assigned a reactivity of 1+. An A/H of 0.1–1 and erythema >21 mm was assigned a reactivity of 2+. An A/H in the range of 1–2 was recorded as a reactivity of 3+. For an A/H of 2–3, reactivity was 4+, and for an A/H >3 reactivity was 5+. A positive responder was defined as a subject who demonstrated a response >2+ on the skin-prick test. This study was reviewed by the institutional review board of Ajou University Medical Center, Suwon, Korea.

Table 1. Primer sequences of inserted PAT and NPT genes.

	Primer sequence
PAT 2587U	5'-TCG TCA ACC ACT ACA TCG AGA-3'
PAT 2893L	5'-ATG ACA GCG ACC ACG CTC TT-3'
NBT 3161U	5'-AGA AAG TAT CCA TCA TGG CTG A-3'
NBT 3574L	5'-ATA CCG TAA AGC ACG AGG AAG-3'

Preparation of Extracts from Wild-Type and GM Potatoes

Extracts were prepared from wild-type and GM potatoes using phosphate-buffered saline (PBS; pH 7.5, 1:10 w/v), kept at 4°C overnight, and then centrifuged at 12,000–15,000 rpm for 20 min. The supernatant was dialyzed (the cutoff molecular weight was 6000 Da; Spectrum Medical Industries, Houston, TX, USA) against 4 l of PBS at 4°C for 72 h, and the resultant fluid was stored at -20°C until ELISA, ELISA inhibition testing, and immunoblot analyses were carried out. For the skin-prick test, extract was mixed with sterile glycerin at a ratio of 1:1. To evaluate the effects of digestive enzymes, simulated gastric fluid (SGF; 471 U pepsin/mg; Sigma Chemical Co., St. Louis, MO, USA) and simulated intestinal fluid (SIF; pancreatin; Sigma) were preincubated with the extracts in the presence or absence of heat.

ELISA for Specific IgE Antibodies to Potato Extracts

The presence of specific IgE antibodies to the two types of extract was determined by ELISA, as previously described [13]. Microtiter plates (Corning, NY) were coated with 100 µl of extract (GM or wild-type, 10 µg/ml)/well and kept overnight at 4°C. Each well was washed three times with 0.05% PBS-Tween (PBST), and the remaining binding sites were blocked by incubation with 10% fetal bovine serum (FBS)-PBS for 1 h at room temperature. The wells were then incubated for 2 h at room temperature with either 50 µl of the patients' sera or the control sera, both at 50% dilution. The control sera were from the 38 healthy controls who had tested negative in the skin-prick tests to the common inhalant and food allergens and to potatoes. The wells were washed three times with PBST, 1:1000 v/v biotin-labeled goat anti-human IgE antibody (Vector Lab, Burlingame, CA, USA) was added to each well, and the plates were incubated for 1 h at room temperature. After another washing, 100 µl of 1:1000 v/v streptavidin-peroxidase (Sigma) was added, the plates were incubated for 30 min, and the wells were washed again.

The colorimetric reaction was developed with a TMB (3,3',5,5'-tetramethyl-benzidine) substrate solution for 15 min at room temperature. The reaction was stopped by the addition of 100 µl of 2 N sulfuric acid, and absorbance was read at 450 nm using an automated microplate reader (Benchmark; Bio-Rad Laboratories, Hercules, CA, USA). All assays were carried out in duplicate. Positive cutoff values were determined from the mean plus two standard deviations of the absorbance values of the 38 healthy controls.

ELISA Inhibition Test

The specificity of IgE binding to the potato extracts was tested and the allergenic potency of the GM and wild-type potato extracts was analyzed by a competitive ELISA inhibition test. Sera from four patients with high levels of specific IgE binding were pooled, preincubated overnight at 4°C with five concentrations (1–100 µg/ml) of either Dermatophagoides pteronyssinus or extract from GM or wild-type potatoes, and then incubated for 12 h in microtiter plates coated with extracts from the two types of potato. The same steps were followed as in the ELISA. As a control, samples were preincubated with equal volumes of PBS (pH 7.5) instead of house dust mite or potato extracts. The percentage of inhibition of serum IgE binding was expressed as: 100 - (absorbance of the samples preincubated with allergens/absorbance of the samples preincubated with PBS) × 100.

SDS-PAGE and Immunoblot Analysis

SDS-PAGE and immunoblot analysis were carried out under reducing conditions according to previously described methods [13]. Extracts from the two potatoes were mixed with sample buffer (Tris-HCl 31 mmol/l, 10% glycerol,

1% SDS, 0.0025% bromophenol blue, 2.5% β-mercaptoethanol, pH 6.8) and heated in boiling water for 5 min. Standard markers (4–250 kDa; Novex, San Diego, CA, USA) and the extracts were loaded on a 4–20% Tris-glycine gel (Novex). Electrophoresis was done with a Novex X cell™ Mini cell for 90 min at 125 V. The gel was fixed and stained with Coomassie brilliant blue. For immunoblotting, proteins were transferred onto a polyvinylidene difluoride membrane (PVDF; Millipore Co., Bedford, MA, USA) in transfer buffer (Tris-base 25 mmol/l, glycine 193 mmol/l, and methanol 20%) using a Bio-Rad transfer apparatus set at 200 mA for 90 min. The blotted PVDF membrane was sliced into 4-mm widths and then incubated in 5% skim milk in Tris-buffered saline (TBS)-Tween (TBST) for 1 h to block nonspecific binding to the membranes. Each membrane slice was incubated overnight at 4°C with patient or control sera diluted 1:2 v/v with 3% skim milk in TBST, and then washed with 0.1% TBST for 30 min. Subsequently, the membranes were incubated with goat anti-human IgE conjugated with alkaline phosphatase (Sigma) for 1 h at room temperature, washed with TBST, and then developed with BCIP/NBT alkaline phosphatase substrate (Sigma).

Effect of Simulated Gastric and Intestinal Fluid on the IgE-Binding Components

Crude extracts were prepared from GM and wild-type potatoes and then heated at 100°C for 5 min. The intrinsic digestibility of the extracts and the digestibility of extracts preheated and incubated in SGF or SIF were examined as previously described [14]. Briefly, SGF digestibility was analyzed by dissolving 680 µg of naïve crude extract or preheated extract in 200 µl of pre-warmed 100 mmol HCl/l (pH 1.2) and 30 mmol NaCl/l containing 0.32% (w/v) pepsin A (Sigma). Digestion was conducted at 37°C with continuous shaking, and aliquots of the digested solution (20 µl) were withdrawn at 0, 0.5, and 60 min. These aliquots were quickly mixed with 26 µl of sample buffer (containing 2.5% 2-mercaptoethanol and 1% SDS) and 6.0 µl of Na_2CO_3 (200 mmol/l). The mixture was then boiled for 5 min and stored at -20°C until further analysis. To evaluate the effects of SIF, 680 µg of the naïve crude extract or the preheated extract were dissolved in 260 µl of prewarmed intestinal control solution (0.05 M KH_2SO_4, pH 6.8) containing 1.0% (w/v) pancreatin (pancreatin USP; Sigma). This solution was incubated at 37°C with continuous shaking. Aliquots (26 µl) were withdrawn at 1, 90, and 240 min, mixed with 26 µl of sample buffer (containing 2.5% 2-mercaptoethanol and 1% SDS), and then boiled for 5 min. SDS-PAGE (12%) and IgE-immunoblot analysis were then carried out as described above.

Results

Allergy Skin-Prick Test and Specific IgE to Wild-Type and GM Potatoes

During the one-year study, 108 (5.7%) of the 1886 patients had an A/H score >2+ in response to the skin-prick test using extracts from GM and wild-type potatoes. All of the subjects who reacted positively to the GM potato also had positive responses to the wild-type potato, as seen in Table 2. House-dust mites are the most common aeroallergen in Korea, followed by weed, tree, and grass pollens [15], but, based on allergy skin-prick testing, no specific antigen showing cross-reactivity with potato proteins has been identified. ELISA using sera from subjects with a positive skin prick test showed that the prevalence of specific IgE to extracts of wild-type and GM potatoes ranged from 0–88%, with 57 and 58% (13 of 23 and 11 of 19, respectively) having a 2+ reaction, 64 and 58% (18 of 28 and 18 of 31, respectively) with a 3+ reaction, 63 and 88% (5 of 8 and 7 of 8, respectively) having a 4+ reaction, and 50 and 60% (2 of 4 and 1 of 2, respectively) with a 5+ reaction, 0 and 50% (0 and 1 of 2, respectively) with a 6+ reaction. (Table 2, Fig. 1).

Table 2. Specific IgE binding, stratified by the results of skin-prick testing, to extracts of wild-type and GM potatoes.

A/H ratio		Wild-type potato					GM potato				
		2+	3+	4+	5+	6+	2+	3+	4+	5+	6+
Skin-prick test result n (%)		38 (35.2)	48 (44.5)	12 (11.1)	7 (6.4)	3 (2.8)	36 (33.3)	51 (47.2)	13 (12)	6 (5.6)	2 (1.9)
sIgE (%) to	Wild	13/23 (57)	18/28 (64)	5/8 (63)	2/4 (50)	0/2 (0)	8/19 (42)	17/31 (55)	6/8 (75)	1/5 (20)	0/2 (0)
	GM	10/23 (44)	16/28 (57)	5/8 (63)	1/4 (25)	0/2 (0)	11/19 (58)	18/31 (58)	7/8 (88)	3/5 (60)	1/2 (50)

Positive skin-prick test response (≥2+): 108/1886 for wild-type and genetically modified (GM) potato. A/H ratio: the ratio of the size of the wheal induced by allergen on allergic skin-prick test to that induced by histamine; N: number of subjects positive to skin-prick test for wild-type and GM potato; sIgE: prevalence of serum specific IgE antibody to wild-type and GM potato (number of subjects with positive serum specific IgE/number of subjects tested in each response group according to the skin-prick test using potato extract).

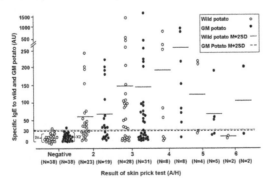

Figure 1. Specific IgE binding to extracts of wild-type and GM potatoes, as determined by ELISA, vs. the A/H ratio of the potato. A/H: the ratio of the size of the wheal induced by allergen on allergic skin-prick test to that induced by histamine. N: number of subjects in each response group according to the skin-prick test using potato extract. Positive cutoff values were determined from the mean plus two standard deviations (M + 2SD) of the absorbance values of the 38 healthy controls.

ELISA Inhibition Test

The ELISA inhibition test showed significant dose-dependent inhibition for wild-type and GM potato extracts. In addition, the two extracts had similar potencies. By contrast, minimal inhibition was noted using D. pteronyssinus. Figure 2 shows the ELISA inhibition test by the addition of extracts from wild-type and GM potato and Dermatophagoides pteronyssinus. The same pooled sera from patients sensitized to potatoes were used (A: wild-type potato; B: GM potato).

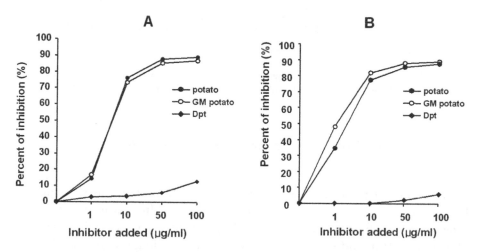

Figure 2. Percent inhibition of IgE-ELISA by the addition of: extract from wild-type potato (●), GM potato extract (O), and Dermatophagoides pteronyssinus (♦). Sera from patients sensitized to potatoes were used. A: wild-type potato; B: GM potato.

SDS-PAGE and IgE-Immunoblot Analysis

The IgE-binding components within the wild-type and GM potatoes were compared using the sera of eight individuals with high specific IgE levels and sera pooled from the controls. The latter was derived from 10 patients who responded negatively to the two potato extracts on the skin-prick test. Three components (45, 34, and 26 kDa) were noted in >50% and 11 components (78, 72, 64, 36, 35, 25, 23, 22, 20, 19, and 14 kDa) in 33% of patients sensitized to wild-type potato. One component (45 kDa) was noted in 88% and eight components (78, 64, 35, 26, 25, 23, 22, and 14 kDa) in <50% of patients sensitized to GM potato. Thus, the 45-kDa band present in serum was the most frequently bound (>80%) by extracts from wild-type and GM potatoes (Fig. 3).

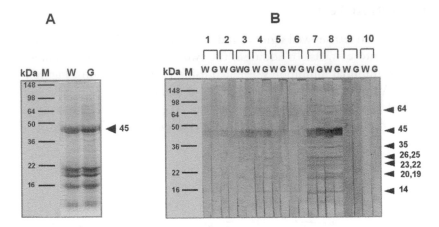

Figure 3. SDS-PAGE of proteins (A) and IgE-binding components (B) present in wild-type and GM potato extracts, as measured in sensitized patients. W: wild-type potato extracts; G: GM potato extracts. M: marker; lanes 1–8: sensitized subjects; lane 9: normal control; lane 10: buffer control.

Effect of Digestive Enzymes on Extracts of GM and Wild-Type Potatoes

Figures 4, 5, and 6 demonstrate the changes both in the proteins, as seen on SDS-PAGE, and in the IgE-binding components of extracts from wild-type and GM potatoes after SGF or SIF treatments in the presence or absence of heating. A combination of SGF and heat treatment resulted in the disappearance of the protein bands and the IgE-binding components in extracts from wild-type and GM potatoes, while minimal changes were noted with SIF treatment alone.

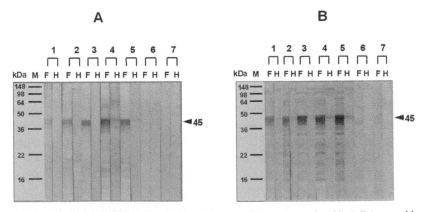

Figure 4. Effect of boiling on wild-type (A) and GM (B) potato allergenicity, analyzed by IgE-immunoblotting. F: Fresh potato extract; H: heated extract. Lanes 1–5: sensitized subjects; lanes 6 and 7: normal control; lane 8: buffer control.

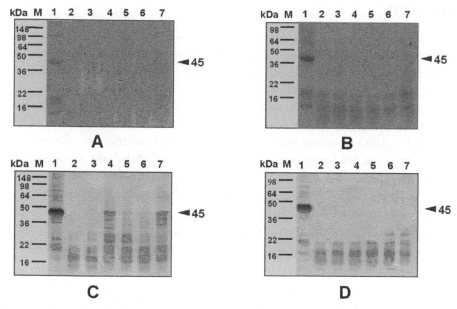

Figure 5. Effect of simulated gastric fluid (SGF) treatment on extracts from wild-type and GM potatoes, as determined by SDS-PAGE (A, B), and on IgE-binding components present in the extracts (C, D). A, C: extracts from wild-type potato; B, D: extracts from GM potato; M: marker; lanes 1–7: incubation for 0 s, 30 s, 1 min, 10 min, 30 min, 1 h, and 1.5 h with sera pooled from five patients. pH of SGF: ca. 1.2.

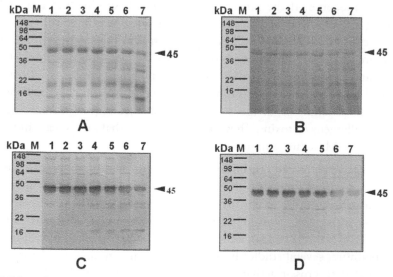

Figure 6. Effect of simulated intestinal fluid (SIF) treatment on extracts from wild-type and GM potatoes, as determined by SDS-PAGE (A, B), and on IgE-binding components present in the extracts (C, D). A, C: extracts from wild-type potato; B, D: extracts from GM potato; M: marker; lanes 1–7: incubation for 0 s, 30 s, 1 min, 10 min, 30 min, 1 h, and 4 h. pH of SIF: ca. 7.5.

Discussion

The results of this study demonstrate that the prevalence of positive skin-prick tests to extracts from wild-type and GM potatoes was 5.7% in our population of adult allergy patients. Serum-specific IgE antibodies were detected among the positive responders. The 45-kDa allergenic band was the most frequently bound (>80%) by the two types of extract, and ELISA inhibition studies suggested that they had similar potencies. Recent investigations to evaluate the allergic risks of GM corn and soybean demonstrated that allergenicity did not increase after genetic manipulation of wild-type corn [11] and soybean [9]. Similarly, our study showed that the genetic manipulation of potatoes, which are one of the most common sources of food allergens in Korea, did not enhance their allergenic risk, as evaluated using in vivo and in vitro methods.

GM crops currently on the market have been thoroughly assessed for safety according to the guidelines developed by the World Health Organization [16] and the Food and Agriculture Organization of the United Nations [17]. In addition, the potential allergenicity of newly introduced proteins must be assessed in all foods produced through agricultural biotechnology, and the FAO [18] and WHO have developed a rigorous approach for this assessment. The NPTII gene introduced in the potato used in this study encodes an enzyme that confers resistance to aminoglycoside antibiotics and was isolated from the prokaryotic transposon Tn5 [19]. The PAT gene was obtained from the aerobic soil bacterium Streptomyces viridochromogenes. If the gene source is bacterial, specific and targeted serum screenings are not necessary because bacterial proteins are rarely allergenic, due to the low exposure levels and lack of allergic sensitization to these proteins [8]. Furthermore, previous studies have confirmed that ingestion of genetically engineered plants expressing NPTII does not pose any safety concerns [20,21]. Herouet et al. [22] found that PAT proteins do not possess the characteristics associated with food toxins or allergens, i.e., they have no sequence homology with any known allergens or toxins. These findings suggest that an increase in the allergenic risk of a GM potato is unlikely.

A few studies have identified IgE-binding components within potato extracts. Wahl et al. [23] reported four (16, 30, 45, and 65) IgE-binding components, which were detected by immunoblotting using sera from 12 patients with IgE-mediated hypersensitivity reactions to potatoes. Four major potato allergens, Sol t 1 (43 kDa), 2 (21 kDa), 3 (21 kDa), and 4 (16 kDa), were identified in children [24,25]. Furthermore, several studies have shown that allergy to natural rubber latex is associated with cross-reactivity to potatoes and tomatoes [7,26]. In our study of the Korean population, only one allergenic protein (45 kDa) could be identified as the major allergen, and it was present in extracts from wild-type and

GM potatoes. Moreover, this protein may be the same one identified in a previous investigation [23].

Proteolytic stability is a useful criterion in assessing the allergenic potential of food allergens. The FAO/WHO [18] decision-tree approach advocates the use of resistance to proteolysis with pepsin as a comparative measure of digestive stability for proteins introduced into food through agricultural biotechnology. In the current study, we demonstrated that SGF and heat treatment substantially suppressed the activity of IgE-binding components in wild-type and GM potato extracts, while minimal changes were noted with SIF treatment alone.

Conclusion

We report that the sensitization rate of patients to wild-type and GM potato extracts was 5.7%, and the prevalence of specific IgE to these extracts in the sera of subjects with a positive skin-prick test was similar. An ELISA inhibition test showed significant dose-dependent inhibition by GM and wild-type potato extracts, and the two extracts had similar potencies. A 45-kDa band was identified as the most frequently bound by both extracts. Our results strongly suggest that genetic manipulation of potatoes using antibiotic resistance and herbicide tolerance genes does not increase their allergenic risk.

Abbreviations

GM = genetically modified

ELISA = enzyme linked immunosorbent assay

SDS-PAGE = sodium dodecyl sulfate-polyacrylamide gel electrophoresis

NPTII = neomycin phosphotransferase II

PAT = phosphinothricin acetyltransferase

SGF = simulated gastric fluid

SIF = simulated intestinal fluid

Authors' Contributions

SKL: ideas, study design and writing

YMY: ideas and writing

SHY: laboratory work

BOL: preparation of GMO and non -GMO potato

SHK: study design and laboratory work

HSP: ideas, study design, laboratory work and writing

Acknowledgements

This study was supported by ARPC (2004070-300). The English in this document has been checked by at least two professional editors, both native speakers of English.

References

1. Sampson HA: Food allergy. Part 1: immunopathogenesis and clinical disorders. J Allergy Clin Immunol 1999, 103:717–728.

2. Castells MC, Pascual C, Esteban MM, Ojeda JA: Allergy to white potato. J Allergy Clin Immunol 1986, 78:1110–1114.

3. De Swert LFA, Cadot P, Ceuppens JL: Allergy to cooked white potatoes in infants and young children: a cause of severe, chronic allergic disease. J Allergy Clin Immunol 2002, 110:524–529.

4. Quirce S, Diez Gomez L, Hinojosa M, Cuevas M, Urena V, Rivas MF, Puyana J, Cuesta J, Losada E: Housewives with raw potato-induced bronchial asthma. Allergy 1989, 44:532–536.

5. Beausolei JL, Spergel JW, Pawlowski NA: Anaphylaxis to raw potato. Ann Allergy Asthma Immunol 2001, 86:68–70.

6. Calkhoven PG, Aalbers M, Koshte L, Pos O, Oei HD, Aalberse RC: Cross-reactivity among birch pollen, vegetables and fruits as detected by IgE antibodies is due to at least three different cross-reactive structures. Allergy 1987, 42:382–390.

7. Reche M, Pascual CY, Vincente J, Caballero T, Martin-Munoz F, Sanchez S, Martin-Esteban M: Tomato allergy in children and young adults: cross-reactivity with latex and potato. Allergy 2001, 56:1197–1201.

8. Taylor ST: Protein allergenicity assessment of foods produced through agricultural biotechnology. Annu Rev Pharmacol Toxicol 2002, 42:99–112.

9. Batisa R, Nunes B, Carmo M, Cardoso C, Jose HS, Almeida AB, Manique A, Bento L, Ricardo CP, Margarida M: Lack of detectable allergenicity of transgenic maize and soya samples. J Allergy Clin Immunol 2005, 116:403–410.

10. Sutton SA, Assa-ad AH, Steinmetz C, Rothenberg ME: A negative, double-blind, placebo-controlled challenge to genetically modified corn. J Allergy Clin Immunol 2003, 112:1011–1012.

11. Yoon SH, Kim HM, Kim SH, Suh CH, Nahm DH, Park HS: IgE sensitization and identification of IgE binding components of the corn allergen in adult allergy patients: Comparison between the wild and the genetically modified corn. J Asthma Allergy Immunol 2005, 25:52–58.

12. European Academy of Allergology and Clinical Immunology: Position paper: Allergen standardization and skin tests. Allergy 1993, 48:48–82.

13. Kim YK, Park HS, Kim HA, Lee MH, Choi JH, Kim SS, Lee SK, Nahm DH, Cho SH, Min KU, Kim YY: Two-spotted spider mite allergy: immunoglobulin E sensitization and characterization of allergenic components. Ann Allergy Asthma Immunol 2002, 89:517–522.

14. Astwood JD, Leach JN, Fuchs RL: Stability of food allergens to digestion in vitro. Nat Biotechnol 1996, 14:1269–1273.

15. Kim TB, Kim KM, Kim SH, Kang HR, Chang YS, Kim CW, Bahn JW, Kim YK, Kang HT, Cho SH, Park HS, Lee JM, Choi IS, Min KU, Kim YY: Sensitization rates for inhalant allergen in Korea: a multi-center study. J Asthma, Allergy Immunol 2003, 23:483–493.

16. World Health Organization: Safety aspects of genetically modified foods of plant origin. Report of a joint FAT/WHO expert consultation; Geneva, Switzerland 2000.

17. Food and Agriculture Organization of the United Nations: Biotechnology and food safety. Report of a joint FAO/WHO consultation: Rome, Italy 1996.

18. Food and Agriculture Organization of the United Nations: Evaluation of the allergenicity of genetically modified foods. Report of a joint FAO/WHO expert consultation: Rome, Italy 2001.

19. Beck E, Ludwig G, Auerswald EA, Reiss B, Schaller H: Nucleotide sequence and exact localization of the neomycin phosphotransferase gene from transposon Tn5. Gene 1982, 19:327–336.

20. Fuchs RL, Heeren RA, Gustafson ME, Rogan GJ, Bartnicki DE, Leimgruber RM, Finn RF, Hershman A, Berberich SA: Purification and characterization of microbially expressed neomycin phosphotransferase II (NPTII) protein and its equivalence to the plant expressed protein. Biotechnology 1993, 11:1537–1542.

21. Fuchs RL, Ream JE, Hammond BG, Naylor MW, Leimgruber RM, Berberich SA: Safety assessment of the neomycin phosphotransferase II (NPTII) protein. Biotechnology 1993, 11:1543–1547.

22. Herouet C, Esdaile DJ, Mallyon BA, Debruyne E, Schulz A, Curtier T, Hendricx K, van der Klis RJ, Rouan D: Safety evaluation of the phosphinothricin acetyltransferase proteins encoded by the pat and bar sequences that confer tolerance to glufosinate-ammonium herbicide in transgenic plants. Regul Toxicol Pharmacol 2005, 41:134–149.

23. Wahl R, Lau S, Maasch HJ, Wahn U: IgE-mediated allergic reactions to potatoes. Int Arch Allergy Appl Immunol 1990, 92:168–174.

24. Seppälä U, Alenius H, Turjanmaa K, Reunala T, Palosuo T, Kalkkinen N: Identification of patatin as a novel allergen for children with positive skin prick test responses to raw potato. J Allergy Clin Immunol 1993, 103:165–171.

25. Seppälä U, Majamaa H, Turjanmaa K, Helin J, Reunala T, Kalkkinen N, Palosuo T: Identification of four novel potato (Solanum tuberosum) allergens belonging to the family of soybean trypsin inhibitors. Allergy 2001, 56:619–626.

26. Seppälä U, Palosuo T, Seppälä U, Kalkkinen N, Ylitalo L, Reunala T, Turjanmaa K, Reunala T: IgE reactivity to patatin-like latex allergen, Hev b 7, and to patatin of potato tuber, Sol t 1, in adults and children allergic to natural rubber latex. Allergy 2000, 55:266–273.

Genetically Modified Parthenocarpic Eggplants: Improved Fruit Productivity Under Both Greenhouse and Open Field Cultivation

Nazzareno Acciarri, Federico Restaino, Gabriele Vitelli,
Domenico Perrone, Michela Zottini, Tiziana Pandolfini,
Angelo Spena and Giuseppe Leonardo Rotino

ABSTRACT

Background

Parthenocarpy, or fruit development in the absence of fertilization, has been genetically engineered in eggplant and in other horticultural species by using the DefH9-iaaM gene. The iaaM gene codes for tryptophan monoxygenase and confers auxin synthesis, while the DefH9 controlling regions drive expression of the gene specifically in the ovules and placenta. A previous greenhouse trial for winter production of genetically engineered (GM) parthenocarpic eggplants demonstrated a significant increase (an average of 33% increase) in fruit production concomitant with a reduction in cultivation costs.

Results

GM parthenocarpic eggplants have been evaluated in three field trials. Two greenhouse spring trials have shown that these plants outyielded the corresponding untransformed genotypes, while a summer trial has shown that improved fruit productivity in GM eggplants can also be achieved in open field cultivation. Since the fruits were always seedless, the quality of GM eggplant fruits was improved as well. RT-PCR analysis demonstrated that the DefH9-iaaM gene is expressed during late stages of fruit development.

Conclusions

The DefH9-iaaM parthenocarpic gene is a biotechnological tool that enhances the agronomic value of all eggplant genotypes tested. The main advantages of DefH9-iaaM eggplants are: i) improved fruit productivity (at least 30–35%) under both greenhouse and open field cultivation; ii) production of good quality (marketable) fruits during different types of cultivation; iii) seedless fruit with improved quality. Such advantages have been achieved without the use of either male or female sterility genes.

Background

Fruit-set and growth of several horticultural plants are negatively affected by adverse environmental conditions. In general, sub and/or supra-optimal temperatures negatively affect reproductive processes and therefore curtail fruit production [1,2]. Under greenhouse cultivation, low temperature, insufficient light intensity, and/or high humidity drastically reduce fruit productivity and quality in eggplant and other species. Moreover, environmental conditions often met in open field cultivation such as drought and high temperatures have a negative effect on fruit productivity and quality in eggplant and other species (e.g. tomato).

Parthenocarpic fruit development (i.e. fruit-set and growth without fertilization) can significantly aid in the resolution of the aforementioned problems. Parthenocarpy can be triggered by exogenous factors, such as plant growth regulators, or it can be achieved by genetic factors. Genes causing parthenocarpic development have been identified in several plant species [3-5], and parthenocarpic eggplant varieties (e.g. Talina, Galine) have been introduced in the production process (e.g. protected cultivation). However, during winter cultivation of egg-plant varieties in unheated greenhouses in the Mediterranean area, the negative effect of suboptimal environmental conditions on fruit production is usually counteracted by treating flower buds with plant growth regulators. Phytohormonal treatments make the production process more expensive due to the cost of both chemicals and labor.

In principle, the genetic engineering of plants allows one to confer a trait of interest to different species and within a species to all the varieties of interest. To confer parthenocarpic fruit development, a chimeric gene has been constructed [6]. Specifically, the DefH9-iaaM gene contains the coding region of the iaaM gene from Pseudomonas syringae pv. savastanoi under the control of the placenta and ovule-specific promoter from the DefH9 gene from Anthirrinum majus[6]. The iaaM gene codes for a tryptophan monoxygenase, which produces indolac-etamide that in turn, is either chemically or enzymatically converted to the auxin indole-3-acetic acid. To date, the DefH9-iaaM chimeric gene has been shown to cause parthenocarpic fruit development in tobacco [6], eggplant [6], tomato [7,8], strawberry and raspberry [9].

We have previously shown that DefH9-iaaM eggplants outperform control plants during protected winter cultivation by an average of 33% [10]. The present manuscript presents data on the agronomic performance of DefH9-iaaM eggplant hybrids during spring and summer cultivation. Seed-derived F1 plants perfectly reflects the real agronomical situation of eggplant production. In addition, they have allowed to demonstrate that the transgene is active after meiosis and give satisfactory results in the hemizygous state. Different types of eggplant hybrids have been evaluated during springtime in unheated greenhouses in two different locations. To evaluate GM parthenocarpic eggplants under optimal environmental conditions, a single trial has been carried out using two different genotypes under standard open field cultivation during summertime. Furthermore, we demonstrate that the DefH9-iaaM gene is expressed during late stages of fruit development.

Results and Discussion

Greenhouse Spring Production

Spring production was evaluated in trials performed in two different locations: Monsampolo and Pontecagnano. Spring production was divided into early production, consisting of the first four harvests, and total production, consisting of sixteen harvests. Early spring production corresponds to the cultivation period from March to the first half of May, with temperatures somewhat low for fruit-set and growth. During this period, the average minimum and maximum temperatures ranged from 7° to 17°C in southern Italy (Pontecagnano) and from 14° to 17°C in central Italy (Monsampolo). The transgenic parthenocarpic hybrids gave a significantly higher yield, on the average a six-fold increase during early production, in comparison to their controls (Table 1). The increment in the number of fruits per plant and the higher average weight of the fruits were the main causes of the increased early spring production of the parthenocarpic hybrids. The suboptimal/adverse environmental conditions did not affect the growth of GM parthenocarpic fruits and the average fruit weight was significantly higher in the GM eggplants in comparison to untransformed controls. The traditional parthenocarpic cultivar Talina produced fruits with an average weight similar to those of GM eggplant fruits (Table 1). However, due to the higher number of fruits per plant, the productivity of the transgenic parthenocarpic hybrids was also increased, by an average of five-fold, when compared to cv Talina (Table 1). During the whole harvesting period, the GM fruits were always seedless (Fig. 1) and were normal in both size and shape.

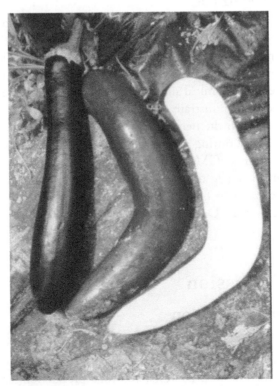

Figure 1. Eggplant fruits of the transgenic parthenocarpic hybrid P1 from the greenhouse spring trial. Left: fruit at the stage of commercial ripeness; middle: an overripe eggplant fruit from a border plant; right: a longitudinally cut fruit showing the complete absence of seeds.

Table 1. Eggplant production of parthenocarpic hybrids and their respective controls at springtime.

Genotype	Early production			Total production		
	Yield/plant (g)	Fruits/plant (n)	Fruit weight (g)	Yield/plant (g)	Fruits/plant (n)	Fruit weight (g)
P1	488 b	1.9 a	268.3 a	2241 a	8.4 a	253.9 a
P2	695 a	2.6 a	290.8 a	2288 a	8.6 a	260.2 a
C1	75 c	0.4 b	227.0 b	1547 b	7.8 a	187.5 b
P5	522 b	2.0 a	281.8 a	2163 a	9.2 a	230.5 a
C2	116c	0.7 b	170.6 b	1574 b	9.3 a	167.3 b
Talina	114c	0.3 b	270.5 a	2360 a	9.4 a	229.9 a

For each trait at least one common letter indicates no significant difference according to the Duncan test (P = 0.05). Mean values of yields per plant, number of fruits per plant, and fruit weight of three transgenic parthenocarpic hybrids (P1, P2 and P5), two controls (C1 and C2) and the commercial cultivar Talina. C1 hybrid plants represent the controls of P1 and P2 transgenic hybrid plants. C2 hybrid plants are the controls of the P5 transgenic hybrids. Data are the average of the Monsampolo and Pontecagnano locations. The experiments were carried out in greenhouse at springtime.

The increased productivity of GM hybrids characterised both the early spring production (i.e. the first four harvests) and the whole spring production cycle (i.e. sixteen harvests). During the whole spring production cycle, the hybrids P1 and P2 gave an average yield that was 46% higher with respect to the corresponding control C1 (Table 1). The hybrid P5 gave a 37% higher yield with respect to its control C2. The average total number of fruits produced per plant in both locations was similar in all the hybrids (8–9 fruits/plant). However, the higher average weight of the GM fruits led to a higher total yield of transgenic hybrids with respect to their controls. When considering the whole spring cultivation cycle, the parthenocarpic cultivar Talina gave a total production that was not significantly different from either of the three GM hybrids (Table 1).

Open Field (Summer) Production

Summer production was evaluated in an open field trial carried out during the optimal period of eggplant cultivation. Plants were transplanted on May 20th and the last harvest took place on September 11th. The early production of the transgenic hybrids, consisting of the first three harvests, was significantly higher than that of the untransformed hybrids (Table 2): P1 and P10 hybrids yielded, respectively, 36% and 76% more than their corresponding controls, C1 and C10. The difference in productivity between P1 and C1 hybrids, which have long-shaped fruits, was caused by the higher average weight of GM fruits. When comparing P10 and C10 hybrids, which have sub-oval fruits, the higher yield obtained with GM plants during the early harvesting period was due to the increased number of fruits per plant.

Table 2. Eggplant production of parthenocarpic hybrids and their respective controls at summertime.

Genotypes	Early production			Total production		
	Yield/plant (g)	Fruits/plant (n)	Fruit weight (g)	Yield/plant (g)	Fruits/plant (n)	Fruit weight (g)
PI	1158a	2.7 a	433 b	3039 a	9.1 a	335 b
CI	846 b	2.4 a	344 c	2211 b	7.7 ab	288 b
P10	1287 a	2.4 a	553 a	2791 ab	6.8 bc	410 a
C10	731 b	1.3 b	541 a	2386 ab	5.7 c	415 a

For each trait, at least one common letter indicates no significant difference according to the Duncan test (P = 0.05). Mean value of yields per plant, number of fruits and fruit weight of two transgenic hybrids (PI and P10) and their corresponding untransformed hybrids (CI and C10), cultivated during summertime under open field conditions.

Consisting of ten harvests, the total production of P1 hybrids was 37% higher than control C1 eggplants (Table 2). The difference in total yield between P1 and control C1 hybrids was statistically significant and due both to the higher number of fruits/plant and to the increased weight of GM fruits. It is noteworthy to point out that when considered individually, neither trait (number of fruits/plant or fruit weight) showed statistically significant differences between the GM and untransformed plants (Table 2). Although higher in P10 hybrids in comparison to its control C10, the total yield (the number and average weight of the fruits) was not significantly different between the two. During the whole harvesting period, fruits from both P1 and P10 parthenocarpic hybrids were always seedless (Fig. 2), whilst control fruits always had seeds. Therefore, under open field cultivation, the DefH9-iaaM gene had a positive influence on fruit quality, as GM DefH9-iaaM fruits were always seedless. Fruit quality affects the economic value of eggplant production.

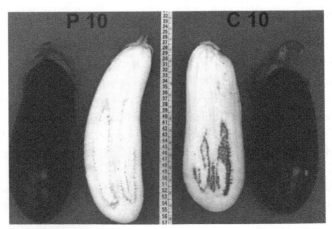

Figure 2. Eggplant fruits from the open field summer trial. Left, uncut and cut fruit of the transgenic hybrid P10; Right, cut and uncut fruit of the C10 control hybrid.

Although the environmental conditions were optimal and consequently did not affect negatively fruit-set, the DefH9-iaaM gene caused both faster development and growth of the fruits as indicated by the increased early-summer production (the first three harvests). In this regard, it is worthwhile to stress that expression of the DefH9-iaaM gene takes place in the placenta and ovules before pollen development. As a consequence, in GM parthenocarpic plants fruits are seedless and fruit development initiates well before non-GM controls [11].

In all trials we have never used homozygous lines because growers mostly employ F1 hybrids. The use of hemizygous primary transformants as pollinator plants allowed us to obtain in rather short time, by in vivo selection for kanamycin resistance, F1 plants transgenic for the DefH9-iaaM gene. Young, healthy and vigorously growing plants did not produce seeds. However, it is possible to obtain seeds from aged DefH9-iaaM transgenic plants both by selfing and crossing. By exploiting the delayed female fertility we have produced the homozygous plants needed as male parents for rapid seed multiplication of F1 eggplant hybrids. Therefore, the female sterility of young and mature plants is not an insuperable hindrance for mass propagation and commercial fruit production.

Expression of the DefH9-iaaM Gene Takes Place During Both Flower and Fruit Development in Transgenic Parthenocarpic Eggplants

The DefH9 gene is expressed specifically in the placenta and ovules during early phases of flower development [6]. To determine whether expression of the DefH9-iaaM gene also occurs during later stages of fruit growth and whether it is influenced by pollen fertilization, mRNA from transgenic flowers and fruits obtained either from emasculated or hand pollinated flowers was analyzed by RT-PCR at different stages of development, until the fruit attained a length of 28 cm. An amplicon of 161 bp corresponding to the spliced DefH9-iaaM mRNA was detected in all stages analyzed (Fig. 3, lanes 2–6). Thus, the presence of the mRNA of the DefH9-iaaM gene and consequently its action is most likely not limited to early stages of flower and fruit development. Pollination did not affect the steady state level of DefH9-iaaM mRNA (Fig. 3, compare lane 5 and lane 6).

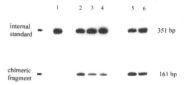

Figure 3. Expression analysis of the DefH9-iaaM gene by competitive RT-PCR during flower and fruit development. Untransformed plant (lane 1); 5, 8, 12 mm long flower buds (lanes 2, 3, 4, respectively); 40 mm long hand-pollinated fruit (lane 5); 280 mm long emasculated fruit (lane 6). An internal standard of 351 bp is present in all lanes.

Treatment with auxin often causes parthenocarpic development in several plant species [for review, see: [12]]. However, in some species and/or varieties, to efficiently sustain fruit growth, the hormonal treatment of the flowers must be repeated [13]. The finding that DefH9-iaaM mRNA is also present during later stages of fruit development is consistent with the interpretation that in DefH9-iaaM parthenocarpic plants, the placenta, the ovules, and the tissues derived therefrom are a source of auxin during the whole growth of the fruit. As a consequence, they efficiently sustain fruit growth.

Conclusions

The data hereby presented show the positive influence of the DefH9-iaaM parthenocarpic gene on eggplant productivity under both greenhouse (spring) and open field (summer) cultivation. Taking into account the data previously obtained under winter greenhouse cultivation [10], we conservatively estimate that the overall increase in productivity is at least 30–35%. The increase in productivity of DefH9-iaaM eggplants is mainly due to a drastically improved fruit-set under sub-optimal temperatures and to an enhanced fruit growth and weight. Fruit quality is also improved because the fruits are seedless and do not show a placental cavity. The qualitative improvement of DefH9-iaaM eggplant fruits is interesting both for the fresh market and for the processing industry. During early spring greenhouse production, DefH9-iaaM parthenocarpic hybrids always gave fruits with an average weight suitable for fresh market commercialization, while untransformed hybrids, under sub-optimal conditions, rarely produced commercial fruits. Thus, the DefH9-iaaM gene quantitatively and qualitatively improves eggplant production under both greenhouse and open field cultivation. In all genotypes tested the DefH9-iaaM gene had a very positive effect on production and quality parameters. Such findings are of paramount importance as the hybrids tested have the same genetic background that the relative controls, except for the presence of the DefH9-iaaM gene. The DefH9-iaaM gene, already known to be expressed in the placenta and ovules during early phases of flower development, is expressed also in mature fruits, most likely in tissues derived from the ovules.

From an economical standpoint, the main advantages conferred to eggplant by the DefH9-iaaM gene are: i) production of marketable fruits under environmental conditions adverse for fruit-set and growth; ii) reduction of cultivation costs (energy, phytohormones and labor) necessary for off-season and open field eggplant cultivation; and iii) enhancement of fruit quality. Last but not least, contrary to conventional wisdom, these advantages have been achieved without the use of either male or female sterility genes.

Materials and Methods

Greenhouse Spring Cultivation

Trials were carried out in central Italy (Monsampolo del Tronto-AP) and in southern Italy (Pontecagnano-SA) (approval of the Italian Ministry of Health N° B/IT/97-29). The greenhouses were rather similar and made of galvanized steel and covered with plastic polyethylene (0.12 mm thick). An apparatus for drip-irrigation was used and the soil was completely mulched. A complete randomized block design with three replicate hybrid genotypes was adopted. Each experimental plot measured 3.12 m2 and contained eight plants in a double row. Transplanting was performed on March 3rd in southern Italy and on March 27th in central Italy. The P1, P2, P5, C1, C2 and the commercial Talina hybrids were employed. Transgenic parthenocarpic hybrids P1, P2 and P5 were obtained by crossing (as male parent) the primary transgenic plant DR2 iaaM #34-1 with the line Tal 1/1 (P1), the primary transgenic plant DR2 iaaM #28-1 with Tal 1/1 (P2) and the transgenic plant Tal 1/1 iaaM #1-1 with the line Tina (P5). The hybrids P1 and P2 are homologous to C1 (DR2 × Tal1/1), except for the presence of the DefH9-iaaM gene integrated in their genome. The transgenic hybrid P5 is homologous to its untransformed control C2 (Tal1/1 × Tina). DR2 and Tina are parental lines obtained through classical breeding, Tal1/1 is a double haploid line derived from anther culture of the F1 commercial cultivar Talina. The segregation of the marker gene nptII was checked by spraying the plants with kanamycin [14] and allowed for the conclusion that the transgenes segregate as a single locus in the backcrossed progenies of the three independent events analyzed (Tal iaaM 1-1: x^2 = 0.01065, P = 0.917; DR2 iaaM 34-1: x^2 = 0.0496 P = 0.824; DR2 iaaM 28-1 x^2 = 0.06467 P = 0.799). Southern blot analysis showed that DR2 iaaM 28-1 and 34-1 had a single copy of the transgene, while Tal iaaM 1-1 had three copies of the transgene (Fig 4). Since the interaction genotype/location was not significant for the yield, the data were computed as average of the two locations and subjected to analysis of variance according to a randomized complete block design. Duncan's Multiple Range Test (P = 0.05) was used for means separations.

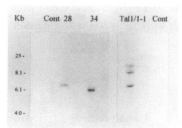

Figure 4. Southern blot analysis of transgenic eggplants. Numbers above the lanes indicate the independent transgenic plant DR2iaaM#28-1 (28), DR2iaaM#34-1 (34) and Tal1/1iaaM#1-1 (Tal1/1-1). Cont indicates untransformed plants, i.e. DR2 and Tal1/1, respectively. The probe used corresponds to the DefH9 regulatory region.

Open Field (Summer) Cultivation

The open field trial was carried out under open field conditions at Monsampolo del Tronto (approval of the Italian Ministry of Health B/IT/99/21). Two transgenic parthenocarpic hybrids were tested: the hybrid P1 (Tal1/1 × DR2 iaaM #34-1) with elongated fruits and the hybrid P10 (UGA × Tal1/1 iaaM #1-1) with sub-oval fruits were compared to their homologous non-transgenic controls C1 (DR2 × Tal1/1) and C10 (UGA × Tal1/1), respectively. The UGA line has oval dark purplish fruits and it has been provided by Dr. S.C. Phatak. A complete randomized block design with the hybrids replicated four times was adopted. Each experimental plot measured 11.7 m2 and contained 30 plants in a double row. Transplanting was performed on May 10th.

Early spring production consisted of the first four harvests (i.e. 4 out of 16 harvests during the whole production cycle), while early summer production, whose cultivation cycle consists of ten harvests, corresponds to the first three harvests. For all trials the number and weight of fruits were recorded. In addition, fruit sample for each harvest and replication was cut to check for the presence of seeds. Data were computed for the early harvesting time and for the whole harvesting season. Analysis of variance was performed according to a randomized complete block design. Duncan's Multiple Range Test (P = 0.05) was used for means separations.

Plant DNA Isolation and Southern Blot Analysis

High-molecular-weight DNA was isolated from the young leaves of transgenic and untransformed eggplants by using Plant DNAzol (Invitrogen), according to the manufacturer's instructions. Ten µg of DNA was digested overnight with 80 units of KpnI (Promega) in a volume of 500 µl, separated on a 0.7% agarose gel and transferred to Hybond N (Amersham Pharmacia Biotech). A 1350 bp fragment of the DefH9 gene was used as a radiolabeled probe. The membrane was hybridized overnight in 5X SSC/50% formamide (Sigma) at 42°C and washed two times for 15 min. in 2 × SSC/0.1% SDS, and two times for 15 min. in 0.1 × SSC/0.1% SDS at 42°C. Signals were detected using Kodak X-OMAT AR5 film (Sigma).

RT-PCR Analysis

Semiquantitative (competitive) PCR analysis was carried out for 38 cycles (annealing temperature 63°C) using as template cDNA (8 ng) obtained by priming poly(A)+ mRNA with an iaaM specific primer (5'-AATAGCTGCCTATGC-

CTCCCGTCAT-3'). The mRNA was extracted from either young flower buds (5,8,11 mm) or eggplant fruits (placental tissue from fruits either 40 or 280 mm long). As an internal standard, 0.5 fg of a 600 bp long DefH9 cDNA fragment was used in the PCR assay. To amplify the 161 bp long amplicon an iaaM specific primer (5'-GGGTGAATTAAAATGGTCATACAT-3') and a DefH9 specific primer (5'-CTTTGGAACTCGTGTTGAGCTCTCA-3') were used. For the internal standard, a 3' primer (5'-TGAGCATTGATCTCCTGAGTGGTGT-3') together with the DefH9 specific primer were used to produce the 351 bp long amplicon. The PCR assays were performed with a thermostable DNA polymerase mixture (Expand High Fidelity PCR system, Roche) in presence of 3 µCi of 32P dCTP. The intensity of the bands was quantified by using an Instant Imager (Packard, Meriden, CT).

Acknowledgements

This research was partially supported by the Consiglio Nazionale delle Ricerce (Progetto Finalizzato Biotecnologie II) and by the Ministero Politiche Agricole e Forestali Progetto "Biotecnologie vegetali". We thank Prof. Phatak, University of Georgia (USA) for the UGA line.

References

1. Nothman J, Koller D: Effects of growth regulators on fruit and seed development in eggplant (S. melongena L.). J Hort Sci 1975, 50:23–27.

2. Romano D, Leonardi C: The responses of tomato and eggplant to different minimum air temperature. Acta Hort 1994, 366:57–66.

3. De Ponti OMB: Breeding parthenocarpic pickling cucumbers (Cucumis sativus L.): necessity, genetical possibilities, environmental influences and selection criteria. Euphytica 1976, 25:29–40.

4. Philouze J: Parthenocarpie naturelle chez tomate. II. Etude d'une collection varietale. Agronomie 1985, 5:47–54.

5. Restaino F, Onofaro-Sanajà V, Mennella G: Facultivative parthenocarpic genotypes of eggplant obtained through induced mutations. XIII Eucarpia Congress, 6–11 July, Angers, France 1992, 297–298.

6. Rotino GL, Perri E, Zottini M, Sommer H, Spena A: Genetic engineering of parthenocarpic plants. Nat Biotechnol 1997, 15:1398–1401.

7. Ficcadenti N, Sestili S, Pandolfini T, Cirillo C, Rotino GL, Spena A: Genetic enginnering of parthenocarpic fruit development in tomato. Mol Breed 1999, 5:463–470.

8. Pandolfini T, Rotino GL, Camerini S, Defez R, Spena A: Optimisation of trans-gene action at the post-transcriptional level: high quality parthenocarpic fruits in industrial tomatoes. BMC Biotechnology 2002, 2:1.

9. Mezzetti B, Landi L, Scortichini L, Rebori A, Spena A, Pandolfini T: Genetic engineering of parthenocarpic fruit development in strawberry. Proc. 4th ISHS Strawberry symposium, Tampere (Finland) 9–14 July 2000. Acta Hortic 2002, 567:101–104.

10. Donzella G, Spena A, Rotino GL: Transgenic parthenocarpic eggplants: superior germplasm for increased winter production. Mol Breed 2000, 6:79–86.

11. Spena A, Rotino GL: Parthenocarpy: state of the art. In: Current trends in the embryology of angiosperm (Edited by: Bhojwani SS, Soh WY). Kluwers Academic Publishers 2001, 435–450.

12. Schwabe WW, Mills JJ: Hormones and parthenocarpic fruit-set: a literature survey. Hort Abstracts 1981, 51:661–698.

13. Saavedra E: Set and growth of Annona cherimolia Mill. Fruit obtained by hand-pollination and chemical treatments. J Am Hort Sci 1979, 104:668–673.

14. Sunseri F, Fiore MC, Mastrovito F, Tramontano E, Rotino GL: In vivo selection and genetic analysis for kanamycin resistance in transgenic eggplant (Solanum melongena L.). J Genet Breed 1993, 47:299–306.

Reconstitution of the Myeloid and Lymphoid Compartments after the Transplantation of Autologous and Genetically Modified CD34+ Bone Marrow Cells, Following Gamma Irradiation in Cynomolgus Macaques

Sonia Derdouch, Wilfried Gay, Didier Nègre, Stéphane Prost, Mikael Le Dantec, Benoît Delache, Gwenaelle Auregan, Thibault Andrieu, Jean-Jacques Leplat, François-Loïc Cosset and Roger Le Grand

ABSTRACT

Background

Prolonged, altered hematopoietic reconstitution is commonly observed in patients undergoing myeloablative conditioning and bone marrow and/or

mobilized peripheral blood-derived stem cell transplantation. We studied the reconstitution of myeloid and lymphoid compartments after the transplantation of autologous CD34⁺ bone marrow cells following gamma irradiation in cynomolgus macaques.

Results

The bone marrow cells were first transduced ex vivo with a lentiviral vector encoding eGFP, with a mean efficiency of 72% ± 4%. The vector used was derived from the simian immunodeficiency lentivirus SIVmac251, VSV-g pseudotyped and encoded eGFP under the control of the phosphoglycerate kinase promoter. After myeloid differentiation, GFP was detected in colony-forming cells (37% ± 10%). A previous study showed that transduction rates did not differ significantly between colony-forming cells and immature cells capable of initiating long-term cultures, indicating that progenitor cells and highly immature hematopoietic cells were transduced with similar efficiency. Blood cells producingeGFP were detected as early as three days after transplantation, and eGFP-producing granulocyte and mononuclear cells persisted for more than one year in the periphery.

Conclusion

The transplantation of CD34⁺ bone marrow cells had beneficial effects for the ex vivo proliferation and differentiation of hematopoietic progenitors, favoring reconstitution of the T- and B-lymphocyte, thrombocyte and red blood cell compartments.

Background

Gene therapy strategies hold promise for the treatment of hematopoietic disorders. All hematopoietic lineages, including polymorphonuclear cells, monocytes, lymphocytes and natural killer cells, and hematopoietic stem cells (HSC) – which are capable of self-renewal and pluripotent differentiation – have been targeted for transduction with therapeutic genes. Most diseases for which gene therapy could be proposed require stable and long-lasting transgene expression for efficacy. Retroviral vectors present the major advantage of integrating the transferred DNA stably into the genome of target cells, which is then passed on to progeny. However, they cannot infect and integrate into non dividing cells [1]. Most HSC are quiescent [2], respond slowly to stimulation [3-7] and tend to differentiate and lose their repopulating capacity upon stimulation [3,8-11]. Lentiviral vectors can be used to transduce cells in growth arrest [12] in vivo and ex vivo[13], thanks to interaction of the preintegration complex – composed of viral VPX and integrase proteins – with the nuclear pore complex[14]. Vectors derived from

HIV-1[15,16], HIV-2[17], FIV[18] and equine infectious anemia virus (EIAV) have been tested[19].

Methods for transferring genes into hematopoietic cells must be tested in relevant animal models before their application to humans [20,21]. Studies in non-human primates (NH)P provide an ideal compromise, because these species are phylogenetically closely related to humans and a high level of nucleotide sequence identity is observed between the genes encoding many hematopoietic growth factors and cytokines in these mammals and their counterparts in humans[22]. Moreover, hematopoiesis in macaques is very similar to that in humans, and the HSC biology of macaques is much more similar to that of humans than is that of rodents, making macaques good candidates for hematopoietic stem cell engraftment studies [23-26]. In addition, testing lentiviral based gene transfer strategies need to be assessed in species that are susceptible to lentivirus induced disease. Or particular interest are the Feline immunodeficiency virus (FIV) infection which causes a clinical disease in cats that is remarkably similar to HIV disease in human [27-30] and experimental infection of macaques with the simian immunodeficiency virus (SIV) reproducing both chronic infection and an AIDS-like disease very similar to those observed in human patients infected with HIV. Despite the theoretical advantages of lentiviral vectors over oncoretroviral vectors, non human primate lentiviruses clearly have pathogenic properties [31]. The use of lentiviral vectors derived from potentially pathogenic primate lentiviruses, such as SIV, therefore continues to raise serious clinical acceptance concerns. SIV-based vectors, such as SIVmac239[31,32] and SIVmac251[33,34], may provide a unique opportunity to test the safety and efficacy of primate lentiviral vectors in vivo.

Recent improvements in the efficiency of gene transfer to NHP repopulating cells[11,35,36] have provided new opportunities to follow the progeny of each primitive progenitor and stem cells directly in vivo, using retroviral marking to track individual progenitor or stem cell clones[37]. Clinically relevant levels (around 10%) of genetically modified cells in the peripheral blood have been achieved by ex vivo gene transfer into HSC and the autologous transplantation of these cells into macaques[37]. Successful and persistent engraftment (up to six months) has also been reported in non human primates with primitive CD34+ progenitors genetically modified with a murine retrovirus vector encoding the murine CD24 gene as a reporter gene[38]. In both trials, marked cells of multiple hematopoietic lineages were identified in the blood: granulocytes, monocytes and B and T cells, including naive T lymphocytes[37,38]. The efficacy of HSC gene transfer could theoretically be improved by the use of newly developed retroviral or lentiviral vectors. Particles bearing an alternative envelope protein, such as that of the feline endogenous virus (RD114), have been shown to be superior to amphotropic vectors for the transduction of NHP stem cells followed by autologous transplantation [39,40].

We report here the results obtained in vitro and in vivo in an experiment assessing the efficacy and safety of a gene transfer protocol based on the transduction of simian CD34+ bone marrow cells with a minimal SIVmac251-derived lentiviral vector. This system is based on the VSVg-pseudotyped SIV vector encoding enhanced green fluorescent protein (eGFP) under control of the phosphoglycerate kinase (PGK) promoter. Most immature CD34+ hematopoietic cells capable of initiating long-term culture (LTC-IC) were efficiently transduced, and eGFP-positive cells were detectable in vivo in all animals more than one year after transplantation.

Methods

Animals

Male cynomolgus macaques (Macaca fascicularis), weighing between 3 and 6 kg were imported from Mauritius and housed in single cages within level 3 biosafety facilities, according to national institutional guidelines (Commission de génie génétique, Paris, France). All experimental procedures were carried out in accordance with European guidelines for primate experiments (Journal Officiel des Communautés Européennes, L358, December 18, 1986).

Immunoselection of Non Human Primate CD34⁺ Bone Marrow Progenitor Cells

Bone marrow mononuclear cells were obtained from the iliac crest or by aspiration from the humerus and isolated by standard Ficoll density-gradient centrifugation (MSL2000, Eurobio, Les Ulis, France). Cells were washed twice in phosphate-buffered saline (PBS, Eurobio, Les Ulis, France) and resuspended in 1% FCS (Fetal Calf Serum; Bio West, France) in PBS. The cellular fraction was then enriched in CD34⁺ cells by positive immuno-magnetic selection, using beads coupled to a specific antibody (clone 561; Dynabeads M-450 CD34, Progenitor Cell Selection System, Dynal, Oslo, Norway), according to the manufacturer' s instructions. Immunoselected CD34⁺ cells were stained with a specific PE-conjugated anti-CD34 antibody (clone 563; Pharmingen, Becton Dickinson, California, USA) and analyzed by flow cytometry (LSR, Becton Dickinson, California, USA) to evaluate the level of enrichment. All preparations contained more than95% CD34⁺ cells, with a mean value of 97% ± 1% (n = 12) for in vitro assays and 96% ± 1% (n = 4) for in vivo assay.

Lentiviral Vector

Two SIV-derived vectors were produced, one for in vitro studies and the other for in vivo studies: 1) pRMES8 is a minimal packaging-competent SIVmac251-based vector[34]. It contains the enhanced green fluorescent protein (eGFP) marker gene under control of the mouse phosphoglycerate kinase (PGK) promoter, placed between the SIVmac251 LTRs and leader sequences. It carries the SIVmac251 RRE region and minimal sequences of the gag and pol genes encompassing central polypurine tract/central termination sequence (cPPT/CTS) regions (figure 1A). pRMES8 was used for in vitro assays investigating the susceptibility of CD34⁺ cells from primate bone marrow to transduction with SIVmac251-derived vectors. 2) For in vivo assays, we used pGASE; this plasmid is an optimised version of pRMES8, with a 3'-SIN-LTR for safety and insertion of an exon splicing enhancer (ESE) upstream the PGK promoter to increase titer [41]

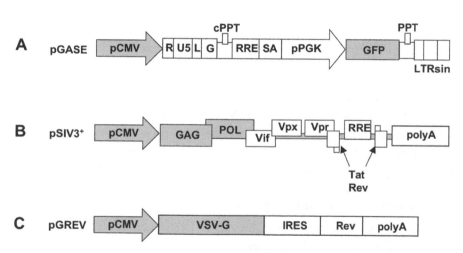

Figure 1. Schematic representation of SIV-derived SIN vector, helper construct and VSV-g encoding plasmid. An SIVmac251-derived vector was produced by cotransfecting 293T cells with three plasmids: A. a plasmid pGASE containing the eGFP gene under control of the PGK promoter; B. a plasmid pSIV3⁺ containing viral genes; C. a plasmid pGREV containing the VSV envelope gene. Cis genetic elements are symbolized with white boxes, whereas promoters and genes are depicted by shadowed boxes. pCMV, early cytomegalovirus promoter; pPGK, mouse phosphoglycerate kinase-1 promoter; RRE, REV-responsible element; SA, SIV Rev/Tat splice acceptor; cPPT and PPT, central and 3' polypurine tracks, respectively; GFP, the gene encoding the enhanced green fluorescent protein; LTRsin, partially U3 deleted 3'LTR; LG, leader and a 5' GAG region.

pSIV3+ is the packaging plasmid derived from the BK28 molecular clone of SIVmac251, as described elsewhere[33]. Briefly, the pSIV3+ gag/pol expression plasmid was obtained by replacing the 5' LTR of SIVmac251 (nucleotides 1 to 506) by the human cytomegalovirus (CMV) early-immediate promoter and enhancer region. The 5' half of the env gene (nt 6582 to 7981) was also removed, leaving the RRE (REV-responsive element) sequence and the 5' and 3' exons of the tat and rev regulatory genes intact. The 3' LTR (nt 9444 to 10249) was replaced by a SV40 polyadenylation sequence, resulting in deletion of the 3' end of the nef gene. Finally, the nef initiation codon was inactivated to prevent translation (figure 1B).

pGREV was used for pseudotyping. It is a bicistronic expression construct encoding the vesicular stomatitis virus glycoprotein (VSV-g) and the REV regulatory protein, linked by an EMCV IRES. Expression of this cassette, which contains the rabbit β-globin intron II and polyadenylation (pA) sequences (figure 1C), is driven by the constitutive CMV promoter.

Production of SIV Vectors

293T cells were plated at a density of 4.0×10^5 cells per well (in 6-well plates) on the day before transfection. Cells were transfected as previously described[42]. SIV vectors were produced by cotransfection with three plasmids: the SIV plasmid vector (pRMES8 or pGASE)(1.7 µg), the helper plasmid, pSIV3+, encoding Gag-Pol and regulatory proteins other than Env and Nef (1.7 µg) and the envelope glycoprotein-encoding plasmid pGREV (2.2 µg). The transfection medium was replaced after 16 hours of incubation. Virus-containing medium was collected 40 hours after transfection, clarified by centrifugation for 5 minutes at 800 g, and passed through a filter with 0.45 µm pores. For high-titer preparations, SIV vectors were concentrated by ultracentrifugation at 110,000 g for 2 hours. The viral pellet was resuspended by incubation for 2 hours at 4°C in phosphate-buffered saline supplemented with 1% glycerol[34].

For determination of the infectious titer, sMAGI cells were seeded at a density of 4×105 cells/ml in six-well plates one day before transduction in DMEM medium (Life Technologies Inc., Berlin, Germany) supplemented with 10% fetal bovine serum (FBS) (Gibco BRL, Grand Island, New York, USA), polybrene (6 µg/ml) (Sigma, Saint Louis, USA) and an antibiotic mixture (5 mg/ml penicillin; 5 mg/ml streptomycin; 10 mg/ml neomycin; Gibco BRL, Grand Island, New York, USA). The cells were cultured for one day, and we then added serial dilutions of virus preparations and incubated the plates for a further four hours. Cells were then washed in DMEM (Life Technologies Inc., Berlin, Germany). Transduction rates was determined 48 hours after infection, as the percentage of GFP-positive sMAGI cells (%GFP+c), by flow cytometry (FACScan, Becton Dickinson, San Jose, Mountain View, California, USA) after transducing 4×105 cells with 1 ml of diluted viral supernatant (dilution factor = d). The infectious titer (IT), expressed as transducing units/ml, was calculated as: IT = %GFP+cells × 4 × 105/100 × d.

Transduction of Immunoselected CD34+ Cells

Following immunoselection, CD34+ cells were cultured in a proliferation medium composed of Iscove's MDM supplemented with 1% bovine serum albumin (BSA), bovine pancreatic insulin (10 µg/ml), human transferrin (200 µg/ml), 2-mercaptoethanol (10-4M) and L-glutamine (2 mM; Stemspan, Stem Cell Technologies, Meylan, France). The medium was supplemented with 50 ng/ml recombinant human (rh) SCF (Stem Cell Technologies, Meylan, France), 50 ng/ml rh Flt3-L (Stem Cell Technologies, Meylan, France), 10 ng/ml rh IL-3 (R&D Systems, Minneapolis, USA),10 ng/ml rh IL-6 (R&D Systems, Minneapolis, USA) and 4 µg/ml polybrene (Sigma, Saint Louis, USA) in plates coated with retronectin (Cambrex Bio Science, Paris, France). The CD34+cells were then transduced by 24 hours of coculture with the vector (multiplicity of infection (MOI) = 100).

Myeloid Differentiation of CD34+ Cells

Following the coculture of CD34+ cells with lentiviral vector, part of the cell culture was fixed in CellFix solution (Becton Dickinson, Erembodegem, Belgium) for evaluation of the rate of transduction of undifferentiated CD34+ cells. Part of the cell culture was cultured for 14 days in 35 mm Petri dishes containing semi-solid medium (Methocult GF H4434, Stem Cell Technologies, Meylan, France) composed of Iscove's MDM medium supplemented with 1% methylcellulose, 30% fetal bovine serum, 10-4 M 2-mercaptoethanol, 2 mM L-glutamine, 50 ng/ml rhSCF, 10 ng/ml rhGM-CSF, 10 ng/ml rhIL-3 and 3 IU/ml rhEPO. Cells were cultured at a density of 104 cells/ml (in triplicate) at 37°C, under an atmosphere containing 5% CO_2, to allow the myeloid differentiation of colony-forming cells (CFC).

The remaining cells were cocultured in 96-well plates for 35 days at 37°C, under an atmosphere containing 5% CO_2, on a layer of stromal cells of the murine fibroblastic cell line M2-10B4, in a medium composed of αMEM supplemented with 12.5% horse serum (HS), 12.5% FBS, 2 mM L-glutamine, 10-4 M 2-mercaptoethanol, 0.16 M I-inositol and 16 µM folic acid (Myelocult H5100, Stem Cell Technologies, Meylan, France) and 10-6 M hydrocortisone. Cells were cultured at a concentration of 103 cells per well (24 wells per condition per monkey), to allow long-term culture-initiating cells (LTC-IC) to undergo myeloid differentiation to generate progenitor cells or CFC. The CFC were cultured for 14 days on semi-solid medium, as described above, to allow their myeloid differentiation into more mature cells.

AZT Pretreatment of Immunoselected CD34⁺ Cells

CD34⁺ cells were treated with AZT before transduction, to inhibit transduction due to reverse transcription of the lentiviral vector genome. Immunoselected CD34+ cells were cultured overnight in the proliferation medium described above, with AZT concentrations of 0, 10^{-7}, 10^{-6} and 10^{-5} molar. The cells were washed twice and transduced with the lentiviral vector, according to the protocol described above. The real percentage of GFP-positive cells resulting from reverse transcription of the lentiviral vector was thus determined by subtracting the percentage of GFP-positive cells obtained after treatment with a saturating dose of AZT, from the percentage of GFP-positive cells obtained in the absence of AZT treatment.

Fluorescence Microscopy

After transduction and myeloid differentiation in semi-solid medium, the colonies formed by AZT-treated CFC were observed by fluorescence microscopy (Axiovert S100, Zeiss) using a magnification factor of 100. Fluorescence microscopy was used to detect GFP in each colony subtype, making it possible to determine the percentage of the colonies positive for GFP. We considered all colonies containing GFP-producing cells to be GFP-positive. Images were analyzed with Adobe Premiere and Adobe Photoshop software (Adobe Systems Inc., San Jose, CA, USA).

Gamma Irradiation

Eight animals were sedated with ketamine (Imalgène; 10 mg/kg, i.m., Rhône-Mérieux, France) and placed in a restraint chair. They received myeloablative conditioning, in the form of total body exposure to 60Co gamma rays with an anterior unilateral direction. A total midline tissue dose of 6 Gy was delivered at a rate of 25.92 cGy/minute. Dosimetry was performed, with 100 µL ionization chambers placed in paraffin wax cylindrical phantoms of a similar size and orientation to the seated animal.

Transplantation of Modified CD34⁺ Bone Marrow Cells

After the coculture of CD34⁺ cells with the lentiviral vector, four animals underwent intramedullary infusion, of whole immunoselected CD34⁺ cells into both humeri (Table 1).

Table 1. Reconstitution with transduced autologous CD34⁺ cells in irradiated cynomolgus macaques

Monkeys	CD34⁺ cells purity	CD34⁺ cells collected	CD34⁺ cells transduced	CD34⁺ cells infused/kg
6653	96.42%	8.8×10^6	76.54%	2.96×10^6
6833	95.85%	8.0×10^6	67.74%	1.50×10^6
6896	95.46%	7.3×10^6	67.76%	1.47×10^6
7036	97.08%	5.5×10^6	74.22%	1.46×10^6

Clinical Support

All animals received clinical support in the form of antibiotics and fresh irradiated whole blood, as required. An prophylactic antibiotic regimen was initiated when leukocyte count fell below $1,000/\mu l$ and continued daily until it exceeded $1,000/\mu l$ for three consecutive days: 1 ml/10 kg/day Bi-Gental® (Schering-Plough Santé Animale) and 1 ml/10 kg Terramycin® (Pfizer). Fresh, irradiated (25 Gy; 137Cs gamma radiation) whole blood (approximately 50 ml/transfusion) from a random donor pool was administered if platelet count fell below $20,000/\mu l$ and hemoglobin concentration was less than 6 g/dl.

Flow Cytometry Analysis

Peripheral blood and bone marrow mononuclear cells were incubated for 30 min at 4°C with 10 μl of selected monoclonal antibodies for single- or triple-color membrane staining. The following antibodies were used: APC-conjugated anti-CD3 (SP34-2, Becton Dickinson), PE-conjugated anti-CD14 (clone M5E2, BD Pharmingen), PE-conjugated anti-CD11b (BEAR-1, Beckman Coulter), PerCP-conjugated anti-CD20 (clone B9E9, Immunotech), PE-conjugated anti CD8 (clone RPA-T8, Becton Dickinson) and PerCP-conjugated antiCD4 (clone L200, BD Pharmingen). Cells were washed twice and fixed in CellFix solution (Becton Dickinson, Erembodegem, Belgium) for 3 days before analysis on a Becton Dickinson FACS apparatus with CellQuest Software (Becton Dickinson). eGFP fluorescence was detected in the isothiocyanate (FITC) channel. Negative controls from normal macaques were run with every experimental sample and were used to establish gates for eGFP quantification.

Polymerase Chain Reaction (PCR) Assays

Cellular DNA was extracted from peripheral blood mononuclear cell (PBMC) samples, using the High Pure PCR Template Preparation Kit according to the manufacturer's instructions (Roche Mannheim, Germany). DNA was quantified by measuring optical density (Spectra Max 190; Molecular Devices, California, USA). The eGFP sequence was analyzed by quantitative real-time PCR on 250 ng of DNA run on an iCycler real-time thermocycler (Bio-Rad, California, USA).

Primers were as follows: forward primer, 5'ACGACGGCAACTACAAGACC3'; reverse primer, 5'GCCATGATATAGACGTTGTGG3'. Amplification was performed in a final volume of 50 μl, with IQ™ SYBR°Green Supermix (Bio-Rad, California, USA), in accordance with the manufacturer's instructions. Amplification was carried out over 40 cycles of denaturation at 95°C, annealing at 59°C and elongation at 72°C. Standard curves for the eGFP sequence were generated by serial 10-fold dilutions of duplicate samples of the eGFP plasmid in DNA from untransduced PBMC, with 250 ng of total DNA in each sample. Samples from animals were run in duplicate, and the values reported correspond to the means for replicate wells.

Statistical Analysis

Paired and unpaired comparisons were performed using non parametric Kruskal Wallis, Wilcoxon rank and Mann & Whitney tests, respectively, both of which can be used for the analysis of small samples when normal distribution is uncertain or not confirmed. Tests were performed using StatView 5.01 software (Abacus Concepts, Berkeley, CA).

Results

Efficient Transduction of Cynomolgus Macaque CD34$^+$ Bone Marrow Cells

We first assessed, in vitro, the efficiency with which a SIVmac251-derived vector transduced CD34$^+$ hematopoietic cells from macaque bone marrow (BM). We harvested BM cells from the iliac crests of 12 different animals. CD34$^+$ cell preparations with a purity of 97% ± 1% were obtained by immunomagnetic purification. The CD34$^+$ cells were then transduced by coculture for 24 h with the lentiviral vector (MOI = 100) in medium supplemented with SCF, Flt3-L, IL-3 and IL-6. The vector used (pRMES8) was derived from SIVmac251 and contains the eGFP reporter gene under control of the phosphoglycerate kinase (pGK) promoter (Figure 1). Transduction efficiency (Figure 2A and 2B), as evaluated by flow cytometry analysis of eGFP expression at 24 h, was 41% ± 9% on average (n = 12). After 24 hours of culture with the lentiviral vector, some of the purified CD34+ cells were cultured for 14 days in semi-solid medium containing SCF, GM-CSF, IL-3 and EPO to allow the myeloid differentiation of colony-forming cells (CFC), whereas some cells were cocultured for 35 days on a layer of murine fibroblasts of the M2-10B4 cell line and were then cultured for 14 days on semi-solid medium containing SCF, GM-CSF, IL-3 and EPO, for the identification of long-term culture-initiating cells (LTC-IC). Transduction had no effect on the clonogenic capacity of CD34+ cells: the mean number of colonies was 41 ± 10 for non transduced cells and 44 ± 12 for pRMES8-transduced cells (12 animals tested, P = 0.60 (Mann & Whitney test)). Similar results were obtained for LTC-IC, with 19 ± 3 colonies obtained for non transduced cells and 19 ± 3 for transduced cells (n = 12; P = 0.79 (Mann & Whitney test)). Transduction rates did not differ significantly between CFC and LTC-IC (P = 0.4884 (Wilcoxon test), n = 12), with 18% ± 7% and 19% ± 7% of colonies, respectively, eGFP-positive. However, in both cases, the percentage of eGFP-positive cells was significantly lower than that observed 24 hours after transduction (P < 0.0001 (Wilcoxon test)). This apparent discrepancy between analyses carried out at 24 h and analyses on CFC or LTC-IC may be due to the eGFP protein present in viral particles and incorporated into the cell cytoplasm during the coculture period. The proportion of cells producing eGFP shortly after transduction was reduced by 25% ± 15% (Figure 2C) if 10-6 M AZT was added to cocultures of CD34+ BM cells and lentiviral vector (MOI = 100). Untreated CFC cultures gave percentages of eGFP-producing cells similar to those observed before differentiation (26% ± 5%) (Figure 2D). No fluorescence was detected after myeloid differentiation of the AZT-treated CFC (n = 3), confirming that eGFP detection resulted from the production of this protein from integrated vector.

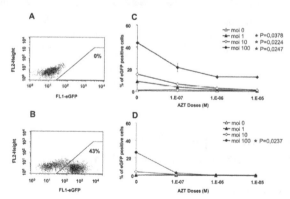

Figure 2. Efficiency of transduction of cynomologus macaque primitive hematopoietic cells with SIV-based lentiviral vectors. A: Non transduced cells were used as a control for each animal. B: Transduction of bone marrow progenitor cells with an SIV-based vector. CD34+ cells were cultured in the presence of cytokines (see materials and methods) and exposed to vector particles at an MOI of 100 for 24 hours before FACS analysis for eGFP production. C: CD34+ cells were cultured overnight in a proliferation medium supplemented with various concentrations of AZT (100 nM, 1 mM, 10 mM). Cells were then washed twice and transduced with various multiplicities of infection (MOI) of the lentiviral vector (0, 1, 10, 100). After 24 hours of coculture with lentiviral vector, some of the CD34+ cells were used to evaluate the rate of transduction of undifferentiated CD34+ cells (C); * indicate statistically significant differences (Kruskal Wallis test) between cultures with and without AZT treatment for MOI = 1 (p = 0,0378), MOI = 10 (p = 0,0224) and MOI = 100 (p = 0,0247). Some of the cells were cultured for 14 days, to allow the myeloid differentiation of CFC. Cells were then resuspended, washed and fixed for three days. They were analyzed by flow cytometry, to evaluate the percentage of eGFP-positive cells and determine the rate of transduction (D); * indicates a statistically significant difference (p = 0,0237(Kruskal Wallis test)) between cultures with and without AZT treatment for MOI = 100. The results shown are the mean values for the three monkeys, each studied in triplicate.

Mosaicism was observed in eGFP gene expression in several colonies (Figure 3). Indeed, eGFP was detected in 56% ± 4% of colonies, whereas only 26% ± 5% of individual cells were eGFP-positive. These results suggest that, on average, only 47% of cells from a single colony contained the SIV vector.

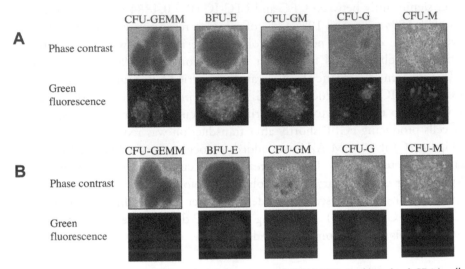

Figure 3. Fluorescence microscopy after myeloid differentiation of CFC (×100). Freshly isolated CD34⁺ cells were transduced or not with the lentiviral vector (24 hours of culture with lentiviral vector at MOI = 100). Cells were then cultured for 14 days in the presence of cytokines, to allow myeloid differentiation of transduced (A) and not transduced (B) CD34⁺ cells. Abbreviations: CFU-GEMM, Colony-Forming Unit-Granulocytes, Erythroid, Macrophage, Megakaryocyte; BFU-E, Burst-Forming Unit-Erythroid; CFU-GM, Colony-Forming Unit-Granulocytes, Macrophage; CFU-G, Colony-Forming Unit-Granulocytes; CFU-M, Colony-Forming Unit-Macrophage.

Transplantation of Autologous BM CD34⁺ Cells Transduced by SIV-Based Vector into Cynomolgus Macaques

We explored the capacity of autologous CD34⁺BM cells transduced ex vivo with a lentiviral vector to engraft efficiently into macaques after total body irradiation (TBI) with a gamma source at the sublethal dose of 6 Gy. Three groups of 4 animals were used: 1) In Group 1, macaque CD34⁺ BM cells (96% ± 1% pure on average) were obtained from the two humeri before gamma irradiation (Table 1). These cells were cocultured, as described above, with pGASE, which is an improved version of pRMES8. Indeed, a mean transduction efficiency of 72% ± 4% was obtained (n = 4) at 24 hours and 37% ± 10% of CFC produced eGFP. Two days after gamma irradiation, 1.4×10^6 to 2.9×10^6 CD34⁺ cells per kg were injected into both humeri of macaques (Table 1); 2) Group 2 included irradiated (6 Gy) macaques that did not undergo cell transplantation: 3) Group 3 included

4 non irradiated animals, which were used as controls, with a similar bleeding frequency.

Reconstitution of Hematopoietic Cells in Vivo

Following total-body irradiation with 6 Gy, transfusion and an antibiotic regimen were required to ensure that all the animals survived. However, one animal from group 1 (7036) died on day 40 due to profound pancytopenia (Figure 4). This macaque received the smallest number of autologous and transduced CD34+ BM cells. All other animals from groups 1 and 2 were studied from days -1 to 471 after gamma irradiation. Controls were followed over the same period.

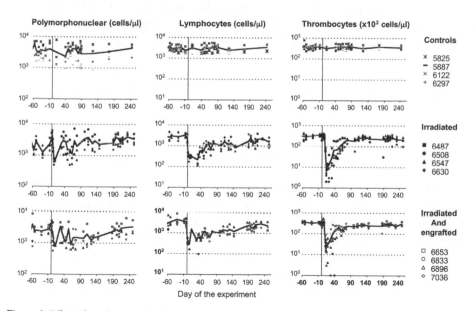

Figure 4. Effect of irradiation and transplantation on polymorphonuclear cell, lymphocyte and thrombocyte counts. All animals were followed during the weeks preceding the study, and for more than 240 days after the irradiation. We carried out hematological analysis including blood cell counts with an automated hemocytometer (Coulter Corporation, Miami, USA).

Radiation rapidly induced severe anemia in all animals (data not shown). A significant decrease in the number of polymorphonuclear cells in the periphery was observed, starting on day 1 after irradiation (Figure 4). No significant difference was observed between the animals of groups 1 and 2 in terms of the minimum number of cells (821 ± 226 cells/µl for group 1 and 658 ± 107 cells/µl for group 2, P = 0.3768 (Mann & Whitney test)) or the time at which that minimum occurred (6 ± 5 days for group 1 and 7 for group 2, P = 0.4795 (Mann &

Whitney test)). Lymphocyte counts also decreased in all macaques by day 1 after gamma irradiation (Figure 4), falling to a minimum of 220 ± 107 lymphocytes/μl on day 18 ± 12 in group 2 and of 347 ± 62/μl on day 11 ± 12 in transplanted animals (group 1). Animals undergoing transplantation tended to display less severe lymphopenia, but no statistical difference was observed between the two groups of irradiated animals in terms of the day on which minimum lymphocyte count was reached (P = 0.1939 (Mann & Whitney test)) or the level of that minimum (P = 0.3805 (Mann & Whitney test)). A significant decrease in platelet counts, beginning by day 10 (Figure 4), was observed in all irradiated animals. Thrombocytopenia (platelet count < 20,000/μl) was characterized in non transplanted animals by a minimum value of 3.75 ± 2.49 × 103 platelets/μl on day 18 ± 3. Thrombocytopenia tended to be less severe in transplanted animals, but this difference was not significant for the minimum number of platelets (10.33 ± 5.25 × 103 platelets/μl; P = 0.1124 (Mann & Whitney test)) or for the day on which that minimum occurred (14.33 ± 0.94; P = 0.3123 (Mann & Whitney test)). This thrombocytopenia required one transfusion in all animals (other than animal 7036, which needed two transfusions) of both groups. However, platelet reconstitution seemed to be correlated with the dose of CD34+ cells infused, the speed of reconstitution increasing with the number of CD34+ cells injected (macaque 6653).

Reconstitution of Bone Marrow Clonogenic Activity

We determined the effects of CD34[+] bone marrow cell transplantation following gamma irradiation on the ex vivo proliferation and differentiation of hematopoietic progenitors. Before gamma irradiation, a mean of 40 ± 9 and 38 ± 6 colonies was observed for groups 1 and 2, respectively (Figure 5). Colony number decreased significantly (P < 0.0001 (Wilcoxon test)) by day 7 in all animals. In both groups, clonogenic activity was detected by day 43 after gamma irradiation with reconstitution significantly better in the animals undergoing transplantation than in those that did not undergo transplantation (P = 0.0009 (Mann & Whitney test)).

Presence of eGFP-Positive Cells in Bone Marrow and Peripheral Blood

Cells with integrated SIV-vector DNA were detected by PCR (Table 2) as early as day 3 after transplantation, in at least two animals (6653 and 6833). These two animals had received the largest numbers of transduced CD34[+] bone marrow cells. Monkey 7036, which died within 40 days of gamma irradiation had very few transduced cells in the bone marrow and SIV-DNA was not detected in peripheral blood cells. In the three remaining animals, vector DNA was detected in peripheral blood cells (up to 500 copies per million cells) and in the bone marrow (up to 6250 copies per million cells) more than one year after transplantation (day 471).

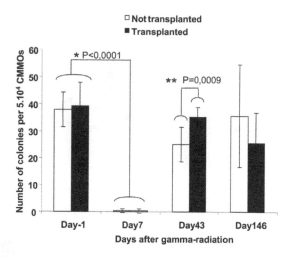

Figure 5. Recovery of bone marrow clonogenic activity. Bone marrow-derived colony-forming units following sublethal irradiation of cynomolgus monkeys transplanted (black bars) or not transplanted with CD34$^+$ cells (white bars). Mean ± SD of CFC number (triplicate). The results of statistical test are indicates; * indicates a statistically significant difference (p < 0,0001 (Wilcoxon test)) between day 0 and day 7 for the both group; ** indicates a statistically significant difference (p = 0,0009 (Mann & Whitney test)) at day 43 between animals undergoing transplantation and those that did not undergo transplantation.

Table 2. Number of DNA copies per million mononuclear cells in peripheral blood (PB) and bone marrow (BM)

					Monkey				
		6653		6833		6896		7036	
Days post transplantation		PB	BM	PB	BM	PB	BM	PB	BM
-3		0	0	0	0	0	0	0	0
3		500	ND	250	ND	0	ND	0	15
5		250	500	ND	250	ND	ND	0	0
108		250	ND	250	ND	1250	ND	*	*
121		750	ND	250	ND	250	ND	*	*
128		250	ND	250	ND	250	ND	*	*
142		250	ND	250	ND	1750	3250	*	*
471		ND	250	250	250	500	6250	*	*

ND: not determined
*: 7036 died on day 40

Flow cytometry analysis demonstrated the presence of eGFP-producing cells among peripheral blood mononuclear cells in myeloid and lymphoid lineges of monkey 6896 (Figure 6). Peripheral blood cells were sorted on the basis of eGFP production, with the aim of characterizing the phenotype of populations of cells expressing the transgene in more detail. We found that 61% of eGFP-positive cells were CD11b-positive,5% of these cells appeared to be CD14+ monocytes, 14% were CD20+ B cells and 10% were CD3+ T cells, 23% of which expressed CD8 and 77% expressed CD4 (data not shown).

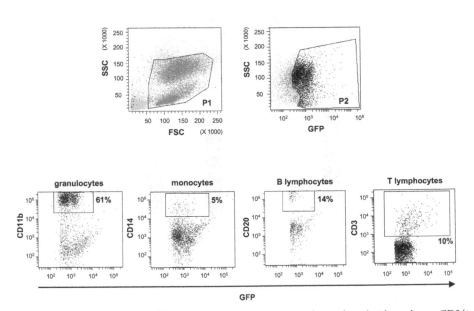

Figure 6. Flow cytometry analysis of hematopoiesis reconstitution. Animal transplanted with autologous CD34⁺ bone marrow cells transduced with an SIV-based vector. eGFP-positive cells present in P1 and P2 were analyzed by immuno-staining to identify the subpopulations of eGFP-positive cells in peripheral blood. CD20-PerCP-Cy5, CD14-PE, CD11b-APC and CD3-APC staining were used to identify the B-lymphocyte, monocyte, granulocyte and T-lymphocyte subpopulations.

Discussion

The aim of this work was to study reconstitution of the myeloid and lymphoid compartments after the autologous transplantation of genetically modified CD34⁺ bone marrow cells into cynomolgus macaques previously subjected to gamma irradiation.

We first assessed, in vitro, the efficiency with which a SIVmac251-derived vector transduced macaque CD34+ hematopoietic bone marrow cells. These vectors are similar to those derived from HIV. However, SIV-derived vectors clearly outperform HIV-derived vectors in simian cells. In fact, HIV-1 fails to replicate in simian cells because of an early postentry block [43,44], and Kootstra et., al showed that the viral determinant involved in postentry restriction of HIV-1 replication in simian cells is located at or near the cyclophilin A (CyPA) binding region of the capside protein [45]. The hydrophobic pocket of cyclophilin A (CypA) makes direct contact with an exposed, proline-rich loop on HIV-1 capsid (CA) and renders reverse transcription complexes resistant to an antiviral activity in human cells. A CypA fusion with TRIM5 (a member of the tripartite motif family) that is unique to New World owl monkeys also targets HIV-1 CA, but this interaction potently inhibits infection. A similar block to HIV-1 infection in Old World monkeys is attributable to the α isoform of the TRIM5 orthologue in these species and using RNA interference techniques, Berthoux et., al demonstrated that CypA inhibits HIV-1 replication in these cells because it is required for CA recognition by TRIM5α [46]. SIV vectors can also efficiently transduce human cells[33,47], and may therefore prove a useful alternative to HIV-1-based vectors, at least in the early phase of preclinical testing of lentivirus vectors. We found that the proportion of eGFP-positive cells obtained before myeloid differentiation (mean value of 30%) was similar to that obtained with CD34+ cells from human donors transduced with lentiviral [48-51], retroviral [52-54], AAV[55], or adenovirus/AAV-derived [56] vectors. However, it is possible to increase the transduction rate, such that 90% transduced human CD34+ cells are obtained from cord blood, 80% from bone marrow and 75% from G-CSF mobilized peripheral blood [57]. We analyzed transduction in two types of assay, based on committed (CFC) and primitive (LTC-IC) hematopoietic progenitors, as analyses of the transduction of committed progenitors only bear little relation to the transduction efficiency for stem cells and less differentiated cells in the long term. After myeloid differentiation, eGFP+ cells were detected, in similar proportions, in CFC on day 15 and LTC-IC on day 50 after transduction, indicating that the vector was able to transduce progenitor cells and most immature hematopoietic cells with a similar efficiency. Similar results have been reported for stimulated human CFC and LTC-IC, which were found to be transduced with similar efficiency by a lentiviral vector based on HIV-1[58]. In this previous study, significant resistance to lentiviral transduction was reported in unstimulated primitive human cells. These results may explain why, in our study, the use of cytokines during transduction made possible the genetic modification of LTC-IC, which are quiescent. Cytokine treatment may have led to these cells entering the cell cycle, facilitating transduction. This result confirms the greater efficiency of lentiviral vectors than of retroviral vectors for the transduction of CD34+ cells. Nevertheless, in our study, only half as many eGFP-positive cells were obtained after differentiation as were obtained from undifferentiated CD34+ cells. Similar observations have been made with MLV-transduced progenitor cells from human donors[59]. We demonstrate here that these differences may be accounted for by the pseudotransduction detected at 24 h of incubation with the vector, confirming the results reported with CD34+ cells in studies using VSVg-pseudotyped MLV-derived[60] or lentivirus-derived vectors[51]. It has been suggested that pseudotransduction may result from VSVg-pseudotyping due to membrane fusion efficiency being higher than the rate of integration of the transgene[61]. Nevertheless, most lentiviral vectors have been generated with VSV-G, as this glycoprotein makes it easy to recover and concentrate the pseudotyped vectors [62].

We also showed that eGFP was produced in all colony subtypes. Clusters of eGFP production were observed on fluorescence microscopy, indicating that not all the cells of a given positive colony – theoretically derived from a single cell – produced eGFP. This result is consistent with those of Mikkola et al. concerning murine HSC transduction by a VSVg-pseudotyped lentiviral vector, in which a mismatch was reported between the transduction rate of cells (almost 25%) and the transduction rate of myeloid colonies (almost 60%). These authors highlighted the occurrence of mosaicism in GFP gene expression in colonies obtained following the myeloid differentiation of CD34+ cells[63], possibly due to a delay in the integration of the transgene during differentiation, resulting in the formation of clusters of GFP-positive cells within a single myeloid colony.

In our in vivo study, autologous HSC were injected into the bone marrow, whereas intravenous injection is currently the most frequently used transplantation method. We aimed to increase seeding efficiency and homing, as only a limited number of stem cells were theoretically available. However, 2.5×10^6 to 5.0×10^6 CD34+ cells is generally sufficient to ensure engraftment, and we found that less than 2.0×10^6 cells were sufficient for long-term reconstitution in macaques. As predicted[64,65], total-body gamma irradiation leads to a drastic decrease in the number of hematopoietic progenitors, preventing the development of mature cells [66]. Despite the occurrence of severe pancytopenia, a positive correlation has been found between the number of CD34+ cells infused and time required for immune reconstitution [42,67,68]. However, hematopoietic recovery may take longer if fewer than 2.0×10^6 CD34+ cells/kg are infused. This notion is consistent with our observation that CD34+ cell transplantation decreases both the severity and duration of irradiation-induced cytopenia. Clonogenic activity also reappeared more strongly in transplanted animals. We also showed that the animals recovered B cells, T cells, monocytes and granulocytes. Nevertheless, the functional activity of these cells requires confirmation, particularly for lymphocytes. However, although we observed long-term reconstitution with lentiviral vector-transduced cells of different lineages, its proportion remained below 1%. Hanawa et al., provided the first evidence that SIV-based vectors can successfully transduce rhesus macaque repopulating hematopoietic stem cells, with an average of 16% of peripheral blood leukocytes containing the SIV vector genome. However, this study was carried out with HSC from mobilized peripheral blood cells, making it possible to obtain larger numbers of HSC than can be harvested from bone marrow. Nevertheless theoretically, these cells contained more progenitors that were already committed and fewer pluripotent stem cells capable of long-term reconstitution than medullary HSC[69]. The small numbers of eGFP-producing cells observed in our study may be due to an anti-eGFP immune response. Some reports have suggested that such reactions do not generally occur after irradiation[70], but two reports described the induction of cytotoxic T-lymphocyte responses to enhanced green (GFP) or yellow (YFP) fluorescent proteins after myeloablative conditioning. One of these reports concerned baboons that had received primitive hematopoietic cells transduced with HIV-1-based lentiviral vectors[71] and the other concerned rhesus macaques that had received CD34+ stem cells transduced with a retroviral vector[72].

Lentiviruses, like retroviruses, can be used to integrate transgenes into the host genome. Two severe adverse events occurred in two patients in the SCID-X1 gene therapy trial 30 to 34 months after injection of the autologous CD34+ cells corrected using a retroviral vector. In these patients, an uncontrolled clonal T lymphoproliferative syndrome, similar to acute lymphoblastic leukemia, was observed [73,74]. This study highlights the risk of insertional mutagenesis restricted to retroviral and lentiviral gene transfer. In the future, additional safety measures could be considered, such as the use of self-inactivating LTRs (as in our study) to reduce enhancer activity, the addition of insulators to reduce the risk further, and the insertion of a second transgene encoding a "suicide" product, such as herpes thymidine kinase, making it possible to kill the transduced cells with ganciclovir. Unlike studies in mice, in which the follow-up period is necessarily limited, studies in large animals, with a longer life span, are compatible with more extensive follow-up. The development of linear amplification-mediated PCR (LAM-PCR), a sensitive and robust approach to molecular clonal analysis, has made it possible to identify and analyze the contribution of individual transduced clones to hematopoiesis. Clonal analysis may provide information about the dominance of transduced clones, potentially predicting possible progression or the propensity to develop clonal hematopoiesis and leukemia. Moreover, replication-competent retrovirus (RCRs), recombinant retrovirus and interaction with endogenous retroviruses should also be investigated, when evaluating the biosafety of retrovirus and lentivirus.

Conclusion

The results reported here provide the first evidence that gene transfer into medullary hematopoietic stem cells and long-term expression of the transgene are possible, using an SIV-based lentiviral vector in non human primates, which provide the best clinical models for in vivo evaluation of the feasibility and safety of gene therapy strategies.

Competing Interests

The authors never received reimbursements, fees, funding, or salary from an organization that may in any way gain or lose financially from the publication of this paper. The authors never have any stocks or shares in an organization that may in any way gain or lose financially from the publication of this paper. The authors have no competing interests to declare in relation to this paper.

Authors' Contributions

SD was the main contributor to this paper. This work is part of her PhD project. She carried out transduction of CD34⁺ cells, transplantation of animals, PCR for identification of cells expressing the transgene in vivo, flow cytometry analysis, WG Have improved assays for transduction of macaque bone marrow CD34⁺ cells with SIV derived vector, DN constructed and produced the SIV derived vector, SP technical assistance to cell sorting, MLD technical assistance to transplantation, BD technical assistance to cell culture, flow cytometry and irradiadion of NHP, GA technical assistance to molecular biology, TA technical assistance to flow cytometry and cell sorting, JLL irradiation of animals and dosimetry, FLC supervises vector design and production, RLG supervisor of SD.

Acknowledgements

We would like to thank M. Ripaux, A. Fort, S. Jacquin, D. Mérigard, P. Pochard, D. Renault, J. C. Wilks and R. Rioux for excellent technical assistance. This work was supported by the Agence Nationale de Recherches sur le SIDA (ANRS, Paris, France), the Centre de Recherches du Service de Santé des Armées Emile Pardé (CRSSA, La Tronche, France), and the Commissariat à l'Energie Atomique (CEA, Fontenay aux Roses, France).

References

1. Miller DG, Adam MA, Miller AD: Gene transfer by retrovirus vectors occurs only in cells that are actively replicating at the time of infection. Mol Cell Biol 1990, 10:4239–4242.

2. Jones RJ, Wagner JE, Celano P, Zicha MS, Sharkis SJ: Separation of pluripotent haematopoietic stem cells from spleen colony-forming cells. Nature 1990, 347:188–189.

3. Gothot A, Loo JC, Clapp DW, Srour EF: Cell cycle-related changes in repopulating capacity of human mobilized peripheral blood CD34(+) cells in non-obese diabetic/severe combined immune-deficient mice. Blood 1998, 92:2641–2649.

4. Hao QL, Thiemann FT, Petersen D, Smogorzewska EM, Crooks GM: Extended long-term culture reveals a highly quiescent and primitive human hematopoietic progenitor population. Blood 1996, 88:3306–3313.

5. Ploemacher RE, Sluijs JP, Voerman JS, Brons NH: An in vitro limiting-dilution assay of long-term repopulating hematopoietic stem cells in the mouse. Blood 1989, 74:2755–2763.

6. Ploemacher RE, Sluijs JP, van Beurden CA, Baert MR, Chan PL: Use of limiting-dilution type long-term marrow cultures in frequency analysis of marrow-repopulating and spleen colony-forming hematopoietic stem cells in the mouse. Blood 1991, 78:2527–2533.

7. Traycoff CM, Kosak ST, Grigsby S, Srour EF: Evaluation of ex vivo expansion potential of cord blood and bone marrow hematopoietic progenitor cells using cell tracking and limiting dilution analysis. Blood 1995, 85:2059–2068.

8. Bodine DM, Crosier PS, Clark SC: Effects of hematopoietic growth factors on the survival of primitive stem cells in liquid suspension culture. Blood 1991, 78:914–920.

9. Larochelle A, Vormoor J, Hanenberg H, Wang JC, Bhatia M, Lapidot T, Moritz T, Murdoch B, Xiao XL, Kato I, et al.: Identification of primitive human hematopoietic cells capable of repopulating NOD/SCID mouse bone marrow: implications for gene therapy. Nat Med 1996, 2:1329–1337.

10. Peters SO, Kittler EL, Ramshaw HS, Quesenberry PJ: Ex vivo expansion of murine marrow cells with interleukin-3 (IL-3), IL-6, IL-11, and stem cell factor leads to impaired engraftment in irradiated hosts. Blood 1996, 87:30–37.

11. Tisdale JF, Hanazono Y, Sellers SE, Agricola BA, Metzger ME, Donahue RE, Dunbar CE: Ex vivo expansion of genetically marked rhesus peripheral blood progenitor cells results in diminished long-term repopulating ability. Blood 1998, 92:1131–1141.

12. Lewis PF, Emerman M: Passage through mitosis is required for oncoretroviruses but not for the human immunodeficiency virus. J Virol 1994, 68:510–516.

13. Naldini L: Lentiviruses as gene transfer agents for delivery to non-dividing cells. Curr Opin Biotechnol 1998, 9:457–463.

14. Fouchier RA, Malim MH: Nuclear import of human immunodeficiency virus type-1 preintegration complexes. Adv Virus Res 1999, 52:275–299.

15. Naldini L, Blomer U, Gallay P, Ory D, Mulligan R, Gage FH, Verma IM, Trono D: In vivo gene delivery and stable transduction of nondividing cells by a lentiviral vector. Science 1996, 272:263–267.

16. Richardson JH, Kaye JF, Child LA, Lever AM: Helper virus-free transfer of human immunodeficiency virus type 1 vectors. J Gen Virol 1995, 76(Pt 3):691–696.

17. Arya SK, Zamani M, Kundra P: Human immunodeficiency virus type 2 lentivirus vectors for gene transfer: expression and potential for helper virus-free packaging. Hum Gene Ther 1998, 9:1371–1380.

18. Poeschla EM, Wong-Staal F, Looney DJ: Efficient transduction of nondividing human cells by feline immunodeficiency virus lentiviral vectors. Nat Med 1998, 4:354–357.

19. Mitrophanous K, Yoon S, Rohll J, Patil D, Wilkes F, Kim V, Kingsman S, Kingsman A, Mazarakis N: Stable gene transfer to the nervous system using a non-primate lentiviral vector. Gene Ther 1999, 6:1808–1818.

20. Hu J, Dunbar CE: Update on hematopoietic stem cell gene transfer using non-human primate models. Curr Opin Mol Ther 2002, 4:482–490.

21. Donahue RE, Dunbar CE: Update on the use of nonhuman primate models for preclinical testing of gene therapy approaches targeting hematopoietic cells. Hum Gene Ther 2001, 12:607–617.

22. Villinger F, Brar SS, Mayne A, Chikkala N, Ansari AA: Comparative sequence analysis of cytokine genes from human and nonhuman primates. J Immunol 1995, 155:3946–3954.

23. Farese AM, MacVittie TJ, Roskos L, Stead RB: Hematopoietic recovery following autologous bone marrow transplantation in a nonhuman primate: effect of variation in treatment schedule with PEG-rHuMGDF. Stem Cells 2003, 21:79–89.

24. Dunbar CE, Takatoku M, Donahue RE: The impact of ex vivo cytokine stimulation on engraftment of primitive hematopoietic cells in a non-human primate model. Ann N Y Acad Sci 2001, 938:236–244. discussion 244–235.

25. Wagemaker G, Neelis KJ, Hartong SCC, Wognum AW, Thomas GR, Fielder PJ, Eaton DL: The efficacy of recombinant TPO in murine And nonhuman primate models for myelosuppression and stem cell transplantation. Stem Cells 1998, 16(Suppl 2):127–141.

26. Shi PA, Hematti P, von Kalle C, Dunbar CE: Genetic marking as an approach to studying in vivo hematopoiesis: progress in the non-human primate model. Oncogene 2002, 21:3274–3283.

27. Burkhard MJ, Dean GA: Transmission and immunopathogenesis of FIV in cats as a model for HIV. Curr HIV Res 2003, 1:15–29.

28. Willett BJ, Flynn JN, Hosie MJ: FIV infection of the domestic cat: an animal model for AIDS. Immunol Today 1997, 18:182–189.

29. Talbott RL, Sparger EE, Lovelace KM, Fitch WM, Pedersen NC, Luciw PA, Elder JH: Nucleotide sequence and genomic organization of feline immunodeficiency virus. Proc Natl Acad Sci USA 1989, 86:5743–5747.

30. Olmsted RA, Barnes AK, Yamamoto JK, Hirsch VM, Purcell RH, Johnson PR: Molecular cloning of feline immunodeficiency virus. Proc Natl Acad Sci USA 1989, 86:2448–2452.

31. Schnell T, Foley P, Wirth M, Munch J, Uberla K: Development of a self-inactivating, minimal lentivirus vector based on simian immunodeficiency virus. Hum Gene Ther 2000, 11:439–447.

32. Wagner R, Graf M, Bieler K, Wolf H, Grunwald T, Foley P, Uberla K: Rev-independent expression of synthetic gag-pol genes of human immunodeficiency virus type 1 and simian immunodeficiency virus: implications for the safety of lentiviral vectors. Hum Gene Ther 2000, 11:2403–2413.

33. Negre D, Mangeot PE, Duisit G, Blanchard S, Vidalain PO, Leissner P, Winter AJ, Rabourdin-Combe C, Mehtali M, Moullier P, et al.: Characterization of novel safe lentiviral vectors derived from simian immunodeficiency virus (SIVmac251) that efficiently transduce mature human dendritic cells. Gene Ther 2000, 7:1613–1623.

34. Mangeot PE, Negre D, Dubois B, Winter AJ, Leissner P, Mehtali M, Kaiserlian D, Cosset FL, Darlix JL: Development of minimal lentivirus vectors derived from simian immunodeficiency virus (SIVmac251) and their use for gene transfer into human dendritic cells. J Virol 2000, 74:8307–8315.

35. Kiem HP, Andrews RG, Morris J, Peterson L, Heyward S, Allen JM, Rasko JE, Potter J, Miller AD: Improved gene transfer into baboon marrow repopulating cells using recombinant human fibronectin fragment CH-296 in combination with interleukin-6, stem cell factor, FLT-3 ligand, and megakaryocyte growth and development factor. Blood 1998, 92:1878–1886.

36. Wu T, Kim HJ, Sellers SE, Meade KE, Agricola BA, Metzger ME, Kato I, Donahue RE, Dunbar CE, Tisdale JF: Prolonged high-level detection of retrovirally marked hematopoietic cells in nonhuman primates after transduction of CD34+ progenitors using clinically feasible methods. Mol Ther 2000, 1:285–293.

37. Kim HJ, Tisdale JF, Wu T, Takatoku M, Sellers SE, Zickler P, Metzger ME, Agricola BA, Malley JD, Kato I, et al.: Many multipotential gene-marked

progenitor or stem cell clones contribute to hematopoiesis in nonhuman primates. Blood 2000, 96:1–8.

38. Rosenzweig M, MacVittie TJ, Harper D, Hempel D, Glickman RL, Johnson RP, Farese AM, Whiting-Theobald N, Linton GF, Yamasaki G, et al.: Efficient and durable gene marking of hematopoietic progenitor cells in nonhuman primates after nonablative conditioning. Blood 1999, 94:2271–2286.

39. Kelly PF, Donahue RE, Vandergriff JA, Takatoku M, Bonifacino AC, Agricola BA, Metzger ME, Dunbar CE, Nienhuis AW, Vanin EF: Prolonged multilineage clonal hematopoiesis in a rhesus recipient of CD34 positive cells marked with a RD114 pseudotyped oncoretroviral vector. Blood Cells Mol Dis 2003, 30:132–143.

40. Hu J, Kelly P, Bonifacino A, Agricola B, Donahue R, Vanin E, Dunbar CE: Direct comparison of RD114-pseudotyped versus amphotropic-pseudotyped retroviral vectors for transduction of rhesus macaque long-term repopulating cells. Mol Ther 2003, 8:611–617.

41. Mangeot PE, Duperrier K, Negre D, Boson B, Rigal D, Cosset FL, Darlix JL: High levels of transduction of human dendritic cells with optimized SIV vectors. Mol Ther 2002, 5:283–290.

42. Chen C, Okayama H: High-efficiency transformation of mammalian cells by plasmid DNA. Mol Cell Biol 1987, 7:2745–2752.

43. Munk C, Brandt SM, Lucero G, Landau NR: A dominant block to HIV-1 replication at reverse transcription in simian cells. Proc Natl Acad Sci USA 2002, 99:13843–13848.

44. Besnier C, Takeuchi Y, Towers G: Restriction of lentivirus in monkeys. Proc Natl Acad Sci USA 2002, 99:11920–11925.

45. Kootstra NA, Munk C, Tonnu N, Landau NR, Verma IM: Abrogation of postentry restriction of HIV-1-based lentiviral vector transduction in simian cells. Proc Natl Acad Sci USA 2003, 100:1298–1303.

46. Berthoux L, Sebastian S, Sokolskaja E, Luban J: Cyclophilin A is required for TRIM5{alpha}-mediated resistance to HIV-1 in Old World monkey cells. Proc Natl Acad Sci USA 2005, 102:14849–14853.

47. Sandrin V, Boson B, Salmon P, Gay W, Negre D, Le Grand R, Trono D, Cosset FL: Lentiviral vectors pseudotyped with a modified RD114 envelope glycoprotein show increased stability in sera and augmented transduction of primary lymphocytes and CD34+ cells derived from human and nonhuman primates. Blood 2002, 100:823–832.

48. Sutton RE, Wu HT, Rigg R, Bohnlein E, Brown PO: Human immunodeficiency virus type 1 vectors efficiently transduce human hematopoietic stem cells. J Virol 1998, 72:5781–5788.

49. Evans JT, Kelly PF, O'Neill E, Garcia JV: Human cord blood CD34+CD38- cell transduction via lentivirus-based gene transfer vectors. Hum Gene Ther 1999, 10:1479–1489.

50. Piconi S, Trabattoni D, Fusi ML, Milazzo F, Dix LP, Rizzardini G, Colombo F, Bray D, Clerici M: Effect of two different combinations of antiretrovirals (AZT+ddI and AZT+3TC) on cytokine production and apoptosis in asymptomatic HIV infection. Antiviral Res 2000, 46:171–179.

51. Case SS, Price MA, Jordan CT, Yu XJ, Wang L, Bauer G, Haas DL, Xu D, Stripecke R, Naldini L, et al.: Stable transduction of quiescent CD34(+)CD38(-) human hematopoietic cells by HIV-1-based lentiviral vectors. Proc Natl Acad Sci USA 1999, 96:2988–2993.

52. Poznansky MC, La Vecchio J, Silva-Arietta S, Porter-Brooks J, Brody K, Olszak IT, Adams GB, Ramstedt U, Marasco WA, Scadden DT: Inhibition of human immunodeficiency virus replication and growth advantage of CD4+ T cells and monocytes derived from CD34+ cells transduced with an intracellular antibody directed against human immunodeficiency virus type 1 Tat. Hum Gene Ther 1999, 10:2505–2514.

53. Rosenzweig M, Marks DF, Hempel D, Lisziewicz J, Johnson RP: Transduction of CD34+ hematopoietic progenitor cells with an antitat gene protects T-cell and macrophage progeny from AIDS virus infection. J Virol 1997, 71:2740–2746.

54. Davis BR, Saitta FP, Bauer G, Bunnell BA, Morgan RA, Schwartz DH: Targeted transduction of CD34+ cells by transdominant negative Rev-expressing retrovirus yields partial anti-HIV protection of progeny macrophages. Hum Gene Ther 1998, 9:1197–1207.

55. Chatterjee S, Li W, Wong CA, Fisher-Adams G, Lu D, Guha M, Macer JA, Forman SJ, Wong KK Jr: Transduction of primitive human marrow and cord blood-derived hematopoietic progenitor cells with adeno-associated virus vectors. Blood 1999, 93:1882–1894.

56. Shayakhmetov DM, Carlson CA, Stecher H, Li Q, Stamatoyannopoulos G, Lieber A: A high-capacity, capsid-modified hybrid adenovirus/adeno-associated virus vector for stable transduction of human hematopoietic cells. J Virol 2002, 76:1135–1143.

57. Amsellem S, Ravet E, Fichelson S, Pflumio F, Dubart-Kupperschmitt A: Maximal lentivirus-mediated gene transfer and sustained transgene expression in human hematopoietic primitive cells and their progeny. Mol Ther 2002, 6:673–677.

58. Zielske SP, Gerson SL: Cytokines, including stem cell factor alone, enhance lentiviral transduction in nondividing human LTCIC and NOD/SCID repopulating cells. Mol Ther 2003, 7:325–333.

59. Yam PY, Yee JK, Ito JI, Sniecinski I, Doroshow JH, Forman SJ, Zaia JA: Comparison of amphotropic and pseudotyped VSV-G retroviral transduction in human CD34+ peripheral blood progenitor cells from adult donors with HIV-1 infection or cancer. Exp Hematol 1998, 26:962–968.

60. Liu ML, Winther BL, Kay MA: Pseudotransduction of hepatocytes by using concentrated pseudotyped vesicular stomatitis virus G glycoprotein (VSV-G)-Moloney murine leukemia virus-derived retrovirus vectors: comparison of VSV-G and amphotropic vectors for hepatic gene transfer. J Virol 1996, 70:2497–2502.

61. Gallardo HF, Tan C, Ory D, Sadelain M: Recombinant retroviruses pseudotyped with the vesicular stomatitis virus G glycoprotein mediate both stable gene transfer and pseudotransduction in human peripheral blood lymphocytes. Blood 1997, 90:952–957.

62. Burns JC, Friedmann T, Driever W, Burrascano M, Yee JK: Vesicular stomatitis virus G glycoprotein pseudotyped retroviral vectors: concentration to very high titer and efficient gene transfer into mammalian and nonmammalian cells. Proc Natl Acad Sci USA 1993, 90:8033–8037.

63. Mikkola H, Woods NB, Sjogren M, Helgadottir H, Hamaguchi I, Jacobsen SE, Trono D, Karlsson S: Lentivirus gene transfer in murine hematopoietic progenitor cells is compromised by a delay in proviral integration and results in transduction mosaicism and heterogeneous gene expression in progeny cells. J Virol 2000, 74:11911–11918.

64. Bolus NE: Basic review of radiation biology and terminology. J Nucl Med Technol 2001, 29:67–73. test 76–67.

65. MacVittie TJ, Farese AM, Herodin F, Grab LB, Baum CM, McKearn JP: Combination therapy for radiation-induced bone marrow aplasia in nonhuman primates using synthokine SC-55494 and recombinant human granulocyte colony-stimulating factor. Blood 1996, 87:4129–4135.

66. Dainiak N: Hematologic consequences of exposure to ionizing radiation. Exp Hematol 2002, 30:513–528.

67. Nieboer P, de Vries EG, Vellenga E, Graaf WT, Mulder NH, Sluiter WJ, de Wolf JT: Factors influencing haematological recovery following high-dose chemotherapy and peripheral stem-cell transplantation for haematological malignancies; 1-year analysis. Eur J Cancer 2004, 40:1199–1207.

68. Jillella AP, Ustun C: What is the optimum number of CD34+ peripheral blood stem cells for an autologous transplant? Stem Cells Dev 2004, 13:598–606.

69. Hanawa H, Hematti P, Keyvanfar K, Metzger ME, Krouse A, Donahue RE, Kepes S, Gray J, Dunbar CE, Persons DA, Nienhuis AW: Efficient gene transfer into rhesus repopulating hematopoietic stem cells using a simian immunodeficiency virus-based lentiviral vector system. Blood 2004, 103:4062–4069.

70. Persons DA, Allay JA, Riberdy JM, Wersto RP, Donahue RE, Sorrentino BP, Nienhuis AW: Use of the green fluorescent protein as a marker to identify and track genetically modified hematopoietic cells. Nat Med 1998, 4:1201–1205.

71. Morris JC, Conerly M, Thomasson B, Storek J, Riddell SR, Kiem HP: Induction of cytotoxic T-lymphocyte responses to enhanced green and yellow fluorescent proteins after myeloablative conditioning. Blood 2004, 103:492–499.

72. Rosenzweig M, Connole M, Glickman R, Yue SP, Noren B, DeMaria M, Johnson RP: Induction of cytotoxic T lymphocyte and antibody responses to enhanced green fluorescent protein following transplantation of transduced CD34(+) hematopoietic cells. Blood 2001, 97:1951–1959.

73. Hacein-Bey-Abina S, von Kalle C, Schmidt M, Le Deist F, Wulffraat N, McIntyre E, Radford I, Villeval JL, Fraser CC, Cavazzana-Calvo M, Fischer A: A serious adverse event after successful gene therapy for X-linked severe combined immunodeficiency. N Engl J Med 2003, 348:255–256.

74. Hacein-Bey-Abina S, Von Kalle C, Schmidt M, McCormack MP, Wulffraat N, Leboulch P, Lim A, Osborne CS, Pawliuk R, Morillon E, et al.: LMO2-associated clonal T cell proliferation in two patients after gene therapy for SCID-X1. Science 2003, 302:415–419.

Open Field Trial of Genetically Modified Parthenocarpic Tomato: Seedlessness and Fruit Quality

Giuseppe Leonardo Rotino, Nazareno Acciarri, Emidio Sabatini,
Giuseppe Mennella, Roberto Lo Scalzo, Andrea Maestrelli,
Barbara Molesini, Tiziana Pandolfini, Jessica Scalzo,
Bruno Mezzetti and Angelo Spena

ABSTRACT

Background

Parthenocarpic tomato lines transgenic for the DefH9-RI-iaaM gene have been cultivated under open field conditions to address some aspects of the equivalence of genetically modified (GM) fruit in comparison to controls (non-GM).

Results

Under open field cultivation conditions, two tomato lines (UC 82) transgenic for the DefH9-RI-iaaM gene produced parthenocarpic fruits. DefH9-RI-iaaM fruits were either seedless or contained very few seeds. GM fruit quality, with the exception of a higher β-carotene level, did not show any difference, neither technological (color, firmness, dry matter, °Brix, pH) nor chemical (titratable acidity, organic acids, lycopene, tomatine, total polyphenols and antioxidant capacity – TEAC), when compared to that of fruits from control line. Highly significant differences in quality traits exist between the tomato F1 commercial hybrid Allflesh and the three UC 82 genotypes tested, regardless of whether or not they are GM. Total yield per plant did not differ between GM and parental line UC 82. Fruit number was increased in GM lines, and GM fruit weight was decreased.

Conclusion

The use in the diet of fruits from a new line or variety introduces much greater changes than the consumption of GM fruits in comparison to its genetic background. Parthenocarpic fruits, produced under open field conditions, contained 10-fold less seeds than control fruits. Thus parthenocarpy caused by DefH9-RI-iaaM gene represents also a tool for mitigating GM seeds dispersal in the environment.

Background

The debate on genetically modified (GM) crop plants has been focused on two main uncertainties: 1) whether a GM plant differs from its non-GM progenitors only in the introduced trait of interest, 2) whether a GM plant is safe in the environment with respect to gene flow and seed dispersal. To address these questions, we have chosen parthenocarpy, the development of the fruit in absence of fertilization, to evaluate the equivalence of GM and non-GM fruit and to evaluate the advantages of parthenocarpy produced by genetic engineering compared to traditional methods. In this work, we present an analysis of parthenocarpic tomato fruit obtained from field-grown GM plants to address some aspects of the equivalence of GM fruit.

The trait of parthenocarpy is particularly important for crop plants whose commercial product is their fruit [1,2]. During flowering, adverse environmental conditions may either prevent or reduce pollination and fertilization decreasing fruit yield and quality. Moreover, parthenocarpic fruits are seedless, and seedlessness is highly valued by consumers in some fruit (e.g. table grape, citrus, eggplant, cucumber).

Parthenocarpic fruits have been produced by traditional breeding methods based either on mutant lines or other strategies such as alteration of the ploidy level as in banana and watermelon [2]. However, genetic parthenocarpy has been used only for a limited number of species and varieties. In some species and varieties, seedless fruit production is often achieved by external application of plant growth regulators as in the case of grape, tomato and eggplant [3].

Several methods to genetically engineer parthenocarpic fruit development have been proposed, and some have also been tested experimentally in crop plants [1,2]. Thus, transgenic parthenocarpic plants have been obtained for horticultural crops [4-7]. In particular, the chimeric gene DefH9-iaaM has been used to drive parthenocarpic fruit development in several species belonging to different plant families [4,5,7]. The DefH9-iaaM transgene promotes the synthesis of auxin (IAA) specifically in the placenta, ovules and tissues derived therefrom [4,8]. The agronomical advantages of DefH9-iaaM GM plants have been assessed by greenhouse and field trials of DefH9-iaaM eggplant [9,10], strawberry and raspberry [7].

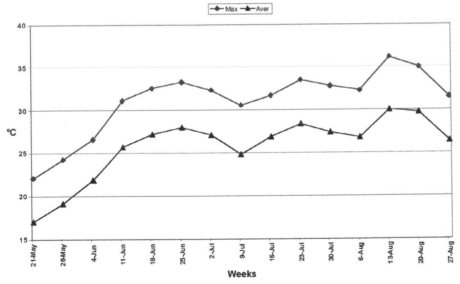

Figure 1. Mean weekly values of maximum and average temperatures from 14 May until 27 August 2003.

The DefH9-RI-iaaM gene construct produces high quality parthenocarpic fruits in the tomato cultivar UC 82, a variety used by the processing industry. The DefH9-iaaM gene construct gave rise to malformed parthenocarpic fruits because of high sensitivity to auxin that is present in the UC 82 genetic background [8].

The DefH9-RI-iaaM gene version is less efficiently translated and has a weaker action in producing IAA than DefH9-iaaM, so its use avoids the malformation of the fruit.

Parthenocarpy represents a useful trait in tomato fruit used for industrial purpose as well. This because parthenocarpic fruits are usually either seedless or contain significantly fewer seeds than non-parthenocarpic varieties. Manipulation of fruit and seed quality, size and number has been recently included among the third-generation traits of GM crop plants [11]. In the tomato sauce industry, seed content is a problem and seeds are removed to obtain sauce of good quality. As far as fruit quality concerns, parthenocarpy may also improve fruit quality through increases in the solid soluble content of the fruit [12]. Productivity may also be increased in some seasons because fruit set and fruit growth are less affected under environmental conditions adverse for pollination and/or fertilization including heavy rain, high humidity, hot and dry wind etc.

Processing tomatoes represent the greatest proportion of tomato production (approximately 113 million tons in 2003 [FAO]). In this paper, data on agronomic performance under open field conditions, technological and biochemical characteristics of two DefH9-RI-iaaM UC 82 tomato lines are presented and compared to untransformed control. We also included in our analysis a modern F1 hybrid tomato, used by the processing industry, to determine the extent to which biochemical and agronomical parameters vary between different non-GM tomato lines. The data indicate that the use in the diet of fruits from a new line or variety introduces much greater changes than the consumption of GM fruits in comparison to its genetic background.

Results

The field trial was performed in 2003. The mean air temperature from the first week of June until the end of the growing season was unusually very high and constantly above 25°C and 30°C average and maximum temperature, respectively (Fig. 1).

Fruit production, as measured by marketable fruit yield, was highest in the modern F1 tomato hybrid Allflesh (Table 1). The transgenic parthenocarpic lines Ri4 and Ri5 gave a fruit yield similar to that of the untransformed control UC 82. However, one transgenic parthenocarpic line (Ri4) gave a marketable fruit production that was not statistically different from all the other three genotypes (Ri5, UC 82 and Allflesh).

Table 1. Mean values (± SE) of marketable and unmarketable yields per plant, number of fruits per plant, and fruit weight for two transgenic parthenocarpic lines (Ri4 and Ri5), the untransformed control (UC 82) and the commercial F1 cultivar Allflesh.

Genotype	Marketable yield			Unmarketable yield		
	Yield/Plant (g)	N° of fruits	Fruit weight (g)	Green fruits		Rotten fruit
				Yield/Plant (g)	N° of fruits	Yield/Plant (g)
Allflesh	1906 ± 362 a	32.3 ± 6.1 ab	60.5 ± 3.4 a	98 ± 22 a	3.4 ± 0.5 a	3.8 ± 0.6 a
UC 82	1380 ± 301 b	21.6 ± 4.3 b	65.5 ± 2.5 a	196 ± 65 a	6.2 ± 1.9 a	5.2 ± 1.2 a
Ri4	1538 ± 159 ab	43.2 ± 5.0 a	37.0 ± 1.1 b	123 ± 15 a	5.6 ± 0.6 a	6.9 ± 1.4 a
Ri5	1227 ± 147 b	33.7 ± 3.6 a	38.0 ± 1.5 b	149 ± 44 a	6.7 ± 1.5 a	5.2 ± 0.9 a

For each trait at least one common letter indicates no significant difference according to the Duncan test (α = 0.05).

The two parthenocarpic lines produced a higher number of fruits with respect to the untransformed control UC 82 (Table 1). The increased fruit number per plant was most likely due to an improved fruit set of GM parthenocarpic plants compared to the untransformed control (UC 82). The two parthenocarpic lines produced fruits of smaller size (Table 1). The reduction of fruit weight observed in the two parthenocarpic lines is most likely due to the increased number of fruits per parthenocarpic plant. The hybrid Allflesh gave a number of fruits that was not different from the numbers of both transgenic lines and of the cultivar UC 82.

The unmarketable yield, represented by green and rotten fruits, was not different between the four genotypes (Table 1).

The shape of the tomatoes, the puffiness fruit index and the number of locules per fruit did not differ between the genotypes tested (Table 2). The two transgenic lines Ri4 and Ri5 produced a significantly lower percentage of seeded fruit and a significantly reduced number of seeds per fruit compared to the untransformed control and the modern F1 hybrid Allflesh (Table 2 and Fig. 2). This shows that the fruits obtained from the transgenic lines were parthenocarpic.

Table 2. Mean values (± SE) of fruit shape index (polar/equatorial ratio), puffiness index (1–3), number of locules, percentage of fruits with seeds and number of seeds per fruit in the transgenic parthenocarpic lines (Ri4 and Ri5), the untransformed control (UC 82) and the commercial F1 cultivar Allflesh.

Genotype	Shape index	Puffiness index	N° of locules	Fruits with seeds (%)	N° of seeds/fruit
Allflesh	1.27 ± 0.02 a	1.20 ± 0.03 a	2.36 ± 0.03 a	86.7 ± 5.4 a	68.8 ± 6.2 a
UC 82	1.25 ± 0.02 a	1.85 ± 0.16 a	2.47 ± 0.05 a	85.0 ± 8.8 a	36.5 ± 6.5 b
Ri4	1.24 ± 0.03 a	1.43 ± 0.12 a	2.38 ± 0.04 a	26.7 ± 5.4 b	18.4 ± 3.8 c
Ri5	1.27 ± 0.03 a	1.62 ± 0.13 a	2.45 ± 0.16 a	20.0 ± 4.7 b	11.4 ± 3.8 c

For each trait at least one common letter indicates no significant difference according to the Duncan test (α = 0.05).

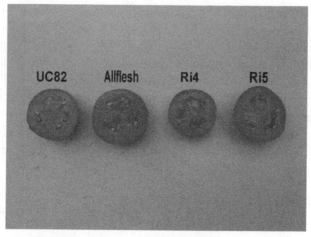

Figure 2. Cut tomato fruits of the four genotypes used: UC 82, Allflesh and GM UC 82 lines Ri4 and Ri5. GM fruits (Ri4 and Ri5) are seedless. UC 82 and Allflesh fruits contain seeds.

Color evaluation of fruits showed that L* values (an index of brightness) did not vary among UC 82 genotypes. A significant higher a* values, representing the red component, was found in non-transformed fruits compared to that of the transgenic fruits (Table 3). The b* value (an index of the yellow color) was different from the UC 82 control only in Ri4 GM line. °Brix values showed a significant higher soluble sugar content in Allflesh tomatoes (5.3). The °Brix value of Ri5 (4.2) did not differ from UC 82 (3.8), while the °Brix value (4.5) of Ri4 was higher than UC 82. However, the °Brix values of the two transgenic lines were not significantly different. The pH values were close to 4.0 in both the cultivar UC 82 and in the transgenic lines derived from it. Allflesh had a slightly higher pH statistically different from all the other genotypes tested (Table 3). Total acidity values showed no significant variation between the genotypes analyzed. The resistance of skin (fruit firmness) was about 0.4 Kg for all samples. Dry matter content was significantly higher in Allflesh with respect to all three UC 82 genotypes tested (Table 3).

Table 3. Mean values (± SE) of color (coordinate L*, a* and b*), °Brix, dry matter (DM), titratable acidity (mEq/100 mL NaOH 0,1 N) and skin resistance (firmness) detected in the four genotypes tested.

Genotypes	Colour L*	a*	b*	°Brix	pH	DM (%)	Titr. ac.	Firmness (kg)
Allflesh	39.2 ± 0.37 b	36.5 ± 0.54 a	25.4 ± 0.48 c	5.3 ± 0.18 a	4.29 ± 0.06 a	6.19 ± 0.18 a	6.12 ± 0.12 a	0.43 ± 0.03 a
UC 82	41.1 ± 0.19 a	35.8 ± 0.12 a	27.9 ± 0.35 a	3.8 ± 0.14 c	4.08 ± 0.04 bc	5.06 ± 0.22 bc	7.23 ± 0.8 a	0.40 ± 0.02 a
Ri4	39.9 ± 0.34 ab	34.1 ± 0.09 b	25.9 ± 1.22 bc	4.5 ± 0.12 b	3.96 ± 0.01 c	4.85 ± 0.01 c	6.12 ± 0.11 a	0.43 ± 0.02 a
Ri5	40.0 ± 0.59 ab	34.0 ± 0.44 b	27.1 ± 0.42 ab	4.2 ± 0.23 bc	4.11 ± 0.01 b	5.24 ± 0.18 b	6.70 ± 0.54 a	0.41 ± 0.03 a

For each trait at least one common letter indicates no significant difference according to the Duncan test (α = 0.05).
L* represents the brightness; a* is an index of red colour (i.e. higher the value, stronger the red colour); b* is an index of yellow colour (i.e. higher the value, more intense the yellow colour).

Analysis of the citric, tartaric and oxalic acid amounts showed that citric acid was the most abundant (Table 4). Citric acid content was significantly higher in Allflesh compared to cv. UC 82. The two lines Ri4 and Ri5 had intermediate amounts not significantly different from each other, although Ri4 had a higher content than UC 82. Tartaric acid content did not show any significant difference between the tomatoes produced by the four genotypes tested. Oxalic acid content differed only in transgenic line Ri5. β-carotene content was significantly higher in the tomatoes of the two transgenic parthenocarpic lines, followed by UC 82 and Allflesh. Vitamin C, lycopene and tomatine did not show any significant difference between the four genotypes tested.

Table 4. Mean values (± SE) of citric acid, tartaric acid, oxalic acid, vitamin C (mg/100 g d.w.), β-carotene and lycopene (μg g-1 d.w.), and tomatine (mg g-1 d.w.) detected in the four genotypes tested.

Genotypes	Citric	Tartaric	Oxalic	Vit C	β-carotene	Lycopene	Tomatine
Allflesh	2194 ± 470 a	40.18 ± 1.4 a	37.30 ± 4.5 a	539.6 ± 44.6 a	491.1 ± 19.7 c	659.5 ± 55.1 a	189.4 ± 34.5 a
UC 82	1434 ± 122 b	32.06 ± 5.8 a	35.97 ± 2.5 a	445.6 ± 69.7 a	566.6 ± 8.6 b	815.9 ± 36.7 a	248.2 ± 52.2 a
Ri4	2271 ± 209 a	30.39 ± 1.7 a	36.03 ± 5.0 a	436.5 ± 59.2 a	698.9 ± 10.1 a	928.5 ± 113 a	202.8 ± 33.9 a
Ri5	1837 ± 419 ab	40.58 ± 3.2 a	22.95 ± 4.5 b	374.3 ± 43.5 a	643.6 ± 15.1 a	811.1 ± 34.1 a	169.3 ± 45.4 a

For each trait at least one common letter indicates no significant difference according to the Duncan test (α = 0.05).

The total antioxidant capacity was rather similar in all four genotypes with Allflesh having the highest value which differed significantly only from that of Ri4 line (Table 5). The difference in the total antioxidant activity was attributable to the hydrophilic phase. In fact, hydrophilic and total TEAC were similar both as values and as trend, whereas the antioxidant activity of the lipophilic matrix showed no significant difference between the four genotypes tested and with values so low that they barely contribute to the overall antioxidant capacity. The content of total polyphenols was not significantly different.

Table 5. Mean values (± SE) of total trolox equivalent antioxidant capacity (TEAC-μmol g-1 f.w.), hydrophilic and lypophilic phases, and total polyphenols (mg gallic acid g-1 f.w.) detected in the four genotypes tested.

Genotypes	TEAC			Total Polyphenols
	Total	Hydrophilic	Lypophilic	
Allflesh	3.01 ± 0.08 a	2.60 ± 0.07 a	0.42 ± 0.02 a	9.03 ± 1.01 a
UC 82	2.79 ± 0.14 ab	2.40 ± 0.06 ab	0.40 ± 0.08 a	7.55 ± 0.34 a
Ri4	2.50 ± 0.07 b	2.11 ± 0.02 c	0.39 ± 0.05 a	9.32 ± 0.37 a
Ri5	2.71 ± 0.07 ab	2.29 ± 0.07 bc	0.43 ± 0.02 a	7.78 ± 0.19 a

For each trait at least one common letter indicates no significant difference according to the Duncan test (α = 0.05).

In conclusion, tomatoes (UC 82) genetically modified for parthenocarpy, grown under open field conditions, show a fruit yield per plant identical to their corresponding control but a 10-fold reduction in the seeds content. All other tested parameters, with the exception of β-carotene, did not show relevant changes between GM and not GM UC 82 tomatoes.

Discussion

The open field trial of two transgenic tomato lines obtained from the processing cultivar UC 82 showed that the DefH9-RI-iaaM gene was able to induce parthenocarpic fruit development under open field cultivation conditions. Most (approximately 75%) of the fruits produced were seedless and, furthermore, seeded fruits contained significantly less amount of seed (on the average the number of seeds per fruit was 60–80% reduced) than both the corresponding untransformed UC 82 control and the modern cultivar F1 Allflesh (Table 2). These results are in accordance with those from field trial of eggplant where most of the transgenic fruits were seedless [10] and confirmed that iaaM-induced parthenocarpy might be used as a tool to reduce transgenic seed dispersal without the need to combine it with male sterility. The significantly fewer number of seeds in DefH9-RI-iaaM tomato (UC 82) fruits may represent a desirable trait for processing tomatoes because the absence of seeds simplifies industrial activities which normally discard the seeds from the paste.

Under the rather high temperature that occurred during flowering, fruit set and growth, the DefH9-RI-iaaM parthenocarpic transgene allowed the production of a significantly larger number of fruits compared to the untransformed cv. UC 82. This indicates, at least under the conditions tested, an improved ability of the DefH9-RI-iaaM plants to set fruits. Poor fruit set and dramatic reduction in fruit size has been reported in tomato plants grown at 26°C under controlled conditions [13]. DefH9-RI-iaaM parthenocarpic fruits had a shape similar to that of the control and no malformations were observed, confirming the data obtained for T0 plants [8]. The reduction of weight observed in parthenocarpic fruit grown under open field conditions, might be due to the increased number of fruits caused by the improved fruit set of parthenocarpic plants. A different agronomical practice (i.e. watering, fertiliser regime, etc) apt for high fruit-set lines (i.e. DefH9-Ri-iaaM GM) might improve fruit weight, and consequently plant productivity.

Except for the color coordinate a* (red), the two transgenic parthenocarpic lines showed no relevant variation of the technological properties (°Brix, pH, dry matter, total acidity and firmness) compared to the untransformed control cv. UC 82. Interestingly, all three genotypes in the UC 82 background differed from the modern F1 hybrid Allflesh for the solid soluble content (°Brix value), which is a very important trait for processing tomato. The presence of the transgene

DefH9-RI-iaaM showed no significant influence on the amount of the biochemical compounds relevant for tomato processing quality. In fact, the organic acids, vitamin C, lycopene and tomatine were not different between the two GM lines and their control. The amount of β-carotene, a beneficial dietary bioactive for humans, was higher in the transgenic lines compared to both the control and the F1 Allflesh. This represents the main difference between transgenic and control fruits. Some natural parthenocarpic variants of tomato have a higher β-carotene level than their original cultivars [12]. No differences were observed for the antioxidant activity and polyphenol content between the transgenic lines and the cv. UC 82. For these traits, few differences were observed between the F1 Allflesh and one UC 82 genotype (Ri4 transgenic). Similar results have been obtained in transgenic tomatoes engineered for other traits [14-16].

To determine the substantial equivalence and identify possible unintended effect in engineered food it has been proposed that biochemical fingerprinting be expanded beyond the comparison between transgenic genotypes and its correspondent untransformed controls to include several non-transgenic lines [17]. This approach can allow the determination of whether any difference originate from metabolic effect associated with the transgenic trait or from expected genetic and/or physiological variation within the species. In the present study the extent of variation was higher between the traditional UC 82 cultivar (transgenic or not) and the last generation F1 hybrid Allflesh.

Conclusion

This trial has demonstrated that, under open field conditions allowing pollination/fertilization, the DefH9-Ri-iaaM transgene was able to sustain parthenocarpic fruit development in the cv. UC 82. Biochemical and technological analyses performed on tomato fruit (GM and non-GM) showed a very little variation that is well within the variability of the species Lycopersicon esculentum. Yield per plant did not differ between GM and non-GM, in GM lines fruit weight was decreased, whilst fruit number was increased.

Methods

Plant Material

The selfed progenies of two single copy transgenic tomato (cv. UC 82) lines carrying DefH9-RI-iaaM (named Ri4 and Ri5), the untransformed control cultivar UC 82 and the commercial F1 hybrid "Allflesh 1000" (Peotec) were compared.

The Ri4 and Ri5 selfed progenies were selected for kanamycin resistance. Consequently, they were either homozygous or hemizygous. The field trial was performed, following authorization by the Italian Ministry of Health (B/IT/02/10 Pomodoro – DefH9-iaaM), at the experimental farm of the Marche Polytechnic University located in Agugliano (Ancona – IT). The experimental design used a latin square with 4 replications, each containing 20 plants at a density of 3 plants/m2. The plants were grown in a single row in soil mulched with black plastic polyethylene film (0.05 mm thick) and standard agronomical techniques were applied throughout the growing season. Plantlets at the third-fourth true leaf were transplanted on May 12, 2003. Harvest was in the last week of August. The following traits were recorded: number and weight of ripe tomatoes (marketable production), number and weight of unripe and rotten fruits (unmarketable production). For each plot, the percentage of fruits containing at least 1 seed, the mean number of seeds/fruit, the fruit shape index (polar/equatorial ratio), puffiness index evaluated by an arbitrary scale from 1 (no puffy fruit) to 3 (deep puffy fruit) and the number of locules was recorded using a representative sample of fruits.

Biochemical and Physical Analyses

All the analyses were performed on a representative sample of fruits from each replicated plot. The physical quality traits (texture, color, firmness, dry matter content), chemical quality traits (soluble solid content-°Brix, pH, titratable acidity, acidic profile, vitamin C levels, total polyphenol levels) and antioxidant activity were determined after 72 hours from the harvest, with 24 hours conditioning at 4°C prior to analysis.

Physical Analyses

Color measurements were performed with a reflectance colorimeter Minolta Chromameter CR 200 (Minolta Co., Osaka, Japan), which measured the central part of the surface of tomatoes. L*, a* and b* values were calculated. L* represents the brightness, in particular a value close to 100 means a very high light; a* is an index of red color (i.e. a high positive value means a strong red color while a high negative value means a green color); b* is an index of yellow color (i.e. a high positive value indicates intense yellow color while a high negative value indicate a blue color).

The firmness of the tomato fruits was determined with a dynamometer Instron model 4301 (Instron Corporation, Canton, MA, USA) by measuring the maximum force (kg) required to make an hole on the skin of the tomato using a

1 mm diameter, cylindrical probe pressed into the pulp of each fruit at a speed of 0.1 m/min.

Chemical Analyses

All chemical assays were once replicated within the same sample. Tomatoes were cut and peeled and immediately homogenized by a Waring blender.

Dry matter and pH determinations were performed as reported in [18]. An aliquot of the homogenized tomato tissues was used to measure the soluble solid content (SSC) with a BS model RFM 81 refractometer.

About 10 g of homogenized flesh was used for the determination of titratable acidity and acidic profile. Titratable acidity was determined according to AOAC methods (AOAC. 1980. Official Methods of Analysis, 13th ed. N° 46024 and N° 22061. Association of Official Analytical Chemists, Washington. DC).

The Total Phenolic Content was determined in tomato extracts by the Folin-Ciocalteu based method [19], using gallic acid (GA) as a standard for the calibration curve. Results were calculated as gallic acid equivalent (GAE) in fresh fruit (mg g-1). A calibration curve (0-500 mg l-1 of GA) was made prior to the determinations. Samples were read in a KONTRON Uvikon 941 Plus spectrophotometer.

Citric, tartaric and oxalic organic acids were determined from an aqueous extract of the homogenized pulp (10 g plus 20 mL H_2O), homogenized for 30 seconds with an Ultra-Turrax, then centrifuged at 5600 × g for 20 min, and filtered through 0.45 μm filter. The extracts were analyzed by HPLC at 20°C, using 0.02 M H_3PO_4 as mobile phase (flow rate: 0.6 mL/min) on an Inertsil ODS-3 column of 0.46 × 25 cm dimension with 5 μm of particle diameter, using detection by a UV-VIS spectrophotometer set at 210 nm. Retention times of standards were: tartaric acid 8.9 min, oxalic acid 9.8 min, citric acid 21.9 min.

Ascorbic acid (vitamin C) was determined from an aqueous extract of the homogenized pulp (10 g plus 20 mL metaphosphoric acid 6% in H_2O), homogenized for 30 seconds in an Ultra-Turrax, then centrifuged at 5600 × g for 20 min, and filtered through a 0.45 μm filter. The extracts were analyzed by HPLC at 20°C, using 0.02 M H_3PO_4 as mobile phase (flow rate: 0.6 mL/min) on an Inertsil ODS-3 column of 0.46 × 25 cm dimension with 5 μm of particle diameter detected by a UV-VIS spectrophotometer set at 254 nm. Under these conditions, ascorbic acid has a retention time of 8.6 minutes.

Compounds were identified following HPLC by comparing their retention times with those of commercial standards. Compounds were quantified by plotting different peak areas against concentrations. Results were expressed as mg/100 g dry weight of whole tomato fruits.

The determinations of tomatine, β-carotene and lycopene were performed on tomato fruit samples frozen in liquid nitrogen at harvest and stored at -80°C. The extraction and HPLC analysis procedures used to determine tomatine levels were as reported in [20]. We employed a Luna C8 5 μm (stainless steel, 150 × 4.6 mm i.d.) (Phenomenex, Torrance, CA) reversed phase column, equipped with the appropriate pre-column, for the separation along a mobile phase of water-acetonitrile-methanol-0.1 M ammonium phosphate buffer, pH 3.5 (using phosphoric acid) at a flow rate of 0,7 mL/min. The injection volume was 5 μL for both standards and samples; detection was at 205 nm and each separation lasted 15 min at room temperature. Peak area was used for quantification.

For β-carotene and lycopene analyses, 10 g of tomato edible portions were homogenized with 2 mL of 0.1% butylated hydroxytoluene (BHT) in methanol (v/v) for 3 minutes, 15 mL of HPLC grade isooctane were added, the samples were then mixed and incubated at 4°C for 1 hour; a 5 mL aliquot of isooctane extract was dried on a rotatory evaporator at 45°C and the dry residue dissolved in 1 mL of mobile phase. Quantification of compounds was based on the response of commercially available standards treated in the same way as the samples by plotting peak areas against carotenoid concentrations in μg per mL.

Twenty microliters of each sample and standard were filtered through a 0.22 μm Millipore filter and injected on a Partisil 5 ODS-3 (stainless steel, 250 × 4,6 mm i.d.) C18 reversed phase column (Whatman) for HPLC analysis. A reversed phase pre-column (Whatman) was also used. Isocratic elution was carried out at 40°C along with a mobile phase of acetonitrile-water-ethyl acetate-ethyl acetate:tetraydrofuran 1:1 (42:10:18:30) containing 0.2% glacial acetic acid. The flow rate was 0,8 mL/min, with detection at 502 nm. Each separation lasted 25 min.

Antioxidant activity (AA) was evaluated according to the TEAC (Trolox equivalent antioxidant capacity), modified assay [21,22]. ABTS, a chromogen colorless substance, is changed into its colored monocationic radical form (ABTS·+) by oxidative agents. The absorption peak of ABTS·+ is 734 nm. Addition of antioxidants reduces the ABTS·+ into the colorless form. AA was expressed as μmoles Trolox (an analogue of vitamin E) equivalents g 1-1 of fresh weight. To assess the antioxidant status of the fruit, two types of extraction (hydrophilic and lipophilic) were carried out. Extraction of the hydrophilic phase (HE) was performed with a T25 Ultra-turrax blender with frozen fruit samples using an ethanol/water (80/20) extraction solution to a final ratio of 1:10 (w/v). Homogenate was centrifuged (2000 × g for 10 min) and the supernatant (HE extract) transferred to vials and stored (-20°C) until assayed. Three extracts were collected per sample.

Extraction of the lipophilic phase (LE) was made on pellet by adding acetone (1:4 w/v), and centrifuging at 2000 × g for 15 min. The supernatant (LE extract) transferred to vials and stored (-20°C) until assayed. Three extracts were collected for each sample.

Each extract (HE and LE) was recovered and the antioxidant activities were measured separately by recording the absorbance at 734 nm in a spectrophotometer (KONTRON Uvikon 941 Plus). Data are expressed as AA induced by the hydrophilic and lipophilic components, and total antioxidant activity by the sum of the two phases

Statistical Analysis

All the data were subjected to ANOVA according on a Latin square scheme with 4 replicates, Duncan test ($p < 0.05$) was used for comparison of means.

Authors' Contributions

GLR and NA selected the genetic material and coordinated the field trial; ES and BM were concerned with agronomical work at the field trial; GM, RLS, AM, JS, BM have performed biochemical and technological analysis; NA and ES have done the statistical analysis; GLR wrote the manuscript; BM, TP and AS performed the molecular analysis of the plants and the gene constructs used.

Acknowledgements

This work is part of the FIRB project RBAUOIJTHS [23] of the MIUR (Italian Ministry of University and Research).

References

1. Varoquaux F, Blanvillain R, Delseny M, Gallois P: Less is better: new approaches for seedless fruit production. Trends Biotec 2000, 18:233–242.

2. Spena A, Rotino GL: Parthenocarpy: State of the Art. In Current Trends in the Embryology of Angiosperms. Edited by: Bhojwani SS, Soh WY. The Netherlands: Kluwer Academic Publishers; 2001:435–450.

3. Schwabe WW, Mills JJ: Hormones and parthenocarpic fruit set: a literature survey. Hort Abstracts 1981, 51:661–698.

4. Rotino GL, Perri E, Zottini M, Sommer H, Spena A: Genetic engineering of parthenocarpic plants. Nat Biotechnol 1997, 15:1398–1401.

5. Ficcadenti N, Sestili S, Pandolfini T, Cirillo C, Rotino GL, Spena A: Genetic engineering of parthenocarpic fruit development in tomato. Molecular Breed 1999, 5:463–470.

6. Carmi N, Salts Y, Dedicova B, Shabtai S, Barg R: Induction of parthenocarpy in tomato via specific expression of the rolb gene in the ovary. Planta 2003, 217:726–735.

7. Mezzetti B, Landi L, Pandolfini T, Spena A: The DefH9-iaaM auxin-synthesizing gene increases plant fecundity and fruit production in strawberry and raspberry. BMC Biotechnology 2004, 4:4.

8. Pandolfini T, Rotino GL, Camerini S, Defez R, Spena A: Optimisation of transgene action at the post-transcriptional level: high quality parthenocarpic fruits in industrial tomatoes. BMC Biotechnology 2002, 2:1.

9. Donzella G, Spena A, Rotino GL: Transgenic parthenocarpic eggplants: superior germplasm for increased winter production. Molecular Breed 2000, 6:79–86.

10. Acciarri N, Restaino F, Vitelli G, Perrone D, Zottini M, Pandolfini T, Spena A, Rotino GL: Genetically modified parthenocarpic eggplants: improved fruit productivity under both greenhouse and open field cultivation. BMC Biotechnology 2002, 2:4.

11. Vasil IK: The science and politics of plant biotechnology – a personal perspectives. Nat Biotechnol 2003, 21:849–851.

12. Lukyanenko AN: Parthenocarpy in tomato. In Monographs on Theoretical and Applied Genetics 14. Genetic improvement of tomato. Edited by: Kalloo G. Berlin Heidelberg: Springer-Verlag; 1991:167–177.

13. Adams SR, Cockshull KE, Cave CRJ: Effect of the temperature on the growth and development of tomato fruits. Ann Bot 2001, 88:869–877.

14. Furui H, Inakuma T, Ishiguro Y, Kiso M: Tomatine content in host and transgenic tomatoes by absorptiometric measurement. Biosci Biotechnol Biochem 1998, 62:556–557.

15. Novak WK, Haslberger AG: Substantial equivalence of antinutrients and inherent plant toxins in genetically modified novel foods. Food Chem Toxicol 2000, 38:473–483.

16. Le Gall GL, Colquhoun IJ, Davis AL, Collins GJ, Verhoeyen ME: Metabolite profiling of tomato (Lycopersicon esculentum) using 1H NMR spectroscopy as a tool to detect potential unintended effects following a genetic modification. J Agric Food Chem 2003, 51:2447–2456.

17. Noteborn HPJM, Lommen A, van der Jagt RC, Weseman JM: Chemical fingerprinting for the evaluation of unintended secondary metabolic changes in transgenic food crops. J Biotech 2000, 77:103–114.

18. Eccher Zerbini P, Gorini F, Polesello A: Measurement of the quality of tomatoes: recommendations of an EEC working group. In Edited by: IVTPA, Milan. 1991, 23–29.

19. Slimkard K, Singleton VL: Total phenol analysis: automation and comparison with manual methods. Am J Enol Vitic 1997, 28:49.

20. Bushway RJ, Perkins LB, Paredis LR, and Vanderpan S: High-Performance liquid chromatographic determination of the tomato glycoalkaloid, tomatine, in green and red tomatoes. J Agric Food Chem 1994, 42:2824–2829.

21. Pellegrini N, Re R, Yang M, Rice-Evans CA: Screening of dietary carotenoids and carotenoid-rich fruit extracts for antioxidant activities applying the ABTS.+ radical cation decolorization assay. Meth Enzymol 1999, 299:589–603.

22. Re R, Pellegrini N, Proteggente A, Pannala A, Yang M, Rice-Evans C: Antioxidant activity applying an improved ABTS radical cation decolorization assay. Free Radic Biol Med 1999, 26:1231.

23. Improved Quality and Productivity in Horticulture (IQPH) [http://www.bioinformatica.unito.it/bioinformatics/spena]

Design and Construction of a Double Inversion Recombination Switch for Heritable Sequential Genetic Memory

Timothy S. Ham, Sung K. Lee, Jay D. Keasling
and Adam P. Arkin

ABSTRACT

Background

Inversion recombination elements present unique opportunities for computing and information encoding in biological systems. They provide distinct binary states that are encoded into the DNA sequence itself, allowing us to overcome limitations posed by other biological memory or logic gate systems.

Further, it is in theory possible to create complex sequential logics by careful positioning of recombinase recognition sites in the sequence.

Methodology/Principal Findings

In this work, we describe the design and synthesis of an inversion switch using the fim and hin inversion recombination systems to create a heritable sequential memory switch. We have integrated the two inversion systems in an overlapping manner, creating a switch that can have multiple states. The switch is capable of transitioning from state to state in a manner analogous to a finite state machine, while encoding the state information into DNA. This switch does not require protein expression to maintain its state, and "remembers" its state even upon cell death. We were able to demonstrate transition into three out of the five possible states showing the feasibility of such a switch.

Conclusions/Significance

We demonstrate that a heritable memory system that encodes its state into DNA is possible, and that inversion recombination system could be a starting point for more complex memory circuits. Although the circuit did not fully behave as expected, we showed that a multi-state, temporal memory is achievable.

Introduction

Synthetic Biology aims to make the implantation of new, complex biological function in cells more of an engineering science than a biology research project. A central effort of this emerging field is the design of modular biological parts that facilitate both the ease of manufacture and the design of predictable function in cells. The hope is that this will allow applications at larger scale and with more sophistication than is currently feasible with the more ad hoc genetic engineering approaches commonly applied today. If successful, a great deal is to be gained from this approach in supporting classical applications such as industrial expression of useful proteins, creation of pathways for biological synthesis of organic molecules for pharmaceuticals and commodity natural products, or to confer a simple phenotype such as pest or drought resistance on a host.

The field, however, looks forward to challenges in global health, energy and the environment that could be well-served by carefully engineered microbes with more complex "programming" to sense and respond to variable and uncertain environments found beyond the bioreactor. Such applications might include, for example, controlled, safety-assured deployment of engineered microorganisms to remediate water and soils, support crop growth in marginal soil, treat disease, or even serve as cheap emergency blood substitutes as one recent undergraduate iGEM team imagined [1].

These applications will require more sophisticated sensing, actuating and regulatory logic than chemical production pathways and may depend on the temporal history of events experienced by the cell. That is, it may be necessary to "remember" past inputs rather than simply compute and actuate on the present ones. This biological circuitry would be more akin to a sequential logical system, a prerequisite for complex computation, than a combinational one such as simple Boolean logic. Regulatory networks with such memory are relatively common in those natural systems that control, for example, development. Memory of past environments can also confer a fitness advantage in some situations[2], and synthetic biological two-state memory switches have been built in both prokaryotes[3]and eukaryotes [4]. It is clear from the complexity of both natural cellular networks and the oncoming synthetic biological applications that regulatory circuitry of some sophistication is required to sense and survive in the outside world.

Systems whose output depends not only on the current inputs but the input history are necessary for sophisticated computation and information storage. While it is unlikely we will use bacteria as super-computers for a number of reasons, it is interesting to note that the chemical networks that underlie their behavior are formally capable of Turing-machine like computations [5], [6], [7], [8]. In these systems, the inputs are generally chemical "inducers", the machine is a set of chemical reactions, and the state is generally the stationary-state concentrations of the internal chemical species. For synthetic biological applications in the foreseeable future, we will not likely need to design networks with extremely large computational power. But scaling to larger applications, with more states and deeper sequential logics is certainly a future need.

In synthetic biological applications to date, the logic of the designed regulatory systems is generally simple and implemented by a handful of genetic "parts" [9]. In most cases, these elements are the standard set of less than five work-horse promoters such as the tetracycline and lactose inducible promoters and their cognate transcription factors. These promoters have been studied and even modified slightly for years rendering them useful for gene expression control. There is an ever increasing bestiary of naturally occurring bacterial promoters and some efforts to engineer synthetic promoters and transcription factors for more complex logical functions. However, the heterogeneity in behavior of these devices and undeveloped rules for their composition make designing larger networks with these components unpredictable. In almost all applications there are unreasonable number of cycles of design and testing before the system works as desired.

Design of parts families for which it is relatively simple to construct a new functionally-independent member from knowledge of the current members has become a key focus of foundational synthetic biology. Examples include the rational design of RNA-based mRNA translation regulation and metabolic sensing [10], [11], modular systems for controlling mRNA stability [12], and a designable way of implementing RNAi-based combinations circuits in mammalian cells [13]. While there are still subtleties and a variety of case-by-case issues, these all exploit the relatively simple structural rules in RNA, Watson-Crick complimentarity rules, and the relative ease of RNA in vitro evolution to design families of parts with more or less predictable function. There has also been progress in engineering protein-based parts families. It is becoming ever more possible to, for example, design zinc finger proteins as transcription factors [14] or to gain control of eukaryotic scaffolding proteins to design signaling switches [15] and more.

Even with the increasing sophistication of synthetic biological parts families, challenges remain in the scalable design of complex regulatory circuitry. Because of the lack of spatial addressing of "signals" between components, like wires between physically separated components in an electronic circuit, it is generally not possible to reuse the same biological component in the same way as one could reuse a transistor (exceptions include such things as ribosome binding sites to some degree [16]). This leads to heterogeneity of "device physics" across the circuit, as every part is somewhat different than every other and a necessity to actually have a chemically different part for every elementary operation of the circuit. The properties of these parts are often very complex, thereby making abstraction into useful mathematics for design, such as Boolean logics, difficult. Further, implementing all these different parts in a cell, were they available, might require fairly large outlays of DNA real estate and place large energetic loads on the cell. Finally, in most current synthetic biological applications, the "states" of the regulatory circuitry are encoded in transient chemical concentrations that require energy outlay to hold and can't be maintained after cell death or transmitted easily from one cell to another.

For all these reasons above, circuits that might operate by changing DNA sequence using, for example, recombinases are attractive complements to gene expression and protein interaction networks. Operations on DNA tend to change state (sequence) in a discrete, almost Boolean, fashion. The state may be maintained without constant energetic input, persists after cell death and even if rare in a cell population a particular sequence may be amplified out from the background by PCR. Further, since the "state" is encoded in DNA, it is possible by the various mechanisms of inter-cell DNA transfer, such as conjugation to pass state-output among cells as a complex form of communication as one iGEM team recently suggested [17]. More subtly, as shown below, the ability to decide the spatial arrangement of recombination sites provides the ability to create circuits with large sequential state spaces accessible from relatively few recombinase inputs with efficient use of DNA.

Motivated by these advantages, in this work, we have explored the use of bacterial inversion recombination systems to construct genetic switches with the above properties and show how more complicated finite-state machine-like devices could be constructed out of such systems. For our purposes invertases are a nice starting point since they act very much as simple switches that may invert a sequence of DNA in place. To summarize briefly, inversion recombination happens between two short inverted repeated DNA sequences, typically less than 30 basepairs (bp) long. The recombinases bind to these inverted repeated sequences, which are recombinase specific. A DNA loop formation, assisted by DNA bending proteins, brings the two repeat sites together, at which point DNA cleavage and ligation occur. This reaction is ATP independent, but requires super-coiled DNA. The end result of the recombination is that the stretch of DNA between the repeated sites inverts. That is, the stretch of DNA switches orientation—what was the coding strand is now the non-coding strand, and vice versa. In this reaction, the DNA is conserved, and no gain or loss of DNA occurs. Additionally, there is great flexibility in the distance between the inverted repeats, ranging from only a few hundred up to 5 kilobases (kb) of additional DNA between the repeats possible (refer to the work of Johnson [18] and Blomfield [19] for a detailed treatment of inversion systems). The advantages of site-specific inversion are the binary dynamics, the sensitivity of output, the efficiency of DNA usage, and its persistent DNA encoding, even after cell death. The disadvantages are the possible interference between multiple recombinases, DNA loss by excision, and reversibility of the reaction.

Previously, we have described a tightly regulated expression switch using the FimE protein of the fim system of E. coli [20] which demonstrated the leak-less properties of this system and persistence of state after removal of recombinase input. Here, we have constructed an artificial overlapped inversion switch by integrating two recombination systems (using the FimB protein of the fim system from E. coli [19] and the hin system from Salmonella [21]) to form an intercalated double inversion system that implements a heritable memory with finite state machine-like behavior with four states dependent on the sequence of invertase activity inputs. We expand on how this works below.

There exist only a few known examples of natural systems that utilize multiple overlapping DNA inversions for diverse gene expression. Some examples are the R64 plasmid shufflon [22], which uses inversion to select among different versions of PilV gene, and the Min system from the p15B plasmid [23], which can make 240 different isomeric forms of a phage protein. The natural example most parallel to our own is the nested inversion system of Campylobacter fetus, where a promoter is moved around via inversion near various S-layer protein genes for expression [24]. All of these systems are thought to be involved in extending the host range of their host pathogen. To date, our system is the first artificial system to incorporate two inversion systems into a single circuit for controlled DNA rearrangement. It is, also, perhaps, one of the first biological finite-state machine encoding more than two sequential states, along with the hin based inversion system used to solve a version of the burning pancake problem [25].

Results and Discussion

An Informative Idealization of Invertase Based Recombinatorics

Before delving into the experimental realization of our circuit we demonstrate an aspect of the possible power of this approach through a quick series of calculations of scalability of designs with invertase activities as input and resulting DNA sequence as the formal output or "state" of the system. In some cases it is possible to have a particular DNA configuration encode the expression of RNA and this could then be an output as well. How the actual physical realization of such circuits can affect the predictions of this idealization will be discussed briefly. In fact, it is critical to the experimental results of our circuit described below. But our goal in this work is not to create a biological computer but only to suggest the power of using DNA read/write as a "stateful" element in synthetic biological regulatory circuitry. Interestingly, an iGEM team from Davidson University has already considered using the possible computational power of a single invertible system as a means of solving a combinatorial problem [25]. As will become clear below, the nature of recombination operations in DNA result in combinatorial equations that describe both their configurations and operations. Thus, we call the theory of design with recombinases "Recombinatorics".

A more complete theory would include, among other things, the wide variety of recombinase activities including excision, insertion and inversion of DNA segments into a target region of a replicon. Here we limit ourselves to invertases like those in our experimental implementation. For the purposes of our arguments here, we assume the following idealizations of invertase circuit dynamics, nearly all of which are violated in some way by our own circuit but also all of which are not beyond the ken of natural engineered inversion systems. First, we assume that there is a single copy DNA target for the recombinases whose activity serves as input to our system. Second, recombinases can only invert a target region once. That is, they are irreversible flippers. This also limits the possible computational power of the device quite a bit. Third, there is no interference among the recombinase inputs such that the ability of one recombinase to flip the region of DNA between its target pair of sites is unaffected by the presence of other recombinases or by the state of the surrounding DNA. Fourth, the length of DNA between a pair of inversion sites is such that the invertases can flip it. Fifth and finally, for simplicity, our input alphabet, which is the set of different invertases, is assumed to be presented as a sequence of single activities, each element of which is well separated in time and on long enough time-scale to effect a flip. No invertases are present more than once in a sequence. Thus, for N invertases there are N! possible ordered input sequences to our device.

Our device is defined by an arrangement of the N pairs of inversion sites on DNA with no pair appearing more than once in the device. While there might be multiple devices encoded on a single duplex of DNA, we call a single (fully connected) device a set of pairs of sites for which every pair brackets a region of DNA that overlaps another region bracketed by at least one other pair of the device. Figure 1 shows possible arrangements of sites for devices accepting one (Figure 1A), two (Figure 1B&C) and three recombinase (Figure 1D) inputs. The number of such possible configurations increases rapidly with number of invertases. Assuming that configurations that are identical under shuffling of site identity are equivalent (that is, x-y-x-y is equivalent to y-x-y-x), an enumeration of all possible devices with n inputs a(n) suggests that with n = N-1, $a(0) = 1; for\ n > 0, a(n) = (2n-1)!! - \sum_{k=1}^{n-1} (2k-1)!!\ a(n-k)$. (The inference of formulae from sequence was provided by the Online Encyclopedia of Integer Sequences.) By the time one has ten invertases there are more than 1010 possible arrangements of sites. The graph of possible configurations as a function of number of invertases, Figure 1E, shows the better than exponential increase in number of configurations as a function of N. If we assume we need at least 500 basepairs between sites for flipping to occur and each site is about 30 bp, a device with N inputs has a minimal size of around 30*2*N+500 basepairs (overlapping regions can decrease this slightly). For a device with 10 inputs then, apart from the DNA encoding the expression of the recombinases, 1.1 kilobases is all that is required to encode any of ten billion machines. This is the length of an average sized gene.

Each of these devices behaves differently under the N! possible inputs. Figure 1A–C shows the state transition graphs for all configurations of 1 and 2 invertase input devices. Each transition shows the transformation of one DNA state to another for each allowed sequence of inputs (see caption). Theoretically, for certain configurations, starting from an initial state of the device (state 0), it is possible that every possible history of input is recordable in the state of the DNA. That is, it is possible to determine which even partial sequence of inputs the device has seen by sequencing the DNA between its outermost sites. Simply counting the internal nodes of the state transition graphs like those in Figure 1C shows that the number of states (excluding state 0) for such devices is the number of permutations of non-empty subsets of {1,...,N} or $S(N) = \sum_{k=1}^{N} k!C(N, k)$. A graph of this function is shown in Figure 1E. Devices accepting input from 10 invertases have maximal state-spaces of nearly 107. This is a large space made accessible by addition of a region of DNA on the order of one gene in size (not counting the recombinases). Of course, not every configuration has this full state space. For example, figure 1C is the only configuration of two pairs of sites that has the full state. The other configuration (Figure 1B) has a cycle with the final states of the input sequences {A,B} and {B,A} being identical. Figure 1D shows all the configurations of three pairs of sites with the number of distinguishable states available to each shown to its right. Only two of the ten configurations show the full rank state-space. The others can "remember" only subsets of the input sequences. Nonetheless, as a system to remember which of a sequence of N inputs occurred, these types of devices could be immensely efficient in terms of size and operation.

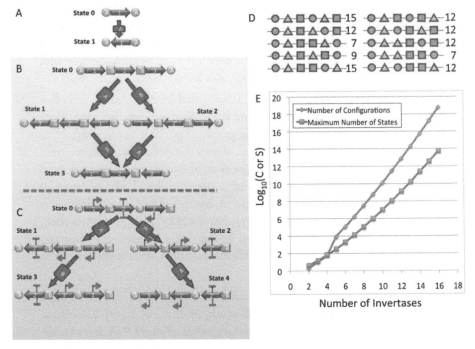

Figure 1. How invertase site configuration defines different "machine behaviors". A) A single invertase can only flip one region ON or OFF. B and C) With two invertases there are two possible intercalated configurations of sites. In B, when one pair of sites is fully contained by another the resultant number of possible "states" of the system is only four, whereas in C, when the site pairs are staggered, five states are accessible and any input sequence results in a different output of the system. C also shows a configuration of promoters and terminators such that states 3 and 4 have promoters pointing outside the machine regions. This configuration is the one experimentally created. D) Possible configurations of three different pairs of recognition sites and the number of possible states. E) How both the number of such configurations and the maximal number of states grows with number of pairs.

These devices may be more than memory even under these restrictive assumptions. With proper placement of "active" elements such as promoters, terminators, and even genes within the device, the system can have active output at chosen states of the device. While a full treatment of how to place such elements within a device is beyond this paper, Figure 1C shows an example that we implement experimentally below. In this case, two promoters and a bi-directional terminator have each been placed in a separate DNA region. Under the operation of the two input recombinases, A and B, the regions containing these elements are rearranged. Only the two "end" states of the graph have an arrangement such that the promoters point away from the terminator and transcription may proceed outside the device boundaries. While the RNA produced might itself be an input to a downstream device, more classically genes may be placed to the left and right of the device and be differentially expressed based on whether the sequence A and then B occurred or the reverse. These genes could be recombinases that drive this device or other devices encoded into other regions of DNA or could be hooks into other more classical synthetic biological circuits. If we allow genes to be placed internal to the device, then all four states could have different output activity. How this ability scales to larger devices remains to be seen and affects how powerful a computer one could build with such a system if one were so motivated.

All this analysis concerns ideal systems however. If flipping is reversible, for example, our device in Figure 1C has an addition state with DNA configuration $(3{\rightarrow}2{\rightarrow}1)$, that is reachable from the end states by action of either invertase A or B. While our fifth assumption above prevents us from reaching this state, reversibility means that the ordered input sequence {A,B} leads first to a probability of being in state 0 or state 1 when A is input, then to a probability of being in state 0, 1, 2 or 3 when B is input. What the ratio is of these probabilities is dependent on the kinetics of the forward and reverse flipping rates of the two invertases and the time they are allowed to be active. In this case, one only gets a probability of state 3 if A and then B and one only gets a probability of state 4 if B and then A so it is still possible, statistically, to determine the order in which inputs were seen from DNA sequence. Each member of a population of cells exposed to conditions that activated invertases A or B in a given order will hold one state of the DNA sequence, and a particular configuration can be read out by PCR or by a screenable output from the device. Violation of the other assumptions above leads to other interesting phenomena which may be either good or bad for specific applications. All the implications of such violations are beyond the scope of the paper other than the fact that in our experimental construct the effects end up being important.

An Experimental Implementation of a Two Invertase, Full Rank System: Basic Design and Construction of the Device

To begin to understand the physical constraints on the building of invertases-based memory devices and switching elements, we chose to construct a device like that shown in Figure 1C. While we could have designed for states 1-4 of our device each to have a separate active output, to simplify our design and to make the measurements realizable we settled on a design with fluorescent proteins to the left and right of the device such that only states 3 and 4 would show output. However, as will be discussed below, the invertases we chose show reversible flipping and so the fifth state mentioned above also becomes accessible. If seen, this state would output both the left and right fluorescent proteins.

The hin and fim systems were chosen for constructing this new genetic switch for the following reasons, in addition to the ones that are obtained for any recombination circuit outlined above. 1) They have very specific recognition sites. These sites are well known and their DNA-protein interactions have been described. 2) They are independent of each other. Unlike inversion recombination systems that are from the same family, the hin and the fim systems are completely orthogonal with different mechanisms. 3) They are inducible. Only in the presence of the Hin and FimB proteins can the system invert. 4) Their mechanisms have been well studied. There is some twenty years of literature exploring the recombination mechanisms. 5) They are known to be flexible. The distance between the recombination sites can be varied greatly, from a few hundred base pairs to several kilobases. This allows the possibility of adding additional promoters or genes in between the recombination sites. 6) They have very low rates of excision, even when the inversion sites are arranged in direct repeats. Other inversion recombination systems (for example Cre/Lox) will excise the region between direct repeats.

Each inversion reaction was initiated by the expression of one of the recombinases, FimB or Hin. The gene encoding FimB was expressed from the arabinose-inducible araBAD promoter (PBAD) [26], and the gene encoding Hin was expressed from the aTc-inducible tet promoter (PTet). Fim and Hin are expressed from a plasmid (pZB) containing the two promoters [27]. The recombinase genes were harbored on a separate plasmid from the switch to facilitate testing. Thus, the system has two inputs and two outputs. Although the two recombinases were expressed from inducible promoters, it would be possible to express them from a different input, for example environmental or metabolic sensors.

The target DNA regions were harbored on a separate plasmid. The switch region was synthesized de-novo and cloned into pPROBE-gfp. The red fluorescence gene was cloned in last. The fluorescent proteins GFP and RFP are placed to the left and right of the device respectively. The resulting plasmid is multicopy, meaning that there are many copies of our device in the cell such that the state of the device could possibly be different for each copy of the plasmid. Since Hin and FimB are reversible enzymes, each cell will harbor a mixture of states of the device as described below. The plasmids for the input and DNA response elements of the device are shown in Figure 2A.

A

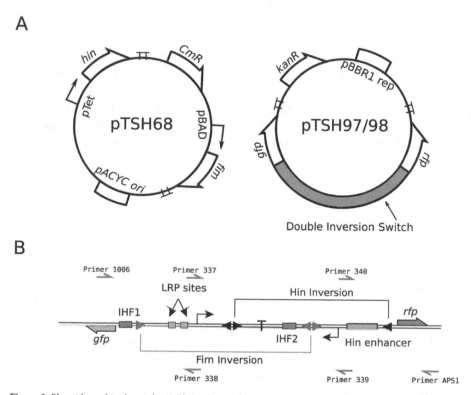

B

Figure 2. Plasmids used in the study. A) Plasmid maps for constructs containing the invertases and the memory switch. B) Detail of structure of the double inversion switch annotated with the primers used to diagnose state as described in the text.

Recombinase Optimization

The double inversion circuit requires two independently induced recombinases, FimB and Hin, to be expressed. In addition, they have to be repressed independently; that is, ideally, one can be induced while the other is repressed. Our switch was sensitive to leaky expression of Hin and FimB, so we required a very tightly controlled expression system. Currently, there are only a few tightly-controlled, regulatable expression systems available. One of the best is the P_{BAD} system, but this system cannot be used with a lac promoter, as they have significant crosstalk

[28]. Thus, Ptet was used along with the P_{BAD} system on a single vector containing both (Fig. 2A), to express the recombinases independently. However, P_{Tet} did exhibit leaky expression and required remedy prior to use with our sensitive switch.

In addition to promoter optimization, translational efficiency was optimized. Seven different ribosome binding sites (RBSs) across a range of translational efficiencies were tested with the hin and fimB genes [16]. The optimization criteria were level of expression (does it cause inversion when expressed) and leakiness (does it cause inversion when not expressed). As our PCR based reporting method was very sensitive, the selection was biased against leakiness rather than towards high expression. In addition, degradation tags (LVA) or LacI binding sites were tested for their effects and possible efficacy at reducing background leaky expression (data not shown). The pTSH68 vector (containing hin with RBS sequence "AGGGACAGGATA" plus the LVA tag driven by PTet and fimB with RBS sequence "GAAGGTTCCTCA" driven by PBAD) was chosen as the recombinase expression vector.

Target Device Optimization

The two recombinase recognition sites in the middle flanking the terminator needed to be constructed in such a way that they form an inverted repeat even after a recombination event. As such, they had to be mirrored; that is, the binding site is repeated immediately adjacent, but in the opposite strand orientation, to maintain the inverted repeat orientation (Fig. 2B). The design for the switch is finalized by placing the necessary elements of both the fim inversion (2 IHF elements, LRP sites, the inverted repeats), and hin inversion (hin enhancer site, inverted repeats) along with generated sequences that separate the components (Fig. 2B). The enhancer elements are somewhat flexible in their location [29], [30], [31], so they were placed in similar places as the original switches. The entire switch, excluding gfp and rfp, is approximately 1 kbp.

Predicted and Measured Function of the Switch

The heritable switch functions as shown in (Fig. 1C) and briefly outlined above. Figure 2B shows the actually constructed state 0. Hin plays the role of invertase B and FimB plays the role of invertase A. State 3 is only reached and GFP expressed when Hin is induced following FimB. State 4 is only reached, and RFP expressed when FimB is induced following Hin.

Because our inversion recombinases are reversible, two deviations from ideality are immediately expected. First given a long enough induction time, the invertases will switch back and forth between the two states its activity connects. Thus, the populations of these two DNA states in the plasmid population will reach some steady-state ratio. This is important to understand because when state 0 is induced to state 1, both state 0 and 1 are present in the population (either as mixed plasmids or cells). Ideally, if the reaction is allowed to go to steady state, there will be a 50/50 mixture of the initial and final states, as FimB and Hin are known to have the same forward and reverse rates. In our case, if FimB is expressed at state 0, we will end up with a 50/50 mixture of states 0 and 1. Subsequently, if FimB is removed, so that there are no further transitions between state 0 and 1, then Hin is expressed and allowed to equilibrate, ½ of state 1 will transition to state 3, and ½ of state 0 will transition to state 2. Thus, the overall population will have ¼ state 0, ¼ state 1, ¼ state 2, and ¼ state 3. A similar result ensues if we apply Hin and then FimB: only states 0, 1, 2, and 4 are populated. Second, because the invertases are reversible it is possible to apply them more than once in sequence. The sequences {Hin, FimB, Hin} and (FimB, Hin, FimB} both result in 1/8 of the population reaching the fifth state of DNA ($3 \rightarrow 2 \rightarrow 1$) mentioned above. Once state 5 is reached, the order information is lost, and the system no longer remembers which states it had been, but rather simply that it had seen both inputs.

Such mixture of states is not entirely desirable, but as there are no known unidirectional, re-settable switches available so we can restart our system, and having only one unidirectional switch (FimE) would unbalance our population distribution, the mixtures were accepted as a compromise for demonstrating our proof-of-principle circuit.

As should be clear from above, our device resembles an AND gate switch in electronic and logic circuitry. An AND gate only outputs when both inputs are present. This circuit, however, behaves differently than a regular AND gate. Instead of requiring that both inputs be present in order to have output, it remembers the input order—that is, it remembers that inputs had been present at some time in the past. The switch then outputs a different fluorescent protein, conditioned on the sequence of inputs it had observed. A temporal memory switch like this has many potential uses, such as in environmental sensors that can track two different conditions occurring one after another, or as in vivo biosensors investigating development or other temporally sensitive assays.

Testing for the Inversions

As many of the inversion states did not have fluorescent output, and because as discussed below flipping events were rarer than expected, the inversion state was assayed using "culture PCR" as described in the methods section. A culture PCR is similar to colony PCR, except the template is not a single colony, but part of the induced culture, likely containing a mixture of inverted genotypes. Six different primers in four permutations were used to probe for the presence or the absence of PCR products (Tables 1, 2 and Fig. 2B). The assay was chosen because, as the orientation of the DNA fragments change, different permutations of primers will amplify different length segments or generate no product at all. This method allowed us to detect very sensitively the presence or absence of a certain orientation of DNA, based on the presence or absence of PCR amplification product of different lengths. Culture PCR was much more accurate than detection by fluorescence, as first generation inversion (states 1 and 2) would show no fluorescence and, it turns out, production of the end states is excessively rare. The sensitive nature of PCR amplification allowed the detection of rare inversions. The inversion state was also verified by sequencing of the PCR product bands and by direct sequencing of the vector.

Table 1. Culture PCR primer sets and lengths of their PCR products. See Fig. 2B for their locations on the switch.

Primer Set	Description	Primers Used	PCR Product Length
E	State 0	337, 339	Short
F	States 2 and 3	337, 1006	State 2: Short, State 3: Long
G	States 1 and 4	339, APS1	State 1: Short, State 4: Long
H	State 5	338, 339	Short

doi:10.1371/journal.pone.0002815.t001

Table 2. Sequences of primers used to interrogate the inversion products.

Primer Name	Sequence
337	cgagccacagaaacgttagctttacatatagcg
338	cgctatatgtaagctaacgtttctgtggctcg
339	cgcgacacgtggcgagtatatgatg
340	catcatatactcgccacgtgtcgcg
APS1	cgaattggggatcggaag
1006	caagaattgggacaactcc

doi:10.1371/journal.pone.0002815.t002

Figure 3 shows the results of our experiments. In these experiments cells were grown in LB at 37°C and exposed to either arabinose or tetracycline. The hin mirrored pair was replaced with hixC, a variant inversion sequence which allows inversion in both directions without impediment [13], as the hin mirrored pair behaved unpredictably due to its native mirrored-pair like structure (data not shown). HixC, however, does have a side effect of resulting in a higher rate of self-excision of the DNA when in a directly repeated orientation.

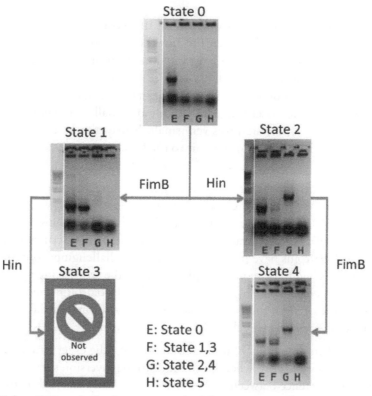

Figure 3. Culture PCR results for the response to the different input sequences. {FimB}, {HinB}, {FimB, HinB} and {HinB, FimB}. Each lane probes for the existence of a state or states, as shown in the legend. Because of the possible arrangements of DNA, lanes E and F could result in either short or long PCR products, indicating states {1, 3}, and {2, 4} respectively. See Figure 2 for relative location of the primers. No State 5 was observed when Hin and FimB were simultaneously expressed.

The fim inversions performed as expected, except with a much lower inversion efficiency than hin. By culture PCR, we determined that some fraction of the cells were inverting, resulting in strong PCR product bands. However, of the 50 isolates sequenced, none of them had the inverted fim orientation, even though sequencing of the PCR product showed clean fim inversion. It would be possible to screen more perhaps by doing a serial dilution and a binary tree search. It appears that the intercalated switch construct has low fim inversion efficiency, with an unknown rate.

In all inversions, inducing the culture at low OD rather than stationary phase resulted in a greater number of inversions. For hin inversion, this is consistent with the fact that Fis production is eliminated during stationary phase [32]. However, it does not explain the increase in fim inversion. Perhaps IHF or LRP is regulated in a similar fashion as Fis, or there is an unknown protein involved in fim inversion.

The rarity of fim inversion had consequences for the second generation (states 3 and 4) inversions. State 4, the hin then fim inversion, although rare, was observed by PCR product sequencing, as a large fraction of hin inversion occurred, and a small fraction of fim inversions then occurred. However, state 3, the fim then hin inversion, was never observed. This is perhaps the result of only having a small fraction of fim inversion occurring, and among the smaller pool of inverted fims, the occurrence of hin inversion was very small. It may also be that the hin inversion efficiency was lower than normal due to the DNA rearrangement via the fim inversion, further suppressing the occurrence of state 3 plasmids.

Many of the problems encountered in creating a functional double inversion switch stemmed from two principal causes. First, the necessity for the mirrored pair introduced a very long hairpin structure that is both difficult to construct and maintain within E. coli. Perhaps strains more tolerant of such structures would be useful. Also, such hairpins have made sequencing very challenging: constructs containing the mirrors had to be sequenced piecemeal, as the sequencing reaction would stop at the very edges of the hairpin. Additionally, the hairpins acted as terminators of unknown strength for the internal constitutive promoters, although this had a positive effect on our switch design.

The second problem is our lack of understanding about the exact structure of protein-DNA complex that forms when FimB binds to the inversion sites. The assumption of non-interference between recombinases was clearly violated. Perhaps a better understanding of the structure of FimB, including its reaction with the enhancer and DNA bending proteins, would reveal insight into why the switch did not perform as expected. As a whole, however, the switch did demonstrate history dependent configurations with states 0, 1, 2, and 4 visited as expected although not with the frequency desired.

Conclusions

A heritable inversion switch was designed and constructed using both the fim and the hin inversion recombination systems. The design allowed for encoding of state information to the DNA, which would be inherited generation after generation. The design also displayed finite state machine-like behavior, as the reporter would transition to and from different states, recording path traversal as it went along.

The switch as constructed was able to transition into three out of the five end states (excluding initial state 0). Thus, we could detect the following sequences: Hin alone, FimB alone, or Hin followed by FimB. It was unable to transition to the other two states, FimB followed by Hin and (Hin:FimB:Hin, FimB:Hin:FimB), probably due to poor transition rates of the fim inversion. The mechanism for the poor transition rate was not determined.

In order to construct a fully functioning heritable switch as envisioned in this paper, robust inversion systems are a necessity. The hin system, using hixC, is quite robust to introduction of exogenous sequences, including strong hairpins and other recombination sites. The fim system, however, seems to be less robust, perhaps because it is more sensitive to the positioning of the IHF and LRP binding sites. Or it is possible that there are yet to be discovered key mechanisms not considered in the design. A greater understanding of the fim inversion system is necessary for the development a robust system. In this study there was not an opportunity to construct a variety of fim inversion systems with many different arrangements of IHF, LRP, mirrored pairs, terminators, in different permutations to concretely discover the exact cause of the inversion repression. With the existing synthesis technology, a study that would rigorously test the necessary elements for fim inversion would have been very costly and time consuming. But perhaps with advanced DNA synthesis and assembly technologies, such tests may be feasible in the near future. Because of the apparent power and theoretical efficacy of these devices for encoding states it seems a useful program on which to embark.

Despite its limitations, this work creates a proof of principle mechanism for production of finite-state DNA read/write systems and has uncovered key challenges that might not have been clear without the construction. With the large variety of enzymes that edit DNA, successful harnessing of DNA recombination systems could lead to powerful applications in biological control and sensing.

Materials and Methods

Construction

All standard molecular techniques were performed using established protocols [33]. The double inversion switch was synthesized de novo in multiple steps. First, a simplified version of the switch, which had the fim and hin mirrored pairs and the middle terminator replaced by restriction sites, was synthesized using the protocol of Rouillard and colleagues [34]. The product was cloned into the pPROBE vector [35], which harbors gfp encoding the green fluorescent protein along with flanking terminators, resulting in pTSH49. Once the sequence was verified, the mirrored pairs and the terminator were cloned in via double strand oligo ligation. Four oligos were synthesized, which when annealed together left a sticky end corresponding to the desired restriction site. The oligos were diluted to 20 µl, and complementary sets were annealed together by mixing, heating to 95°C for 10 minutes, then cooling to room temp gradually. The annealed oligos were kinased, and finally mixed with the vector in a ligation reaction, producing pTSH117, with 10 bp within the fim and hin mirrored inverted repeats, and pTSH118, with 30 bp within the fim mirrored inverted repeats and 10 bp within the hin mirrored repeat.

The strong hairpins formed by the mirrored pairs prevented sequencing through them, so their successful insertion was verified by checking the insert length via PCR and by the presence of an un-sequenceable hairpin. A construct containing hixC in place of the hix mirrored pairs was also made (pTSH89, 10-bp gap fim mirrored repeat, pTSH90, 30-bp gap fim mirrored repeat). The completed switch, which includes hixC, the mirrored fim repeats, the terminator, gfp and rfp, were named pTSH97 (10 bp gap fim mirrored pair) and pTSH98 (30 bp gap fim mirrored pair), respectively.

Testing

The testing of the double inversion switch was performed in vivo, in E. coli DH10B. The vector containing the switch, along with any reporter fluorescent genes, was co-transformed with another vector containing the recombinases fimB and hin. The expression of FimB and Hin was performed through induced expression via P_{BAD} (fimB) and P_{Tet} (hin). The ribosome binding sites (RBS) of the genes were mutated to adjust the expression level. This was done via addition of overhanging sequences during PCR.

Induction experiments were performed as follows. The strain was grown at 30°C overnight in LB medium containing kanamycin (50 µg/ml), chloramphenicol (30 µg/ml), and dextrose (5% w/v). Dextrose was added to prevent spurious PBAD expression. The overnight culture (50 µl) was inoculated into 2 ml LB medium containing kanamycin and chloramphenicol and incubated at 37°C with shaking. After reaching an OD600 of 0.2 to 0.5, the inducer (10 mM arabinose for fimB induction and/or 100 nM anhydrous tetracycline (aTc) for hin) was added. The culture was incubated at 37°C with shaking for 6 hours or overnight. Control cultures were inoculated into non-inducing medium. To test for states 3 and 4, the cultures in states 1 and 2 were inoculated into fresh medium (50 µl of culture into 2 ml of LB) without an inducer, grown overnight, and induced with the complement inducers.

After the prescribed induction time, a 1 ml sample of the culture was centrifuged (15,000×g, 1 minute), and the pelleted culture was used as the template for four PCR reactions ("Culture PCR"). Four primer sets were used: 337 and 339 (which checks for state 0); 337 and 1006 (checks for states 2 and 3); 339 and APS1 (checks for states 1 and 4); and 338 and 340 (checks for state 5). They are labeled as primer sets E, F, G and H, respectively (Tables 1 and 2). The culture PCR was performed as follows. A PCR master mix (containing 2.5 µl 10×Taq PCR buffer, 0.2 µl 25 mM dNTP's, 0.2 µl taq polymerase, 0.5 µl 20 uM primers in 20.6 µl nuclease free water) was mixed with 0.5 µl pelleted cells. The PCR cycling parameters were 95°C for 2 minutes, 35 cycles of (95°C for 30 seconds, 55°C for 30 seconds, 72°C for 90 seconds), then 72°C for 2 minutes, finally, hold at 4°C.

The PCR products were run on 1% agarose gels, and visualized by ethidium bromide stain under UV light. Sometimes the bands were cut and gel extracted for sequencing. The vector itself also was sequenced for state identification. Because different copies of a multi-copy vector might harbor different inversion switch states, which would make sequencing impossible, plasmids were mini-prepped after the induction and transformed into E. coli to isolate individual plasmids for sequencing.

Acknowledgements

APA would like to thank Tom Knight of M.I.T. for pointing out the amazing Online Encyclopedia of Integer Sequences which allowed us to be lazy about proofs.

Author Contributions

Conceived and designed the experiments: TSH. Performed the experiments: TSH SKL. Analyzed the data: TSH SKL JDK APA. Contributed reagents/materials/analysis tools: TSH SKL JDK APA. Wrote the paper: TSH SKL JDK APA.

References

1. Day A, Cole H, Doan K, Fuller K, Liang S, et al. (2007) BactoBlood

2. Wolf DM, Fontaine-Bodin L, Bischofs I, Price G, Keasling J, et al. (2008) Memory in microbes: quantifying history-dependent behavior in a bacterium. PLoS ONE 3: e1700.

3. Gardner TS, Cantor CR, Collins JJ (2000) Construction of a genetic toggle switch in Escherichia coli. Nature 403: 339–342.

4. Ajo-Franklin CM, Drubin DA, Eskin JA, Gee EP, Landgraf D, et al. (2007) Rational design of memory in eukaryotic cells. Genes Dev 21: 2271–2276.

5. Arkin A, Ross J (1994) Computational functions in biochemical reaction networks. Biophys J 67: 560–578.

6. Benenson Y, Gil B, Ben-Dor U, Adar R, Shapiro E (2004) An autonomous molecular computer for logical control of gene expression. Nature 429: 423–429.

7. Hjelmfelt A, Weinberger ED, Ross J (1991) Chemical implementation of neural networks and Turing machines. Proc Natl Acad Sci U S A 88: 10983–10987.

8. Parker J (2003) Computing with DNA. EMBO Rep 4: 7–10.

9. Keasling JD (2008) Synthetic biology for synthetic chemistry. ACS Chem Biol 3: 64–76.

10. Isaacs FJ, Dwyer DJ, Collins JJ (2006) RNA synthetic biology. Nat Biotechnol 24: 545–554.

11. Win MN, Smolke CD (2007) A modular and extensible RNA-based gene-regulatory platform for engineering cellular function. Proc Natl Acad Sci U S A 104: 14283–14288.

12. Carrier TA, Keasling JD (1997) Controlling messenger RNA stability in bacteria: strategies for engineering gene expression. Biotechnol Prog 13: 699–708.

13. Rinaudo K, Bleris L, Maddamsetti R, Subramanian S, Weiss R, et al. (2007) A universal RNAi-based logic evaluator that operates in mammalian cells. Nat Biotechnol 25: 795–801.

14. Mandell JG, Barbas CF 3rd (2006) Zinc Finger Tools: custom DNA-binding domains for transcription factors and nucleases. Nucleic Acids Res 34: W516–523.

15. Bashor CJ, Helman NC, Yan S, Lim WA (2008) Using engineered scaffold interactions to reshape MAP kinase pathway signaling dynamics. Science 319: 1539–1543.

16. Barrick D, Villanueba K, Childs J, Kalil R, Schneider TD, et al. (1994) Quantitative analysis of ribosome binding sites in E.coli. Nucleic Acids Res 22: 1287–1295.

17. Anderson JC, Bosworth W, Davis KA, Dueber JE, Fleming M, et al. (2006) Addressable Conjugation in Bacterial Networks.

18. Johnson RC (2002) Bacterial site-specific DNA inversion systems. Mobile DNA II: ASM Press.

19. Blomfield IC (2001) The regulation of pap and type 1 fimbriation in Escherichia coli. Adv Microb Physiol 45: 1–49.

20. Ham TS, Lee SK, Keasling JD, Arkin AP (2006) A tightly regulated inducible expression system utilizing the fim inversion recombination switch. Biotechnol Bioeng 94: 1–4.

21. van de Putte P, Goosen N (1992) DNA inversions in phages and bacteria. Trends Genet 8: 457–462.

22. Komano T, Kubo A, Nisioka T (1987) Shufflon: multi-inversion of four contiguous DNA segments of plasmid R64 creates seven different open reading frames. Nucleic Acids Res 15: 1165–1172.

23. Sandmeier H, Iida S, Meyer J, Hiestand-Nauer R, Arber W (1990) Site-specific DNA recombination system Min of plasmid p15B: a cluster of overlapping invertible DNA segments. Proc Natl Acad Sci U S A 87: 1109–1113.

24. Dworkin J, Blaser MJ (1997) Nested DNA inversion as a paradigm of programmed gene rearrangement. Proc Natl Acad Sci U S A 94: 985–990.

25. Haynes KA, Broderick ML, Brown AD, Butner TL, Harden L, et al. (2008) Computing with Living Hardware. IET Synthetic Biology 1: 44–47.

26. Guzman LM, Belin D, Carson MJ, Beckwith J (1995) Tight regulation, modulation, and high-level expression by vectors containing the arabinose PBAD promoter. J Bacteriol 177: 4121–4130.

27. Lee SK, Newman JD, Keasling JD (2005) Catabolite repression of the propionate catabolic genes in Escherichia coli and Salmonella enterica: evidence for involvement of the cyclic AMP receptor protein. J Bacteriol 187: 2793–2800.

28. Lee SK, Chou HH, Pfleger BF, Newman JD, Yoshikuni Y, et al. (2007) Directed evolution of AraC for improved compatibility of arabinose- and lactose-inducible promoters. Appl Environ Microbiol 73: 5711–5715.

29. Blomfield IC, Kulasekara DH, Eisenstein BI (1997) Integration host factor stimulates both FimB- and FimE-mediated site-specific DNA inversion that controls phase variation of type 1 fimbriae expression in Escherichia coli. Mol Microbiol 23: 705–717.

30. Gally DL, Rucker TJ, Blomfield IC (1994) The leucine-responsive regulatory protein binds to the fim switch to control phase variation of type 1 fimbrial expression in Escherichia coli K-12. J Bacteriol 176: 5665–5672.

31. Moskowitz IP, Heichman KA, Johnson RC (1991) Alignment of recombination sites in Hin-mediated site-specific DNA recombination. Genes Dev 5: 1635–1645.

32. Osuna R, Lienau D, Hughes KT, Johnson RC (1995) Sequence, regulation, and functions of fis in Salmonella typhimurium. J Bacteriol 177: 2021–2032.

33. Sambrook J, Fritsch EF, MT (1989) Molecular Cloning: A Laboratory Manual. Cold Spring Harbor, New York: Cold Spring Harbor Laboratory.

34. Rouillard JM, Lee W, Truan G, Gao X, Zhou X, et al. (2004) Gene2Oligo: oligonucleotide design for in vitro gene synthesis. Nucleic Acids Res 32: W176–180.

35. Miller WG, Leveau JH, Lindow SE (2000) Improved gfp and inaZ broad-host-range promoter-probe vectors. Mol Plant Microbe Interact 13: 1243–1250.

Transplantation of Genetically Engineered Cardiac Fibroblasts Producing Recombinant Human Erythropoietin to Repair the Infarcted Myocardium

Emil Ruvinov, Orna Sharabani-Yosef, Arnon Nagler,
Tom Einbinder, Micha S Feinberg, Radka Holbova,
Amos Douvdevani and Jonathan Leor

ABSTRACT

Background

Erythropoietin possesses cellular protection properties. The aim of the present study was to test the hypothesis that in situ expression of recombinant human erythropoietin (rhEPO) would improve tissue repair in rat after myocardial infarction (MI).

Methods and Results

RhEPO-producing cardiac fibroblasts were generated ex vivo by transduction with retroviral vector. The anti-apoptotic effect of rhEPO-producing fibroblasts was evaluated by co-culture with rat neonatal cardiomyocytes exposed to H2O2-induced oxidative stress. Annexin V/PI assay and DAPI staining showed that compared with control, rhEPO forced expression markedly attenuated apoptosis and improved survival of cultured cardiomyocytes. To test the effect of rhEPO on the infarcted myocardium, Sprague-Dawley rats were subjected to permanent coronary artery occlusion, and rhEPO-producing fibroblasts, non-transduced fibroblasts, or saline, were injected into the scar tissue seven days after infarction. One month later, immunostaining identified rhEPO expression in the implanted engineered cells but not in controls. Compared with non-transduced fibroblasts or saline injection, implanted rhEPO-producing fibroblasts promoted vascularization in the scar, and prevented cell apoptosis. By two-dimensional echocardiography and postmortem morphometry, transplanted EPO-engineered fibroblasts did not prevent left ventricular (LV) dysfunction and adverse LV remodeling 5 and 9 weeks after MI.

Conclusion

In situ expression of rhEPO enhances vascularization and reduces cell apoptosis in the infarcted myocardium. However, local EPO therapy is insufficient for functional improvement after MI in rat.

Background

Erythropoietin (EPO), a hematopoietic cytokine, has been shown to possess cardioprotective characteristics that can minimize ischemic injury and improve myocardial viability and function [1]. Systemic administration of recombinant human EPO (rhEPO) before or after myocardial ischemia reduces infarct size and improves cardiac function [2,3]. The beneficial effects of EPO are mediated by apoptosis inhibition, driven by Akt/phosphoinositide3-kinase signaling [4,5], and neovascularization by stimulation of endothelial progenitor cells [6,7]. However, systemic administration of EPO to patients with coronary artery disease could lead to adverse effects, such as high viscosity, impaired tissue perfusion, high blood pressure and an increased incidence of thrombosis [8,9]. Hematocrit elevation has been linked to excess mortality in patients with ischemic heart disease [10,11]. Over-expression of rhEPO in transgenic mice resulted in cardiac dysfunction and reduced life span [12,13]. Overall, the potential adverse effects associated with systemic EPO administration suggest that alternative approaches are needed.

A particularly beneficial approach would be a strategy that could provide local rhEPO delivery into the infarcted heart. A recent study has suggested that cardiac fibroblasts are critical effectors for erythropoietin-mediated pro-survival signaling and cardiac protection [14]. In the present study, we aimed to test the hypothesis that cardiac fibroblasts can be reprogrammed ex vivo to produce rhEPO, and that re-implantation of the engineered cells into the infarcted myocardium could promote tissue healing and repair.

Materials and Methods

The study was performed in accordance with the Animal Care and Use Committee guidelines of the Tel Aviv University, Tel Aviv, Israel, which conforms to the policies of the American Heart Association.

Retroviral Vector

Amphotrophic rhEPO (pBabe-EPO) retrovirus-containing packaging cell line PT67 was used for retroviral vector production [15]. Twenty-four hour filtered (0.45 μm) culture virus-containing supernants were stored at -80°C for future experiments. The presence and titer of rhEPO-retroviral vector were assayed by infecting BALB/3T3 (ATCC number CCL-163) mouse fibroblast cell line and a CFU (colony-forming unit) number determination.

Isolation of Cardiac Cells

Cardiac fibroblasts and cardiomyocytes were isolated from 1–2 day old neonatal Sprague-Dawley rats (Harlan Lab, Jerusalem, Israel) as previously described [16]. We used a 30 minute pre-plating procedure to obtain cardiac fibroblasts and to reduce the number of non-myocyte cells in cardiomyocyte culture. The purity of obtained cardiac fibroblast culture was confirmed microscopically by characteristic cell morphology. Neonatal rat cardiomyocytes were used for oxidative stress studies and cardiac fibroblasts were used for transduction and transplantation studies.

Cell Transduction

BALB/3T3 (ATCC number CCL-163) mouse fibroblast cell line and isolated cardiac fibroblasts were used for transduction experiments. RhEPO retrovirus-containing supernants were first incubated at room temperature with a Superfect cationic transfection reagent (QIAGEN, Westburg Ltd., Be'er-Sheva, Israel) to a final concentration of 10 µg/ml for 15 minutes [17]. Finally, the viral stock was added to 70–80% confluent BALB/3T3 or cardiac fibroblast culture. After 24 h incubation at 37°C, puromycin selection was applied (2.5 µg/ml). Transduction efficiency was approximately 80%, as evaluated qualitatively by relative confluence determination (before and after puromycin selection). Twenty-four hour culture supernants were collected for EPO assessment. Supernants were stored at -20°C for future experiments.

Assessment of EPO Production and Activity

The presence of rhEPO in transduced culture supernants was analyzed by a commercial ELISA kit (R&D Systems, Minneapolis, MN, USA). EPO concentration was normalized to 106 cells for 24 h.

EPO biological activity was analyzed using the EPO-sensitive human erythroleukemia cell line TF-1, in which proliferation and metabolic activity is EPO-dependent [18].

TF-1 cells were seeded in 96-well plates (1 × 105 cell per well in 100 µl of RPMI medium). A total of 100 µl of previously collected culture supernants, controls or standards (0.01 to 5 IU/ml rhEPO; Eprex, Beerse, Belgium) were added to each well. On day 5 of the experiment, TF-1 relative cell number was analyzed using colorimetric XTT assay kit (Biological Industries, Beth Haemek, Israel). The XTT assay is based on the ability of metabolically active (live) cells to reduce tetrazolium salt (XTT) to orange-colored compounds of formazan. EPO activity in supernants from rhEPO-transduced fibroblast culture was calculated from XTT rhEPO standard calibration curve. EPO relative biological activity was determined by EPO activity level/EPO protein level (measured by ELISA) ratio. This amount-normalized protein ratio was used for comparison of biological activity between expressed EPO (produced from transduced cells) and recombinant protein.

Oxidative Stress

After 6 days in culture, isolated rat neonatal cardiomyocytes were pretreated with supernants collected from rhEPO-transduced or non-transduced (for control) cardiac fibroblasts for 24 h, after which H_2O_2 was directly added to culture medium to a 150 µM final concentration for 60 minutes, and the degree of apoptosis was tested.

Apoptosis Assays

Apoptosis assessment was performed using Annexin-V-FLUOS (fluorescein)/propidium iodide (PI) staining kit (Roche Diagnostics GmbH, Penzberg, Germany). The samples were prepared according to the manufacturer's instructions and were analyzed by flow cytometry. Annexin-V-positive and PI-negative cells were scored as apoptotic cells.

Apoptotic morphological changes were evaluated by DAPI (4',6'-diamidino-2-phenylindole) staining. After oxidative stress induction, cells were gently washed with phosphate-buffered saline (PBS) and fixed with cold methanol for 10 minutes. The plates were washed twice with PBS and incubated for 5 minutes with 300 μM DAPI (catalogue number D-3571, Molecular Probes, Eugene, Oregon, USA) at 37°C. Finally, the cells were washed three times with PBS and examined by fluorescence microscopy. Apoptotic cells were identified by condensation and fragmentation of the nuclei, and were quantified by counting a total of 200 nuclei from each well and calculating the percentage of apoptotic nuclei.

Rat Model of Acute Myocardial Infarction

Male Sprague-Dawley rats (approximately 250 g; Harlan Lab, Jerusalem, Israel) were subjected to acute myocardial infarction (AMI) induction by permanent coronary artery occlusion, as previously described [19].

RhEPO-transduced and non-transduced fibroblast transplantations were performed 7 days after AMI. After 10 days in culture (7 days after transduction, 3 passages), cardiac fibroblasts were prepared for transplantation by trypsinization with 0.25% Trypsin-EDTA solution for 1 minute at 37°C. The rhEPO expression level in prepared cells was 7.0 $\pm$ 0.23 mIU/24 h/106 cells. Following centrifugation (1,500 rpm, 10 minutes at room temperature) the cells were re-suspended in saline. The final cell density for implantation was 1 $\times$ 106 cells/100 μl. Rats were anesthetized and the chest was opened under sterile conditions. The infarcted area was identified visually by the surface scar and wall motion akinesis. Rats were randomized and selected either for injection of 1 $\times$ 106 rhEPO-transduced cells, 1 $\times$ 106 non-transduced cells, or saline (n = 5 in each group). All injections were made into the center of the scar. After the injection, the air was expelled from the chest, and the surgical incision was sutured closed [19,20].

Echocardiography

Transthoracic echocardiography was performed on all animals before transplantation (baseline echocardiogram) and 5 and 9 weeks later (4 and 8 weeks post-transplantation, respectively). Echocardiograms were performed with a commercially available echocardiography system (Sonos 7500; Philips) equipped with 12.5 MHz phased-array transducer (Hewlett-Packard, Andover, MA, USA) as previously described [21,22]. All measurements were averaged over three consecutive cardiac cycles and were performed by an experienced technician who was blinded to the treatment group.

Hematocrit Measurements

For evaluation of possible systemic effects of rhEPO produced from transplanted cells, hematocrit measurements were performed before or 1 and 4 weeks after transplantation.

Immunohistochemical Analysis

Four weeks after transplantation, hearts were arrested with 15% KCl, perfused with formaldehyde 4% (15 mmHg) for 20 minutes, sectioned into 3–4 transverse slices, and parallel to the atrioventricular ring. Each slice was fixed with 10% buffered formalin, embedded in paraffin, and sectioned into 5 μm slices. Serial sections were stained with the following agents: anti-α-smooth muscle actin monoclonal antibodies (Sigma-Aldrich, Rehovot, Israel) to localize pericytes and arterioles for neovascularization evaluation; anti-active caspase-3 polyclonal antibodies (Biocare Medical, Walnut Creek, CA, USA) for apoptosis detection; anti-rhEPO monoclonal antibodies (R&D Systems, Minneapolis, MN, USA) to localize rhEPO production and distribution; lectin (Bandeirea simplicifolia agglutinin BS-1; Sigma-Aldrich, Rehovot, Israel) to localize vascular endothelium; and terminal deoxynucleotidyl transferase dUTP nick end labeling (TUNEL) using Apoptag Peroxidase In situ Apoptosis Kit (S7101, Chemicon, Temecula, CA, USA) to assess apoptosis. The degree of neovascularization was evaluated by capillary blood vessel and arteriole counts in the infarct border zone (at least three different fields were counted). Apoptosis evaluation was made by active caspase-3 or TUNEL-positive cell counts in the infarct border zone (at least five different fields were counted). Co-staining was made with haematoxylin or methyl green for TUNEL.

Morphometric Analysis

After perfusion pressure fixation (15 mmHg), slides stained with Masson's trichrome (Dexmore, Israel) were used to assess left ventricular (LV) remodeling by morhpometric analysis, as previously described [23]. The following parameters were measured: LV maximal diameter (mm), average wall thickness (mm; averaged from three measurements of septum thickness), average scar thickness (mm; averaged from three measurements of scar thickness), relative scar thickness (average scar thickness/average wall thickness), LV muscle area (mm_2; including the septum), LV cavity area (mm^2), whole LV area (mm2), infarct expansion index ([LV cavity area/Whole LV area]/Relative scar thickness), epicardial scar length (mm), and endocardial scar length (mm).

Statistical Analysis

All values are given as means ± standard error of the mean (SEM) from at least three independent experiments. RhEPO activity results were compared with Student's t-test for unpaired data. Cardiomyocyte or general cell apoptosis results and blood vessel counts were compared by one-way ANOVA with Bonferroni post test.

In the echocardiography study, because each animal was used as its own control, changes between baseline and 4 weeks in the control and treated groups were assessed with paired t-tests. Comparisons of the changes from baseline to 5 and 9 weeks in the control and treatment groups were made with repeated-measure two-way ANOVA. The ANOVA model included control versus treatment and baseline versus 5 and 9 weeks as factors, as well as the interaction between the two factors. GraphPad Prism version 4.00 for Windows (GraphPad Software, San Diego, California, USA) was used for analysis.

Comparison of morphometry parameters and hematocrit measurements among treatment groups was analyzed by one-way ANOVA with Bonferroni post test. All tests were performed using GraphPad Prism Version 4.0. $P < 0.05$ was considered statistically significant.

Results

Ex Vivo Retroviral Gene Transfer and Rhepo Expression

To test retroviral ability for stable expression of biologically active rhEPO, the initial transduction experiments were carried out using BALB/3T3 mouse fibroblast cell line. Table 1 describes the activity of EPO produced from transduced cells. ELISA was used to evaluate EPO concentrations, and its biological activity was determined by proliferation of the EPO-sensitive human erythroleukemia cell line TF-1. The activity induced by recombinant human EPO (Eprex) was used as a standard in this assay. Produced EPO protein was found to be 35 times more biologically active than recombinant protein (Eprex; Table 1, p = 0.0006). The rhEPO production level in transduced cultures was 11.5 ± 0.8 mIU/24 h/10^6 cells.

Table 1. EPO biological activity in vitro

Culture	EPO ELISA (mIU/ml)	EPO activity assay (TF-1) (mIU/ml)	Activity ratio (activity/protein)
Mock BALB/3T3	0	0	-
Transduced BALB/3T3	10.74 ± 0.78	373.3 ± 37.12*	×35
Mock neonatal fibroblasts	0	0	-
Transduced neonatal fibroblasts	51.24 ± 11.17	576.7 ± 186.7*	×12

RhEPO protein levels: the presence of rhEPO in transduced culture supernants was analyzed by a commercial ELISA kit. RhEPO biologic activity (TF-1 cell proliferation assay): EPO biological activity was analyzed using the EPO-sensitive human erythroleukemia cell line TF-1, in which proliferation and metabolic activity is EPO-dependent. *Significantly more active than rhEPO standard (Eprex), p < 0.05.

To generate viable cardiac cell culture producing rhEPO for subsequent transplantation, we used rat neonatal cardiac fibroblasts. EPO activity parameters of transduced cardiac fibroblasts are presented in Table 1. The produced protein was found to be 12 times more active than recombinant protein (Eprex; p = 0.03). The rhEPO production level in transduced cells was 14.2 ± 3.1 mIU/24 h/106 cells.

Anti-Apoptotic Effect of EPO in Isolated Cardiomyocytes

The rate of apoptosis was significantly lower in cultured cardiomyocytes treated with supernants derived from rhEPO-tranduced fibroblasts. To determine the protective effect of rhEPO produced from genetically modified cardiac fibroblasts on cardiomyocyte apoptosis, we used an H_2O_2-induced oxidative stress model [24]. Isolated rat neonatal cardiomyocytes were treated with 150 μM H_2O_2 for 60 minutes, with or without 24 h preconditioning with supernants containing rhEPO (12 mIU/ml) from transduced cardiac fibroblasts. The degree of apoptosis was assayed by Annexin V/PI staining and morphologically evaluated using a DAPI nuclear stain. FACS analysis of Annexin V/PI staining showed that the percentage of apoptotic cells was significantly lower in cardiomyocytes pretreated with rhEPO-containing supernants compared with control (Figure 1A): 17.8 ± 2.3% versus 32.1 ± 3.7% apoptotic cells (p < 0.05).

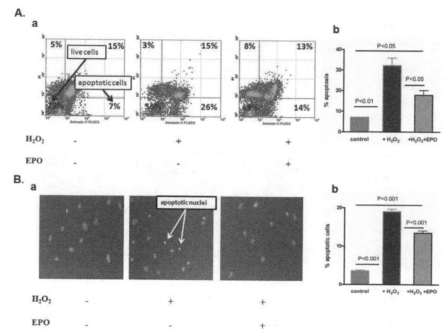

Figure 1. RhEPO inhibits H_2O_2-induced apoptosis in isolated cardiomyocytes. Neonatal rat cardiomyocytes were treated with PBS as a control or with H_2O_2 (150 μM) for 60 minutes in the 24 h presence or absence of rhEPO produced from transduced fibroblasts (12 mIU/ml). (A) Annexin V/PI staining: (a) Representative density profile (percentage of cells in each quadrant is indicated); (b) Quantitative analysis, showing that EPO attenuated cardiomyocyte apoptosis. (B) DAPI staining: (a) representative photomicrographs (arrows show apoptotic cardiomyocyte nuclei chromatin condensation and nuclei fragmentation); (b) quantitative analysis, as represented by relative apoptotic nuclei counts.

Morphological changes in cells that were exposed to oxidative stress were evaluated after DAPI staining. Apoptotic nuclei counts showed a significant reduction in cell death in cells treated with rhEPO-containing supernants, compared to untreated cells: 13.5 ± 0.6% versus 19.0 ± 0.7%, p < 0.001 (Figure 1B).

In Situ Expression of Erythropoietin Promotes Cytoprotection and Vascularization

There was no difference among the groups in the levels of hematocrit at all time points (Table 2). Four weeks after cell transplantation, heart tissue sections were stained with anti-rhEPO monoclonal antibodies to detect rhEPO expression from implanted cells. Human kidney tissue was used as positive control (Figure 2A). RhEPO was detected only in hearts treated with transduced cardiac fibroblasts 5 weeks post MI (Figure 2B–D). The staining was identified in all animals treated with transduced cells (n = 5), towards the center of the infarct zone.

Table 2. Serial hematocrit measurements in rats after MI, showing that EPO-transduced cell injection does not increase systemic hematocrit levels

Treatment time	Saline (n = 9)	EPO-transduced cells (n = 9)	Non-transduced cells (n = 9)
Baseline (before injection; %)	47.5 ± 0.3	47.2 ± 0.2	47.2 ± 0.1
I week post-injection (%)	47.4 ± 0.2	47.0 ± 0.1	47.39 ± 0.2
4 weeks post-injection (%)	47.4 ± 0.18	46.8 ± 0.2	47.15 ± 0.2

$p = 1.00$.

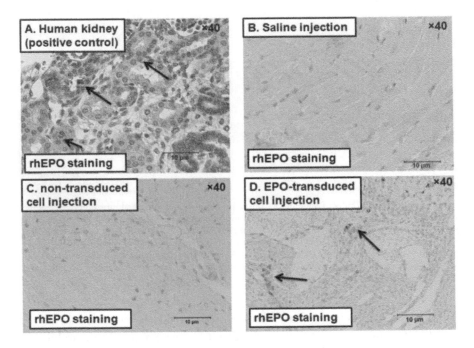

Figure 2. Photomicrographs of heart sections, immunostained with anti-human erythropoietin antibodies (brown), 4 weeks after cell transplantation. (A) Human kidney positive control. (B) saline-injected heart. (C) Non-transduced fibroblast-transplanted heart. (D) RhEPO in transduced fibroblast-transplanted heart (infarct zone). Arrows show rhEPO expression.

The degree of vascularization was evaluated by counting capillary blood vessels in the infarct border zones using α-smooth muscle actin and lectin stainings for vascular endothelium localization. Compared with non-transduced fibroblasts and saline, implanted rhEPO-producing fibroblasts enhanced angiogenesis, as shown in both α-smooth muscle actin (Figure 3A; 15.27 ± 0.37 versus 9.93 ± 0.97 versus 9.57 ± 0.72 vessels/mm2, respectively, $p < 0.05$), and lectin stained sections (Figure 3B; 70.1 ± 8.5 versus 47.5 ± 5.9 versus 20.7 ± 3.6 vessels/mm2, respectively, $p < 0.05$).

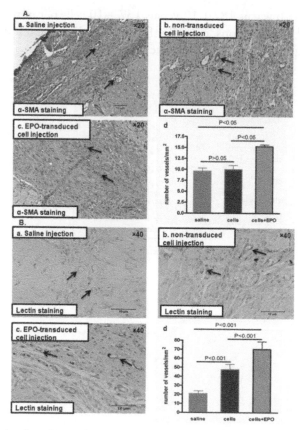

Figure 3. Four weeks after cell transplantation, rhEPO expression promotes neo-angiogenesis in vivo. (A) α-Smooth muscle actin staining (brown) in infarcted hearts (infarct border zones). Arrows show stained blood vessels. (a) Saline injection; (b) non-transduced cell injection; (c) rhEPO-transduced cell injection; (d) quantitative analysis. (B) Lectin (BS-1) staining (brown) in infarcted hearts (infarct border zones). Arrows show stained blood vessels. (a) Saline injection; (b) non-transduced cell injection; (c) rhEPO-transduced cell injection; (d) quantitative analysis.

Apoptosis in infarcted hearts was detected by active caspase-3 staining with polyclonal antibodies detecting only the active (effector) form of caspase-3 (Figure 4A) [25,26]. Human tonsil tissue was used as a positive control (Figure 4A, a). RhEPO-transduced cell transplantation resulted in a significant decrease in the number of active-caspase-3 positive cells (24 ± 6 versus 176 ± 20 versus 110 ± 27 cells per mm2 in saline and non-transduced cell injections, $p < 0.05$; Figure 4C, a). The staining revealed mainly nuclear localization of active caspase-3 (Figure 4A) [26]. Apoptosis was also evaluated by commercially available TUNEL kit (Figure 4B). Rat mammary gland tissue served as a positive control (Figure 4B, a). Similar to active caspase-3 staining, TUNEL staining showed that rhEPO-transduced cell transplantation resulted in a significant decrease in the number of apoptotic cells (118 ± 6 versus 182 ± 12 versus 157 ± 10 cells per mm2 in saline and non-transduced cell injections, p < 0.05; Figure 4C, b). These results suggest that local rhEPO expression prevented cardiac cell apoptosis in vivo.

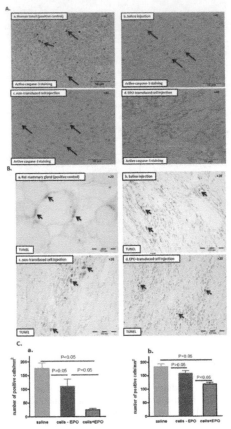

Figure 4. Four weeks after cell transplantation, rhEPO expression attenuates cardiac cell apoptosis in vivo. (A) Representative photomicrographs of active caspase-3 staining (brown) in infarcted hearts (infarct border zone). Arrows show nuclear active caspase-3 staining. (a) Human tonsil tissue (positive control); (b) saline injection; (c) non-transduced cell injection; (d) rhEPO-transduced cell injection. (B) Representative photomicrographs of TUNEL staining (brown) in infarcted hearts (infarct border zone). (a) Rat mammary gland tissue (positive control); (b) saline injection; (c) non-transduced cell injection; (d) rhEPO-transduced cell injection. (C) Quantitative analysis of apoptosis in infarcted hearts. (a) active caspase-3 staining; (b) TUNEL.

Epo-Transduced Cell Injection Did Not Improve Lv Remodeling and Function

To examine potential cardioprotective effects of rhEPO-producing fibroblasts in vivo, we used a rat model of AMI, which resulted in progressive LV remodeling and dysfunction [19]. Overall, 45 rats were included in the final transplantation study. Rats were randomized into three treatment groups (n = 15 in each group): transduced rhEPO-producing cells, non-transduced cells, and saline injection. Ten animals from each group (five additional animals in each group were sacrificed one month after transplantation for immunohistochemical examination) were assessed with echocardiography.

Serial echocardiography tests showed that EPO-transduced and non-transduced cell injection did not reduce LV remodeling and dysfunction 5 and 9 weeks post MI (Table 3). At 5 weeks, there was a trend towards more scar thinning in the EPO-treated group (anterior wall diastolic thickness, p = 0.07; LV end diastolic dimension, p = 0.08).

Table 3. Comparison of LV remodeling and function between EPO-transduced, non-transduced cell or saline injections by two-dimensional echocardiography, before (baseline), or 5 or 9 weeks after MI

	EPO-transduced cells (n = 10)	Non-transduced cells (n = 9)	Saline (n = 9)	p†
5-week follow-up				
Heart rate				0.17
Baseline	308.8 ± 12.6	273.8 ± 11.7	273.0 ± 11.6	
5 weeks	252.7 ± 9.5	252.5 ± 14.0	262.8 ± 13.5	
p*	0.0068	0.32	0.55	
AW d (cm)				0.07
Baseline	0.13 ± 0.01	0.17 ± 0.007	0.13 ± 0.01	
5 weeks	0.09 ± 0.006	0.10 ± 0.01	0.10 ± 0.01	
p*	0.006	0.0002	0.04	
LVEDD				0.08
Baseline	0.76 ± 0.02	0.72 ± 0.02	0.82 ± 0.02	
5 weeks	1.06 ± 0.02	0.98 ± 0.03	1.00 ± 0.03	
p*	<0.0001	<0.0001	0.0009	
LVESD				0.22
Baseline	0.60 ± 0.03	0.56 ± 0.02	0.64 ± 0.03	
5 weeks	0.88 ± 0.03	0.81 ± 0.04	0.80 ± 0.04	
p*	<0.0001	<0.0001	<0.0001	
LVEDA				0.34
Baseline	0.39 ± 0.03	0.35 ± 0.01	0.43 ± 0.02	
5 weeks	0.73 ± 0.04	0.65 ± 0.04	0.67 ± 0.05	
p*	<0.0001	<0.0001	0.0011	
LVESA				0.36
Baseline	0.24 ± 0.02	0.21 ± 0.01	0.26 ± 0.02	
5 weeks	0.52 ± 0.04	0.42 ± 0.04	0.44 ± 0.05	
p*	<0.0001	<0.0001	0.0004	
FS (%)				0.99
Baseline	20.6 ± 2.9	22.3 ± 1.7	25.07 ± 2.2	
5 weeks	16.2 ± 1.4	17.8 ± 1.6	20.96 ± 2.0	
p*	0.14	0.008	0.002	
9-week follow-up				
Heart rate				0.28
Baseline	303.2 ± 12.7	274.4 ± 13.3	273.0 ± 11.6	
9 weeks	242.0 ± 6.3	235.5 ± 12.2	246.2 ± 7.7	
p*	0.04	0.12	0.0014	
AW d (cm)				0.19
Baseline	0.13 ± 0.01	0.17 ± 0.007	0.13 ± 0.01	
9 weeks	0.09 ± 0.01	0.12 ± 0.01	0.11 ± 0.01	
p*	0.03	0.0002	0.3	
LVEDD				0.22
Baseline	0.76 ± 0.02	0.72 ± 0.02	0.82 ± 0.02	
9 weeks	1.11 ± 0.04	1.03 ± 0.03	1.05 ± 0.03	
p*	<0.0001	<0.0001	0.0003	
LVESD				0.3
Baseline	0.60 ± 0.03	0.56 ± 0.02	0.64 ± 0.03	
9 weeks	0.95 ± 0.05	0.84 ± 0.04	0.86 ± 0.05	
p*	<0.0001	<0.0001	<0.0001	
LVEDA				0.42
Baseline	0.39 ± 0.03	0.35 ± 0.01	0.43 ± 0.02	

Table 3. (Continued)

9 weeks	0.83 ± 0.06	0.71 ± 0.05	0.74 ± 0.06	
p^*	<0.0001	<0.0001	0.0006	
				0.37
LVESA				
Baseline	0.24 ± 0.02	0.21 ± 0.01	0.26 ± 0.02	
9 weeks	0.61 ± 0.06	0.48 ± 0.05	0.51 ± 0.06	
p^*	0.0001	0.0001	0.0003	
				0.09
FS (%)				
Baseline	20.6 ± 2.9	22.3 ± 1.72	25.1 ± 2.2	
9 weeks	14.3 ± 1.8	18.7 ± 1.73	19.8 ± 2.4	
p^*	0.05	0.02	0.0008	

AW d, anterior wall diastolic thickness; LVEDD, LV end diastolic dimension; LVESD, LV end systolic dimension; LVEDA, LV end diastolic area; LVESA, LV end systolic area; FS, LV fractional shortening [(LVIDd – LVIDs)/LVIDd] × 100. Values are mean ± SEM; p^*, p-values derived from paired comparisons between baseline and 5/9 week measurements; $p^\dagger$, p-values for the differences between groups over time (two-way repeated-measures ANOVA, p-value for interaction).

Morhpometric analysis of the infarcted hearts showed that EPO-transduced cell injection also fails to prevent infarct expansion and scar thinning (Table 4). For major parameters (average scar thickness, LV cavity area, whole LV area and expansion index), EPO-transduced cells showed significant adverse remodeling compared to non-transduced cells with a magnitude similar to saline-treated animals. Furthermore, non-transduced cell injection attenuated infarct expansion and preserved LV area, compared to saline-treated animals.

Table 4. Comparison of LV remodeling by postmortem morphometry, 9 weeks after MI

	EPO-transduced cells (n = 10)	Non-transduced cells (n = 9)	Saline (n = 9)	p^* ANOVA
LV maximal diameter (mm)	8.2 ± 0.4	6.2 ± 0.31	8.1 ± 0.33	<0.0001
p (Bonferroni test)	<0.001†	<0.001‡	-	
Average wall thickness (mm)	1.4 ± 0.05	1.2 ± 0.04	1.4 ± 0.03	0.003
p (Bonferroni test)	<0.05†	<0.05‡	-	
Average scar thickness (mm)	0.62 ± 0.03	0.92 ± 0.04	1.0 ± 0.05	<0.0001
p (Bonferroni test)	<0.001§	-	-	
Relative scar thickness	0.46 ± 0.05	0.59 ± 0.03	0.56 ± 0.05	0.22
p (Bonferroni test)	-	-	-	
Muscle area (mm²)	43 ± 1.4	43 ± 4.2	47 ± 2.3	0.55
p (Bonferroni test)	-	-	-	
LV cavity area (mm²)	57 ± 8.0	40 ± 4.9	65 ± 5.3	0.002
p (Bonferroni test)	<0.05†	<0.01‡	-	
Whole LV area (mm²)	110 ± 7.5	77 ± 7.0	110 ± 7.0	<0.0001
p (Bonferroni test)	< 0.001†	< 0.001‡	-	
Expansion index	1.1 ± 0.08	0.89 ± 0.05	1.2 ± 0.13	0.04
p (Bonferroni test)	-	<0.05‡	-	
Epicardial scar length (mm)	7.2 ± 1.2	6.5 ± 1.2	8.1 ± 1.6	0.68
p (Bonferroni test)	-	-	-	
Endocardial scar length (mm)	7.3 ± 1.6	6.2 ± 1.2	6.6 ± 1.0	0.65
p (Bonferroni test)	-	-	-	

Values are means ± SEM. p (Bonferroni test). p-values from Bonferroni multiple comparison test; p^*, p-values for the differences between groups (one-way ANOVA). †Versus non-transduced cells; ‡versus saline; §versus saline and non-transduced cells.

Discussion

To the best of our knowledge, the present study is the first to use reprogrammed cardiac fibroblasts encoding the rhEPO gene to repair infarcted myocardium in situ. In previous experiments, rhEPO was delivered systemically prior to, at different time points and immediately after ischemia/reperfusion or cardiomyopathy induction [27-30]. In our study, rhEPO produced from genetically modified cardiac fibroblasts exhibited a strong anti-apoptotic effect in vitro, promoted angiogenesis and attenuated apoptosis in the infarcted tissue. However, despite positive in vitro results and cytoprotection of EPO-transduced cells in the infarcted hearts, restoration of LV function was not observed, and there was a trend to more adverse remodeling.

Cardiac fibroblasts present an attractive gene carrier into the infarcted heart. Although not an 'ideal' source for transplantation (due to lack of contractility and so on), fibroblasts, constituting the major cell population of the myocardium, can be easily obtained from clinically accessible sites, expanded in vitro without significant limitation, and, unlike cardiomyocytes, can be transduced highly efficiently by retroviral EPO-based gene transfer.

Our findings concur with other studies that show a lack of functional improvement following EPO treatment despite positive cellular effects (apoptosis inhibition and angiogenesis) [31-35], as well as one clinical trial conducted with EPO analogue darbepoetin alfa [36].

The reason for the inconsistency between the cytoprotective effect of EPO in vitro and in situ versus the lack of functional improvement in vivo is unclear. A possible explanation is that an EPO therapeutic effect on LV remodeling and function is mainly mediated by endothelial progenitor cell recruitment, which is achieved by systemic administration [34,37,38]. This notion is supported by the positive correlation found between EPO plasma levels and endothelial progenitor cell recruitment in an observational study of patients after acute MI [39]. Thus, local EPO therapy is insufficient for functional improvement.

In addition, the lack of improvement could be related to the timing of therapy. In most of the positive EPO studies systemic cytokine administration was initiated before or immediately after ischemia/reperfusion induction. We used a more severely injured model with more clinically relevant timing of cell delivery (1 week post MI). However, at this stage, EPO could be less effective due to a relatively low cell apoptotic rate and the lack of stem cell homing signals from the infarct [35].

Study Limitations

Cell engraftment and survival was not quantified. Bromodeoxyuridine (BrdU) or other cellular markers were not used to avoid interference with cell capacity to colonize and proliferate. Such interference could cause a decrease in the proposed EPO effect.

Retroviral transduction itself can potentially alter the phenotype of the transduced cells. However, the expression profile of transduced cells did not exhibit any negative effects when the transduced cell culture supernants were added to cardiomyocyte culture in vitro.

In order to avoid possible systemic interference, we did not apply an immunosuppression protocol. However, EPO detection 1 month after cell transplantation suggests that immune response, if evoked, is limited in magnitude. The EPO tissue level could not be quantified due to technical constraints for detection of predicted low levels of the protein.

Conclusion

In situ expression of rhEPO by genetically engineered cardiac fibroblasts exerts cytoprotective and pro-angiogenic effects. However, these favorable effects were not translated into functional improvement after MI. The results suggest that for successful local EPO-based therapy, several improvements should be made to determine the optimal cell type for gene delivery, the most effective recombinant viral vector, myocardial EPO response and, notably, an accurate therapeutic window for cell and gene transfer.

Competing Interests

The authors declare that they have no competing interests.

Authors' Contributions

ER carried out the in vitro and in vivo experiments, including data acquisition and analysis. OS participated in study design and coordination, and helped in drafting the manuscript. AN participated in conceptual study design and coordination, and helped to draft the manuscript. TE provided help in study design, data acquisition and analysis. MF participated in data acquisition and analysis. RH carried out the in vivo experiments. AD participated in conceptual study design and coordination, and helped to draft the manuscript. JL conceived the study, participated in its design and coordination, and helped to draft the manuscript.

References

1. Maiese K, Li F, Chong ZZ: New avenues of exploration for erythropoietin. JAMA 2005, 293:90–95.

2. Parsa C, Matsumoto A, Kim J, Riel RU, Pascal LS, Walton GB, Thompson RB, Petrofski JA, Annex BH, Stamler JS, Koch WJ: A novel protective effect of erythropoietin in the infracted heart. J Clin Invest 2003, 112:999–1007.

3. Calvillo L, Latini R, Kajstura J, Leri A, Anversa P, Ghezzi P, Salio M, Cerami A, Brines M: Recombinant human erythropoietin protects the myocardium from ischemia-reperfusion injury and promotes beneficial remodeling. Proc Natl Acad Sci USA 2003, 100:4802–4806.

4. Tramontano AF, Muniyappa R, Black AD, Blendea MC, Cohen I, Deng L, Sowers JR, Cutaia MV, El-Sherif N: Erythropoietin protects cardiac myocytes from hypoxia-induced apoptosis through an Akt-dependent pathway. Biochem Biophys Res Comm 2003, 308:990–994.

5. Cai Z, Semenza GL: Phosphatidylinositol-3 kinase signaling is required for erythropoietin-mediated acute protection against myocardial ischemia/reperfuson injury. Circulation 2004, 109:2050–2053.

6. Ribatti D, Vacca A, Roccaro AM, Crivellato E, Presta M: Erythropoietin as an angiogenic factor. Eur J Clin Invest 2003, 33:891–896.

7. Heeschen C, Aicher A, Lehmann R, Fichtlscherer S, Vasa M, Urbich C, Mildner-Rihm C, Martin H, Zeiher A, Dimmeler S: Erythropoietin is a potent physiologic stimulus for endothelial progenitor cell mobilization. Blood 2003, 102:1340–1346.

8. Rao SV, Stamler JS: Erythropoietin, anemia, and orthostatic hypotension: the evidence mounts. Clin Auton Res 2002, 12:141–143.

9. Lipsic E, Schoemaker RG, Meer P, Voors AA, vanVeldhuisen D, vanGilst WH: Protective effects of erythropoietin in cardiac ischemia. From bench to bedside. J Am Coll Cardiol 2006, 48:2161–2167.

10. Besarab A, Bolton WK, Browne KF, Egrie JC, NIssenson AR, Okamoto DM, Schwab SJ, Goodkin DA: The effects of normal as compared with low hematocrit values in patients with cardiac disease who are receiving hemodialysis and epoetin. N Engl J Med 1998, 339:584–590.

11. Hebert PC, Wells G, Blajchman MA, Marshall M, Martin M, Pagliarello G, Tweeddale M, Schweitzer I, Yetishir E: A multicenter, randomized, controlled clinical trial of transfusion requirements in critical care. N Engl J Med 1999, 340:409–417.

12. Briest W, Homagk L, Baba HA, Deten A, Rabler B, Tannapfel A, Wagner KF, Wenger RH, Zimmer H: Cardiac remodeling in erythropoietin-transgenic mice. Cell Physiol Biochem 2004, 14:277–284.

13. Wagner KF, Katschinski DM, Hasegawa J, Schumacher D, Meller B, Gembruch U, Schramm U, Jelkmann W, Gassmann M, Fandrey J: Chronic inborn erythrocytosis leads to cardiac dysfunction and premature death in mice overexpressing erythropoietin. Blood 2001, 97:536–542.

14. Parsa CJ, Kim J, Riel RU, Pascal LS, Walton GB, Thompson RB, Petrofski JA, Annex BH, Stamler JS, Koch WJ: Cardioprotective effects of erythropoietin in the reperfused ischemic heart. A potential role for cardiac fibroblasts. J Biol Chem 2004, 279:20655–20662.

15. Einbinder T, Sufaro Y, Yusim I, Byk G, Passlick-Deetjen J, Chaimovitz C, Douvdevani A: Correction of uremic mice by genetically modified peritoneal mesothelial cells. Kidney Int 2003, 63:2103–2112.

16. Shneyvays V, Nawrath H, Jacobson KA, Shainberg A: Induction of apoptosis in cardiac myocytes by A3 adenosine receptor antagonist. Exp Cell Res 1998, 243:383–397.

17. Porter CD, Lukacs KV, Box G, Takeuchi Y, Collins MKL: Cationic liposomes enhance the rate of transduction by a recombinant retroviral vector in vitro and in vivo. J Virol 1998, 72:4832–4840.

18. Kitamura T, Tange T, Terasawa T, Chiba S, Kuwaki T, Miyagawa K, Piao YF, Miyazono K, Urabe A, Takaku F: Establishment and characterization of a unique human cell line that proliferates dependently on GM-CSF, IL-3, or erythropoietin. J Cell Physiol 1989, 140:323–334.

19. Etzion S, Battler A, Barbash IM, Cagnano E, Zarin P, Granot Y, Kedes LH, Kloner RA, Leor J: Influence of embryonic cardiomyocyte transplantation on the progression of heart failure in a rat model of extensive myocardial infarction. J Mol Cell Cardiol 2001, 33:1321–1330.

20. Leor J, Patterson M, Quinones MJ, Kedes LH, Kloner RA: Transplantation of fetal myocardial tissue into the infarcted myocardium of rat: a potential method for repair of infarcted myocardium? Circulation 1996, 94(9 Suppl):II332–II336.

21. Amsalem Y, Mardor Y, Feinberg MS, Landa N, Miller L, Daniels D, Ocherashvilli A, Holbova R, Yosef O, Barbash IM, Leor J: Iron-oxide labeling and

outcome of transplanted mesenchymal stem cells in the infarcted myocardium. Circulation 2007, 116:I38–45.

22. Landa N, Miller L, Feinberg MS, Holbova R, Shachar M, Freeman I, Cohen S, Leor J: Effect of injectable alginate implant on cardiac remodeling and function after recent and old infarcts in rat. Circulation 2008, 117:1388–1396.

23. Hale SL, Kloner RA: Left ventricular topographic alterations in the completely healed rat infarct caused by early and late coronary artery reperfusion. Am Heart J 1988, 116:1508–1513.

24. Hoek TL, Becker LB, Shao Z, Li C, Schimaker PT: Reactive oxygen species released from mitochondria during brief hypoxia induce preconditioning in cardiomyocytes. J Biol Chem 1998, 273:18092–18098.

25. Rodriguez M, Schaper J: Apoptosis: measurement and technical issues. J Mol Cell Cardiol 2005, 38:15–20.

26. Kamada S, Kikkawa U, Tsujimoto Y, Hunter T: Nuclear translocation of caspase-3 is dependent on its proteolytic activation and recognition of a substrate-like protein(s). J Biol Chem 2005, 280:857–860.

27. Nishiya D, Omura T, Shimada K, Matsumoto R, Kusuyama T, Enomoto S, Iwao H, Takeuchi K, Yoshikawa J, Yoshiyama M: Effects of erythropoietin on cardiac remodeling after myocardial infarction. J Pharmacol Sci 2006, 101:31–39.

28. Meer P, Lipsic E, Henning RH, Boddeus K, Velden J, Voors AA, vanVeldhuisen DJ, vanGilst WH, Schoemaker RG: Erythropoietin induces neovascularization and improves cardiac function in rats with heart failure after myocardial infarction. J Am Coll Cardiol 2005, 46:125–133.

29. Liu X, Xie W, Liu P, Duan M, Jia Z, Li W, Xu J: Mechanism of the cardioprotection of rhEPO pretreatment on suppressing the inflammatory response in ischemia-reperfusion. Life Sci 2006, 78:2255–2264.

30. Li L, Takemura G, Li Y, Miyata S, Esaki M, Okada H, Kanamori H, Khai NC, Maruyama R, Ogino A, Minatoguchi S, Fujiwara T, Fujiwara H: Preventive effect of erythropoietin on cardiac dysfunction in doxorubicin-induced cardiomyopathy. Circulation 2006, 113:535–543.

31. Hale SL, Sesti C, Kloner RA: Administration of erythropoietin fails to improve long-term healing or cardiac function after myocardial infarction in the rat. J Cardiovasc Pharmacol 2005, 46:211–215.

32. Olea FD, Janavel GV, DeLorenzi A, Cuniberti L, Yannarelli G, Meckert PC, Cearras M, Laguens R, Crottogini A: High-dose erythropoietin has no

long-term protective effects in sheep with reperfused myocardial infarction. J Cardiovasc Pharmacol 2006, 47:736–741.

33. Kristensen J, Maeng M, Rehling M, Berg JS, Mortensen UM, Nielsen SS, Nielsen TT: Lack of acute cardioprotective effect from preischaemic erythropoietin administration in a porcine coronary occlusion model. Clin Physiol Funct Imaging 2005, 25:305–310.

34. Prunier F, Pfister O, Hadri L, Liang L, Del Monte F, Liao R, Hajjar RJ: Delayed erythropoietin therapy reduces post-MI cardiac remodeling only at a dose that mobilizes endothelial progenitor cells. Am J Physiol Heart Circ Physiol 2007, 292:H522–529.

35. Chanseaume S, Azarnoush K, Maurel A, Bellamy V, Peyrard S, Bruneval P, Hagege AA, Menasche P: Can erythropoietin improve skeletal myoblast engraftment in infarcted myocardium? Interact Cardiovasc Thorac Surg 2007, 6:293–297.

36. Lipsic E, Meer P, Voors AA, Westenbrink BD, de Boer HC, van Zonneveld AJ, Schoemaker RG, van Gilst WH, Zijlstra F, van Veldhuisen DJ: A single bolus of a long-acting erythropoietin analogue darbepoetin alfa in patients with acute myocardial infarction: a randomized feasibility and safety study. Cardiovasc Drug Ther 2006, 20:135–141.

37. Jorgensen E, Bindslev L, Ripa RS, Kastrup J: Epo 'cytokine-doping' of heart disease patients, will it work? Eur Heart J 2006, 27:1767–1768.

38. Westenbrink BD, Lipsic E, Meer P, Harst P, Oeseburg H, Du Marchie Sarvaas GJ, Koster J, Voors AA, van Veldhuisen DJ, van Gilst WH, Schoemaker RG: Erythropoietin improves cardiac function through endothelial progenitor cell and vascular endothelial growth factor mediated neovascularization. Eur Heart J 2007, 28:2018–2027.

39. Ferrario M, Massa M, Rosti V, Campanelli R, Ferlini M, Marinoni B, De Ferrari GM, Meli V, De Amici M, Repetto A, Verri A, Bramucci E, Tavazzi L: Early haemoglobin-independent increase of plasma erythropoietin levels in patients with acute myocardial infarction. Eur Heart J 2007, 28:1805–1813.

Anti-Angiogenesis Therapy Based on the Bone Marrow-Derived Stromal Cells Genetically Engineered to Express Sflt-1 in Mouse Tumor Model

M. Hu, J-L Yang, H. Teng, Y-Q Jia, R. Wang, X-W Zhang, Y. Wu, Y. Luo, X-C Chen, R. Zhang, L. Tian, X. Zhao and Y-Q Wei

ABSTRACT

Background

Bone marrow-derived stromal cells (BMSCs) are important for development, tissue cell replenishment, and wound healing in physiological and pathological conditions. BMSCs were found to preferably reach sites undergoing the

process of cell proliferation, such as wound and tumor, suggesting that BM-SCs may be used as a vehicle for gene therapy of tumor.

Methods

Mouse BMSCs were loaded with recombinant adenoviruses which express soluble Vascular Endothelial Growth Factor Receptor-1 (sFlt-1). The anti-angiogenesis of sFlt-1 in BMSCs was determined using endothelial cells proliferation inhibition assay and alginate encapsulation assay. The anti-tumor effects of BMSCs expressing sFlt-1 through tail-vein infusion were evaluated in two mouse tumor metastases models.

Results

BMSCs genetically modified with Adv-GFP-sFlt-1 could effectively express and secret sFlt-1. BMSCs loaded with sFlt-1 gene could preferentially home to tumor loci and decrease lung metastases and prolong lifespan in mouse tumor model through inducing anti-angiogenesis and apoptosis in tumors.

Conclusion

We demonstrated that BMSCs might be employed as a promising vehicle for tumor gene therapy which can effectively not only improve the concentration of anticancer therapeutics in tumors, but also modify the tumor microenvironment.

Background

Bone marrow-derived stromal cells (BMSCs), also known as mesenchymal stem cells or nonhematopoietic progenitor cells, are precursors that can be differentiated into chondrocytes, osteoblasts, adipocytes, neurons and other cell types [1]. They are important for development and cell replenishment of active proliferating tissues in physiological conditions, such as blood, skin and gut. In pathological conditions, they are involved in the process of wound healing and tissue regeneration. In animal and clinical experiments, BMSCs have been used for tissue damage repair and functional reconstruction of organs, such as myocardial infarction, osteogenesis imperfecta, syndrome of multiple organ failure and rebuilding of hematopoietic system after chemotherapy of breast cancer [2-5].

It has been reported that the conditions featured by enhanced cell proliferation and tissue remodeling, such as bone fractures, embryo growth and tumorigenesis, offer an appropriate microenvironment for migration, proliferation and differentiation of stem cells delivered systemically [3,6-9]. The formation of tumor stroma is similar to that of wound healing and results in a microenvironment which is suitable for the proliferation of bone marrow-derived stem cells [10]. Therefore, we hypothesize that systemically delivered BMSCs would preferentially home to tumor tissues and participate in the formation of tumor stroma. Therefore, it will be useful to develop a new targeting system based on genetically modified BMSCs for tumor gene therapy. The feasibility of this approach has been demonstrated by Matus Studeny's experimemt [11].

Metastasis is the most common and fundamental characteristics of solid tumors. Most deaths of cancerous patients result from the ruthless growth of metastases that are resistant to conventional remedies. This urges us to develop a novel therapeutic strategy for inhibiting tumor metastasis. The concept that tumor growth and metastasis relies on angiogenesis has been widely proven and accepted. The signal axis of vascular endothelial growth factor (VEGF) and its receptors (VEGFR-1, VEGFR-2 and VEGFR-3) is involved in the formation and progress of tumors. The soluble VEGF receptor-1 (sFlt-1) has been shown to be effective in inhibition of cancerous angiogenesis [12].

In this report, we developed a novel strategy of tumor gene therapy in which BMSCs loaded with recombinant adenoviruses expressing sFlt-1 could effectively suppress tumor growth through inhibiting angiogenesis and metastases and prolong the lifespan in mouse model, indicating that BMSCs might be employed as an effective carrier for tumor gene therapy.

Methods

BMSCs Isolation and Culture

BMSCs were isolated from femur bone marrow of male BALB/c mouse. Mononuclear cells were separated by centrifugation over a Ficoll gradient (Sigma Chemical Co., St. Louis, MO) of 1.077 g/ml. The cells were cultured in L-DMEM containing 10% fetal bovine serum (Gibco BRL, Inc) at an initial seeding density of 1×10^5 cells/cm2 at 37°C with 5% CO_2 and 95% humidified atmosphere. The nonadherent cells were removed and washed with PBS after 24 hours, and the monolayer of adherent cells were cultured until they reached to confluence. The cells were then trypsinized (0.25% trypsin with 0.1% EDTA), subcultured at densities of 5000–6000 cells/cm2. and used for experiments during passages five to eight. CD34, CD29 and CD44 expressions in BMSCs were determined by flow cytometry. Briefly, the trypsinized BMSCs were adjusted to 1×10^7 cells/ml in media (1% bovine serum albumin, 0.2% sodium azide in PBS), and then stained with FITC-conjugated monoclonal antibodies (BD Biosciences, San Jose, CA, USA) on ice according to the manufacturer's protocol, and followed by detection with flow cytometry.

Production of Adenovirus and Genetic Modification of BMSCs

AdenoVec-GFP-sFlt-1 and AdenoVec-GFP (InvivoGen, San Diego, CA) were transfected into HEK-293 cells using jetPEI (Qbiogene, Nottingham, UK) according to the manufacturer's protocol. Seven to 14 days later, the recombinant adenoviruses were harvested from the cultures and the viruses were further amplified in 293 cells, and then tittered by plaque assay. BMSCs were infected with adenoviruses at a multiplicity of infection (MOI) of 3000 for 2 hours followed by replacing the infection medium. Twenty-four hours later, the virus-infected BMSCs were harvested for subsequent experiments [11].

Western Blot Analysis

Western blot analysis was performed as described previously [13]. Briefly, the secreted proteins in supernatants of the culture were precipitated by TCA-DOC/acetone and the western blot assay was performed using standard method with primary monoclonal antibody of anti-sFlt-1 (1:500 dilution, Santa Cruz Biotechnology, Santa Cruz, CA, USA) and a biotinylated secondary antibody, and the proteins on the membrane were visualized by Vectastain ABC kit (Vector Laboratories, Burlingame, CA, USA).

Mouse Endothelial Cells Proliferation Inhibition Assay

The mouse endothelial cells proliferation inhibition assay was performed as described previously for examining suppression of sFlt-1 in the conditioned media on VEGF-driven proliferation of endothelial cells [14]. The conditioned media were obtained from BMSCs infected with Adv-GFP-sFlt-1, Adv-GFP, or PBS (mock infected), respectively. Endothelial cells were grown in 24-well plates at 37°C in DMEM containing 10% FBS. At 50% confluence, the cells were washed with PBS after removal of the DMEM media, and then 0.5 ml of conditioned media was added. Fifteen minutes later, VEGF was added to each well with a final concentration of 10 ng/ml and the cells were incubated at 37°C. After 72 h, the cells were trypsinized, and the number of viable cells was counted using a trypan blue assay.

Alginate Encapsulation Assay

Alginate-encapsulated tumor cell assays were performed as described previously [13]. Briefly, tumor cells, BMSCs expressing sFlt-1 or mixed cells were resuspended in a 1.5% solution of sodium alginate and added dropwise into a swirling 37°C solution of 250 mM calcium chloride. Alginate beads were formed with about 1 × 105 cells per bead. After mice were anesthetized, 4 beads were implanted subcutaneously into an incision made on the dorsal side. Incisions were closed with surgical clamps. After 21 days, mice were injected intravenously with 100 μl of a 100 mg/kg FITC-dextran solution (Sigma). Beads were surgically removed, and FITC-dextran was quantified against a standard curve of FITC-dextran.

Murine Tumor Metastases Models and Treatment

The colon carcinoma CT26, Lewis lung cancer LLC and fibrosarcoma MethA cell lines were used in this study. Metastasis models of LLC and CT26 were generated as described previously [15]. Briefly, female C57BL/6 and BALB/c mice were received i.m. injections of 2×10^5 LLC in leg or i.v. injection of 2×10^5 CT26 cells from tail vein in 100 μl of PBS, respectively and then the animals were randomly divided into four groups (10 mice/group). After 4 days for CT26 model and 10 days for LLC model of tumor inoculation, 1×10^6 of BMSCs infected with Adv-GFP-sFlt-1, BMSCs infected with Adv-GFP, unifected BMSCs or 100 μl of 0.9% NaCl solution alone were administered via tail vein for four doses at 3 days intervals for CT26 model and 5 days intervals for LLC model. Mice received LLC or CT26 injection were sacrificed on day 32 or day 16 after inoculation, respectively, when control mice became moribund, and the lungs from each mouse were removed and the surface lung metastases (>3 mm) were measured and scored [15]. The lungs of animals were also fixed in 10% buffered formalin followed by histological analysis. All of the mice used in this study were approved by the West China Hospital Cancer Center's Animal Care and Use Committee. In the handling and care of animals, all possible steps were taken to avoid animals' suffering and efforts were made to use the minimum number of animals.

Histological Analysis

The tissues were fixed in 10% neutral buffered formalin solution and embedded in paraffin. Sections of 3–5 μm were stained with hematoxylin and eosin (H.E.). Frozen sections were fixed in acetone, incubated, and stained with anti-CD31 antibody, then visualized with DAKO LSAB kit (DAKO, Carpinteria, CA), as described previously. Vessel density was determined by counting the number of microvessels per high-power field in the sections, as described [16]. Apoptosis in tumor tissues was determined by terminal dUTP nick-end labeling (TUNEL) method using an in situ cell death detection kit (Roche Molecular Biochemicals) following the manufacturer's protocol [17,18]. Sections in H&E staining and immunohistochemical staining were observed by two pathologists in a blinded manner.

Fluorescent in Situ Hybridization (FISH)

Fluorescent in situ hybridization was performed according to the mouse Y chromosome probe manual (STAR*FISH; Cambio, Cambridge, England) with some modifications. Cryostat sections of 5 μm in thickness were fixed in Carnoy's fixative for three times, 10 min each, and the sections were incubated in pepsin solution (1% in 0.1N HCl) for 10 minutes and rinsed in PBS. Serial ethanol dehydration was done (1.5 min each), and the slides were air-dried at room temperature. Sections were denatured at 65°C for 2 min in preheated 70% formamide and 2×SSC buffer, pH 7.0, and were then 'quenched' with ice-cold 70% ethanol for 1.5 min. Serial ethanol dehydration was done again. The mouse Y chromosome probe labeled with Cy3 was denatured at 65°C for 10 min and applied to the sections. The sections were coverslipped and sealed with rubber cement for incubation overnight in a hydrated slide box at 37°C. The next day, the coverslips were carefully removed. The sections were washed twice in preheated 50% formamide in 2×SSC buffer for 5 min each at 45°C and were then gently washed twice in preheated 1×SSC buffer for 5 min each at 45°C.

Statistical Assay

Significance was determined using one-way ANOVA and log-rank test (SPSS 11.0 for windows). Difference between groups were significant at a value of $P < 0.05$.

Results

Bmscs Isolation, Genetic Modification and Sflt-1 Function Validation

After harvest of monolayer cells with spindle-like morphology from mouse bone marrow, three CD markers (CD34, CD29 and CD44) were used for characterizing BMSCs. The cells were positive for CD29, CD44, and negative for CD34, indicating that they didn't belong to hematopoietic progenitor cells, but rather bone marrow-derived stromal cells (BMSCs) or mesenchymal stem cells (MSCs). BMSCs were cultured to reach to 90% confluence and incubated with adenoviruses at a MOI of 3000 for 2 hours. Twenty four hours later, all of the cells were GFP-positive checked by fluorescence microscopy and the cells were ready for tail vein infusion after PBS washing two times (Figure 1A). The secreted sFlt-1 in the supernatants of conditioned media was verified by Western blot (Figure 1B) and the function of sFlt-1 was validated by endothelial cell growth inhibition assay in vitro. The concentrated conditioned media obtained from BMSCs infected with Adv-GFP-sFlt-1 or with Adv-GFP were applied to the mouse endothelial cells grown in 24-well plates, followed by stimulation with a 10 ng/ml of VEGF 15 minutes later. Conditioned media from BMSCs infected with Adv-GFP-sFlt-1 resulted in inhibition of endothelial cells proliferation by about 50% compared with that from BMSCs infected with Adv-GFP or uninfected BMSCs (Figure 1C, P < 0.05), indicating that the BMSCs infected with Adv-GFP-sFlt-1 could effectively express and secret sFlt-1.

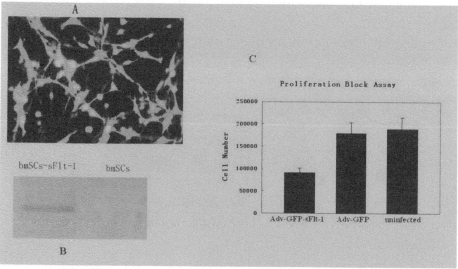

Figure 1. BMSCs infection and verification of secreted sFlt-1. After the BMSCs were confirmed by flow cytometric analysis, BMSCs were grown to 90% confluence and infected with adenoviruses at a multiple of infection of 3000 for 2 hours. 24 hours later, BMSCs were harvested and ready for use. Fluorescence microscope showed up to 100% GFP-positive cells. (Figure 1-A). Supernatants deposits from sFlt-1 bearing BMSCs could be recognized by antibodies reactive to NH2 terminus of mouse Flt-1, but negative staining in supernatants deposits from control BMSCs in Western blot analysis. (Figure 1-B). Conditioned media from Adv-sFlt-1 infected BMSCs was shown to inhibit the VEGF-driven mouse endothelial cells proliferation by about 50% compared with controls (Figure 1-C, P < 0.05).

Anti-Metastasis Effect

The mouse CT26 and LLC metastasis models were used to determine whether BMSCs expressing sFlt-1 suppress the growth of metastatic neoplasm. The metastatic nodules > 3 mm were assessed as an indicator of angiogenesis because the growth of tumors > 3 mm is thought to be vasculature sensitive [19]. Most of the mice systemically administered with control BMSCs or NaCl solution developed macroscopic lung metastases, whereas mice treated with BMSCs expressing sFlt-1 decreased lung metastases. The tumor nodules (> 3 mm) of lung surface were significantly decreased in the mice treated with BMSCs expressing sFlt-1 (P < 0.05) (Figure 2A, B). Lung weight, which correlates with total tumor burden, was also reduced significantly in the team of mice administered with BMSCs expressing sFlt-1 (Figure 2C). Booming proliferation of tumor cells and severe destruction of lung tissues were observed in the H.E. staining for control mice, whereas in the mice treated with BMSCs expressing sFlt-1, cancerous cells were less nourished and lung tissues were relatively normal (Figure 2D). Benefited from the decrease of tumor burden in the lung metastases, the lifespan of the tumor-bearing mice was prolonged significantly (Figure 2E, P < 0.05). The similar results were also observed from mouse LLC metastasis model (Figure 3).

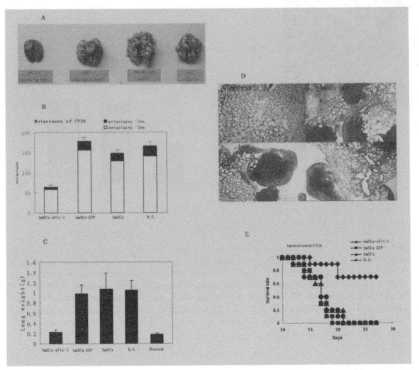

Figure 2. Anti-tumor effect in mouse CT26 colon adenocarcinoma model. Female BALB/c mice received i.v. injection of 2×10^5 CT26 cells in 100 μl of PBS. Treatment with 1×10^6 Adv-GFP-sFlt-1 infected BMSCs (black rhombus), 1×10^6 Adv-GFP infected BMSCs (■, black quadrilateral), 1×106 uninfected BMSCs (black triangle) and 100 μl of 0.9% NaCl solution alone (black circularity) were administered i.v. from tail vein. Treatment with sFlt-1-bearing BMSCs could decrease the number of and growth of surface metastases (Figure 2-A, D) and abolish the tumor burden by rendering the weight of lungs similar to that of normal mice (Figure 2-C). Mean numbers of lung tumor nodules in each group are shown (Figure 2-B). Values are plotted as means ± SEM (P < 0.05). The percentage of metastatic foci > 3 mm are marked as solid bars. A significant increase of survival rate in sFlt-1-bearing BMSCs treated mice, compared with the controls (Figure 2-E, P < 0.05. By log-rank test), was found in the tumor model.

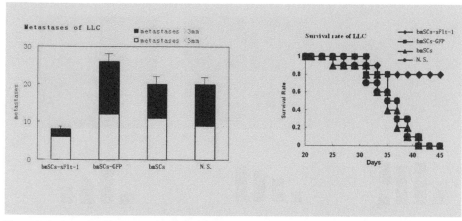

Figure 3. Anti-tumor effect in mouse LLC lung carcinoma model. Female C57BL/6 mice received i.v. injection of 2×10^5 LLC cells in 100 μl of PBS. Treatment with 1×10^6 Adv-GFP-sFlt-1 infected BmSCs (black rhombus), 1×10^6 Adv-GFP infected BmSCs (■, black quadrilateral), 1×106 uninfected BmSCs (black triangle) and 100 μl of 0.9% NaCl solution alone (black circularity) were administered i.v. from tail vein. A significant increase of survival and decrease of metastases in sFlt-1-expressing BMSCs treated mice, compared with the controls ($P < 0.05$. By log-rank test), was found in this model.

Inhibition of Angiogenesis and Induction of Apoptosis

Systemically delivered BMSCs which express sFlt-1 were shown to induce apparent inhibition of angiogenesis in vivo compared with control mice by immunohistochemistry. Angiogenesis within tumor tissue was estimated by counting the number of microvessels on the section staining with an anti-CD31 antibody. Very few newborn microvessels were found around the tumor cells in the BMSCs-sFlt-1 group, whereas abundant microvessels existed in three controls (Figure 4A, B, $P < 0.05$). The inhibition of angiogenesis in mice treated with BMSCs expressing sFlt-1 was further confirmed in alginate encapsulation assay. The surface vessel densities and the FITC-dextran uptake were apparently reduced in beads which contain relatively high ratio of sFlt-1-bearing BMSCs to LLC cells (Figure 4C, D, $P < 0.05$). Furthermore, an another alginate assay showed that the anti-angiogenesis effect of local production of sFlt-1 (came from the alginate beads being filled with sFlt-1-bearing BMSCs and tumor cells, or the BMSCs intravenous injected) was stronger than systemic production of sFlt-1 (came from the alginate beads being filled with sFlt-1-bearing BMSCs and implanted in the distant site) (Figure 4-E and 4F). Being starved of the nourishment by the inhibition of angiogenesis in neoplasm, augmentation of apoptosis of cancerous cells was detected by the TUNEL assay (Figure 5, $P < 0.05$). The inhibition of angiogenesis and the consequent induction of apoptosis underlie the mechanism of the metastases decrease and the life-span prolongation in tumor-bearing mice.

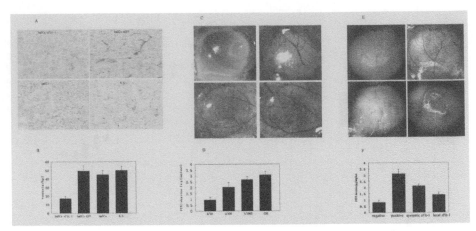

Figure 4. Evaluation of anti-angiogenetic effect by CD31 immunohistochemistry and alginate assay. Sections of frozen CT26 tumor tissues obtained from mice treated with sFlt-1-bearing BMSCs, GFP-expressing BMSCs, unmodified BmSCs and 0.9% NaCl solution were stained with CD31 antibody and Peroxidase-DAB (Figure 4-A). Vessel density was determined by counting the number of microvessels per high-power field (×200) in sections. (Figure4-B). Alginate beads containing different ratio of sFlt-1-bearing BMSCs and LLC (1/10, 1/100, 1/1000, LLC only) were implanted subcutaneously into C57BL/6 mice. Three weeks later, beads were surgically removed, and FITC-dextran was quantified. FITC-dextran uptake decrease and photograph of alginate implants showed the reduction of vascularization in beads containing relatively more sFlt-1-expressing BMSCs. (Figure4-C and D). Alginate beads containing MethA, BMSCs and mixture of both with ratio of 1/10 were prepared. Beads-MethA and beads-mixture were served as positive and negative controls. The third team of mice was implanted with two different beads containing MethA or BMSCs respectively in different sites of mice, which served as systemic sFlt-1 model to show the non-targeted effect of viruses. The fourth team of mice was implanted with beads containing MethA and hence injected intravenously with BMSCs in the same number of the former team, which served as local sFlt-1 model to show the targeted effect of viruses loaded in BMSCs. Compared with the controls, the effect of systemic production of sFlt-1 was weaker than local one. (Figure 4-E and F).

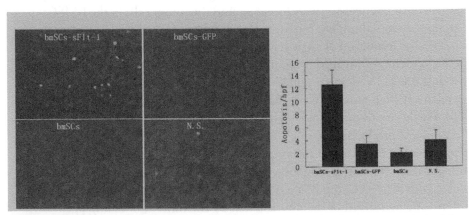

Figure 5. Apoptosis Assay by TUNEL staining. Paraffin sections from CT26 tumor tissues obtained from mice systemically administered with sFlt-1-expressing BMSCs, GFP expressing BMSCs, unmodified BMSCs and NaCl solution were stained using an in situ cell death detection kit (Roche) following the manufacturer's protocol. Apoptosis was determined by counting the number of the positive cells per high-power field in tumor sections. Original magnification, ×200. (Figure 5-A and B, P < 0.05).

Verification of BMSCs' Homing to Tumor

It has been reported that ubiquitous tissue distribution of adipose-derived stem cells in normal mouse was observed when the cells were systematically administrated [20]. We are concerned about the existence of BMSCs in the metastases and the surrounding tissues. To detect the tumor tropism of BMSCs after systemically administration, fluorescent in situ hybridization of Y chromosome was performed on frozen section of CT26 lung metastasis after BMSCs administration one week. Positive signals were observed in the frozen section of lung metastases exclusively, whereas the relatively normal area of lung tissue was totally negative for the fluorescent signals (Figure 6). The tropism of BMSCs to the sites undergoing active cell growth and differentiation makes them a smart delivery system of therapeutic agents, especially in gene therapy of cancer and severe destruction of tissues.

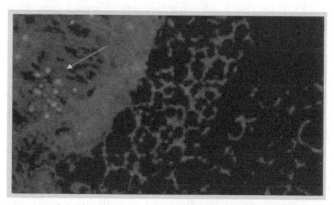

Figure 6. Fluorescent in situ hybridization. Frozen sections of lung metastases were tested for the presence of Y chromosome which is indication of male originated BMSCs by Cy3 labeled Y chromosome probe (STAR*FISH; Cambio, Cambridge, England). Positive result could be seen in the tumor area obviously, whereas the relatively normal parts of the sections were totally negative for fluorescent signals.

Discussion and Conclusion

BMSCs have been used in the therapy of myocardial infarction, osteogenesis imperfecta, syndrome of multiple organ failure and rebuilding of haematopoietic function after chemotherapy of breast cancer in pre-clinical and clinical experiments [2-5]. In this report, we demonstrated that systemically delivered BMSCs loaded with sFlt-1 gene could preferentially home to tumor loci and induce anti-angiogenesis as well as apoptosis of cancerous cells, resulting in the decrease of lung metastases and prolongation of lifespan in mouse tumor model.

The BMSCs' capability of tissue repairing and homing has been attracted great attention of scientists. Grove reported that up to 20% of lung epithelium could be derived from circulating BMSCs after lung injury, and a large portion of lung damage could be repaired by forming entire alveoli. Meantime, these adult stem cells could be effectively used to deliver therapeutic gene to lung while retaining the ability to differentiate into lung cells [21]. Darwin J Prockpop demonstrated that injured tissue create a milieu that enhances homing, proliferation and differentiation of BMSCs [7]. It has been found that there are high concentrations of fibroblast growth factor, transforming growth factor β, platelet-derived growth factor, and epidermal growth factor in injured tissue and in the microenvironment of tumor, in which these facters are needed in the process of homing, proliferation and differentiation of BMSCs [22,23]. Therefore, we hypothesize that tissue rebuilding in tumorigenesis, wound healing and fetal development entitles the microenvironments the capability of attracting the systemically administered BMSCs.

Suppression of VEGF by sFlt-1 adenoviruses administered systemically was proved to contribute to inhibition of tumor growth [12]. In our study, we employed BMSCs as a gene vehicle loaded with sflt-1 recombinant adenovirus for tumor therapy, and found that the BMSCs expressing sFlt-1 preferentially homed to and survived in the tumor loci where its strong anti-angiogenesis and anti-metastasis effect has been observed. We also observed that BMSCs expressing IL-12 induced a stronger antitumor effect than administration with IL-12 adenoviruses alone (data not shown). This vehicle may be useful and effective in tumor gene therapy for four reasons. Firstly, the BMSCs after ex vivo administration may survive for hundreds of days, which enable them to offer a relatively long effect for chronic diseases [24]. Secondly, the capacity of BMSCs homing to tumor sites and protecting viruses against the immune surveillance would strengthen the effect of recombinant adenovirus gene therapy, which is unsatisfactory when systemically administered alone because of nonspecific distribution and serious hepatotoxicity [11,25]. Thirdly, the lower immunogenicity, which is the result of low expression of MHC and costimulatory molecules, enables BMSCs advantageous for allogeneic or xenogeneic applications [26]. Last, BMSCs are easy to be expanded in vitro and to be genetically modified, which means convenience and cost effectiveness in the future preclinical or clinical applications.

However, in the experiments performed by our lab (data not shown) and others, the naive BMSCs seemed to promote growth of tumors. On the other hand, Aarif reported that mesenchymal stem cells injected intravenously homed to neoplasm and potently inhibited tumor growth in Kaposi sarcoma model [27]. The relationship between the stem cells and the cancer cells may depend on the microenvironment which is created by these two kinds of cells and others [28-31]. Besides, the possible transformation and the consequent tumorigenesis of BMSCs themselves remains to be clarified. Jakub and his colleagues reported that bone marrow-derived mesenchymal stem cells showed cytogenetic aberrations after several passages in vitro, which resulted in malignant lesions when infused to secondary recipients [32,33]. All of these contradictions indicated that the fate and effect of BMSCs on tumor growth should be further investigated.

In conclusion, we demonstrated that BMSCs might be employed as a promising vehicle for tumor gene therapy which can effectively not only improve the concentration of anticancer therapeutics in tumors, but also modify the tumor microenvironment. However, the fate and effect on tumor growth of BMSCs systemically administrated should be further evaluated.

Competing Interests

The authors declare that they have no competing interests.

Authors' Contributions

JLY and YQW conceived of the study, and participated in its design and coordination. MH, JLY and HT participated in the design, cell experiment, animal experiment and drafted the manuscript. XZ, YQJ, LT carried out the cell culture and relevant analysis. RW, XWZ carried out the immunochemistry experiment, statistical analysis. YW, YL carried out the mouse alginate test. XCC, RZ participated in the pathologic analysis. All authors read and approved the final manuscript.

Acknowledgements

This work was supported by National Key Basic Research Program of China (2004CB518800), Project of National Natural Sciences Foundation of China, Natural 863 Projects.

References

1. Pittenger MarkF, Mackay AlastairM, Beck StephenC, Jaiswal RamaK, Douglas Robin, Mosca JosephD, Moorman MarkA, Simonetti DonaldW, Craig Stewart, Marshak DanielR: Multilineage potential of adult human mesenchymal stem cells. Science 1999, 284:143–147.

2. Mangi AA, Noiseux N, Kong D, He H, Rezvani M, Ingwall JS, Dzau VJ: Mesenchymal stem cells modified with Akt prevent remodeling and restore performance of infarcted hearts. Nat Med 2003, 9:1195–1201.

3. Horwitz EM, Prockop DJ, Fitzpatrick LA, Koo WW, Gordon PL, Neel M, Sussman M, Orchard P, Marx JC, Pyeritz RE, Brenner MK: Transplantability and therapeutic effects of bone marrow-derived mesenchymal cells in children with osteogenesis imperfecta. Nat Med 1999, 5:309–313.

4. Chapel A, Bertho JM, Bensidhoum M, Fouillard L, Young RG, Frick J, Demarquay C, Cuvelier F, Mathieu E, Trompier F, Dudoignon N, Germain C, Mazurier C, Aigueperse J, Borneman J, Gorin NC, Gourmelon P, Thierry D: Mesenchymal stem cells home to injured tissues when co-infused with hematopoietic cells to treat a radiation-induced multi-organ failure syndrome. J Gene Med 2003, 5(12):1028–38.

5. Koç OmerN, Gerson StantonL, Cooper BrendaW, Dyhouse StephanieM, Haynesworth StephenE, Caplan ArnoldI, Lazarus HillardM: Rapid hematopoietic recovery after coinfusion of autologous-blood stem cells and culture-expanded marrow mesenchymal stem cells in advanced breast cancer patients receiving high-dose chemotherapy. J Clin Oncol 2000, 18:307–316.

6. Liechty KW, MacKenzie TC, Shaaban AF, Radu A, Moseley AM, Deans R, Marshak DR, Flake AW: Human mesenchymal stem cells engraft and demonstrate site-specific differentiation after in utero transplantation in sheep. Nat Med 2000, 6:1282–1286.

7. Hanahan D, Weinberg RA: The hallmarks of cancer. Cell 2000, 100:57–70.

8. MacKenzie TC, Flake AW: Human mesenchymal stem cells persist, demonstrate site-specific multipotential differentiation, and are present in sites of wound healing and tissue regeneration after transplantation into fetal sheep. Blood Cells Mol Dis 2001, 27:601–604.

9. Watt FM, Hogan BL: Out of Eden: Stem Cells and Their Niches. Science 2000, 287:1427–1430.

10. Tumors DvorakHF: wounds that do not heal. Similarities between tumor stroma generation and wound healing. N Engl J Med 1986, 315:1650–1659.

11. Studeny Matus, Marini FrankC, Champlin RichardE, Zompetta Claudia, Isaiah FidlerJ, Andreeff Michael: Bone marrow-derived mesenchymal stem cells as vehicles for interferon- beta delivery into tumors. Can Res 2002, 62:3603–3608.

12. Mori A, Arii S, Furutani M, Mizumoto M, Uchida S, Furuyama H, Kondo Y, Gorrin-Rivas MJ, Furumoto K, Kaneda Y, Imamura M: Soluble Flt-1 gene therapy for peritoneal metastases using HVJ-cationic liposomes. Gene Therapy 2000, 7:1027–1033.

13. Liu Ji-yan, Wei Yu-quan, Yang Li, Zhao Xia, Tian Ling, Hou Jian-mei, Niu Ting, Liu Fen, Jiang Yu, Hu Bing, Wu Yang, Su Jing-mei, Lou Yan-yan, He

Qiu-ming, Wen Yan-jun, Yang Jin-liang, Kan Bing, Mao Yong-qiu, Luo Feng, Peng Feng: Immunotherapy of tumors with vaccine based on quail homologous vascular endothelial growth factor receptor-2. Blood 2003, 102:1815–1823.

14. Mahasreshti ParameshwarJ, Navarro JesusG, Kataram Manjula, Wang MingH, Carey Delicia, Siegal GeneP, Barnes MackN, Nettelbeck DirkM, Alvarez RonaldD, Hemminki Akseli, Curiel DavidT: Adenovirus-mediated Soluble FLT-1 Gene Therapy for Ovarian Carcinoma. Clin Can Res 2001, 7:2057–2066.

15. Su Jing-Mei, Wei Yu-Quan, Tian Ling, Zhao Xia, Yang Li, He Qiu-Ming, Wang Yu, Lu You, Wu Yang, Liu Fen, Liu Ji-Yan, Yang Jin-Liang, Lou Yan-Yan, Hu Bing, Niu Ting, Wen Yan-Jun, Xiao Fei, Deng Hong-Xin, Li Jiong, Kan Bing: Active Immunogene Therapy of Cancer with Vaccine On the Basis of Chicken Homologous Matrix Metalloproteinase-2. Can Res 2003, 63:600–607.

16. Blezinger P, Wang J, Gondo M, Quezada A, Mehrens D, French M: Systemic inhibition of tumor growth and tumor metastases by intramuscular administration of the endostatin gene. Nat Biotechnol 1999, 17:343–348.

17. Kasahara Yasunori, Tuder RubinM, Taraseviciene-Stewart Laimute, Le Cras TimothyD, Abman Steven, Hirth PeterK: Inhibition of VEGF receptors causes lung cell apoptosis and emphysema. J Clin Invest 2000, 106:1311–1319.

18. Xiao F, Wei Y, Yang L, Zhao X, Tian L, Ding Z, Yuan S, Lou Y, Liu F, Wen Y, Li J, Deng H, Kang B, Mao Y, Lei S, He Q, Su J, Lu Y, Niu T, Hou J, Huang MJ: A gene therapy for cancer based on the angiogenesis inhibitor, vasostatin. Gene Ther 2002, 9:1207–1213.

19. Folkman J: What is the evidence that tumors are angiogenesis dependent? J Natl Cancer Ins 1990, 82:4–6.

20. Meyerrose ToddE, De Ugarte DanielA, A Alex Hofling, Herrbrich PhillipE, Cordonnier TaylorD, Shultz LeonardD, Eagon J Chris, Wirthlin Louisa, Sands MarkS, Hedrick MarcA, Nolta JanA: In Vivo Distribution of Human Adipose-Derived Mesenchymal Stem Cells in Novel Xenotransplantation Models. Stem Cells 2007, 25(1):220–227.

21. Theise ND, Henegariu O, Grove J, Jagirdar J, Kao PN, Crawford JM, Badve S, Saxena R, Krause DS: Radiation pneumonitis in mice: a severe injury model for pneumocyte engraftment from bone marrow. Exp Hematol 2002, 30:1333–1338.

22. Prockop DarwinJ, Gregory CarlA, Spees JefferyL: One strategy for cell and gene therapy: Harnessing the power of adult stem cells to repair tissues. PNAS 2003, 30:11917–11923.

23. Sasaki Mikako, Abe Riichiro, Fujita Yasuyuki, Ando Satomi, Inokuma Daisuke, Shimizu Hiroshi: Mesenchymal Stem Cells Are Recruited into Wounded Skin and Contribute to Wound Repair by Transdifferentiation into Multiple Skin Cell Type. J Immunol 2008, 180:2581–2587.

24. Bartholomew A, Patil S, Mackay A, Nelson M, Buyaner D, Hardy W, Mosca J, Sturgeon C, Siatskas M, Mahmud N, Ferrer K, Deans R, Moseley A, Hoffman R, Devine SM: Baboon mesenchymal stem cells can be genetically modified to secrete human erythropoietin in vivo. Hum Gene Ther 2001, 12:1527–1541.

25. Steitz J, Bruck J, Knop J, Tuting T: Adenovirus-transduced dendritic cells stimulate cellular immunity to melanoma via a CD4+ T cell-dependent mechanism. Gene Ther 2001, 8:1255–1263.

26. Saito Takayuki, Kuang Jin-Qiang, Bittira Bindu, Al-Khaldi Abdulaziz, Chiu RayC-J: Xenotransplant cardiac chimera: immune tolerance of adult stem cells. Ann Thorac Surg 2002, 74(1):19–24.

27. Khakoo AarifY, Pati Shibani, Anderson StasiaA, Reid William, Elshal MohamedF, Rovira IlsaI, Nguyen AhnT, Malide Daniela, Combs ChristianA, Hall Gentzon, Zhang Jianhu, Raffeld Mark, Rogers TerryB, Stetler-Stevenson William, Frank JosephA, Reitz Marvin, Finkel Toren: Human mesenchymal stem cells exert potent antitumorigenic effects in a model of Kaposi's sarcoma. J Exp Med 2006, 203(5):1235–1247.

28. Djouad F, Plence P, Bony C, Tropel P, Apparailly F, Sany J, Noël Danièle, Christian Jorgensen: Immunosuppressive effect of mesenchymal stem cells favors tumor growth in allogeneic animals. Blood 2003, 102:3837–3844.

29. Glennie Sarah, Soeiro Inês, Dyson PeterJ, Lam EricW.-F, Dazzi Francesco: Bone marrow mesenchymal stem cells induce division arrest anergy of activated T cells. Blood 2005, 105(7):2821–2827.

30. Haniffa MuzlifahA, Wang Xiao-Nong, Holtick Udo, Rae Michelle, Isaacs JohnD, Dickinson AnneM, Hilkens CatharienMU, Collin MatthewP: Adult Human Fibroblasts Are Potent Immunoregulatory Cells and Functionally Equivalent to Mesenchymal Stem Cells. J Immunol 2007, 179:1595–1604.

31. Blankenstein Thomas: The role of tumor stroma in the interaction between tumor and immune system. Curr Opin Immunol 2005, 17(2):180–186.

32. Tolar Jakub, Nauta AlmaJ, Osborn MarkJ, Mortari AngelaPanoskaltsis, McElmurry RonT, Bell Scott, Xia Lily, Zhou Ning, Riddle Megan, Schroeder TaniaM, Westendorf JenniferJ, McIvor R Scott, Hogendoorn PancrasCW, Szuhai Karoly, Oseth LeAnn, Hirsch Betsy, Yant StephenR, Kay MarkA, Peister Alexandra, Prockop DarwinJ, Fibbe WillemE, Blazar BruceR: Sarcoma Derived from Cultured Mesenchymal Stem Cells. Stem Cells 2007, 25:371–379.

33. Bernardo MariaEster, Zaffaroni Nadia, Novara Francesca, Cometa AngelaMaria, Avanzini MariaAntonietta, Moretta Antonia, Montagna Daniela, Maccario Rita, Villa Raffaella, Daidone MariaGrazia, Zuffardi Orsetta, Locatelli Franco: Human Bone Marrow-Derived Mesenchymal Stem Cells Do Not Undergo Transformation after Long-term In vitro Culture and Do Not Exhibit Telomere Maintenance Mechanisms. Can Res 2007, 67:9142–9149.

A Self-Inactivating Retrovector Incorporating the IL-2 Promoter for Activation-Induced Transgene Expression in Genetically Engineered T-Cells

Diana E. Jaalouk, Laurence Lejeune, Clément Couture and
Jacques Galipeau

ABSTRACT

Background

T-cell activation leads to signaling pathways that ultimately result in induction of gene transcription from the interleukin-2 (IL-2) promoter. We hypothesized that the IL-2 promoter or its synthetic derivatives can lead to T-cell specific, activation-induced transgene expression. Our objective was to develop a retroviral vector for stable and activation-induced transgene expression in T-lymphocytes.

Results

First, we compared the transcriptional potency of the full-length IL-2 promoter with that of a synthetic promoter composed of 3 repeats of the Nuclear Factor of Activated T-Cells (NFAT) element following activation of transfected Jurkat T-cells expressing the large SV40 T antigen (Jurkat TAg). Although the NFAT3 promoter resulted in a stronger induction of luciferase reporter expression post stimulation, the basal levels of the IL-2 promoter-driven reporter expression were much lower indicating that the IL-2 promoter can serve as a more stringent activation-dependent promoter in T-cells. Based on this data, we generated a self-inactivating retroviral vector with the full-length human IL-2 promoter, namely SINIL-2pr that incorporated the enhanced green fluorescent protein (EGFP) fused to herpes simplex virus thymidine kinase as a reporter/suicide "bifunctional" gene. Subsequently, Vesicular Stomatitis Virus-G Protein pseudotyped retroparticles were generated for SINIL-2pr and used to transduce the Jurkat T-cell line and the ZAP-70-deficient P116 cell line. Flow cytometry analysis showed that EGFP expression was markedly enhanced post co-stimulation of the gene-modified cells with 1 μM ionomycin and 10 ng/ml phorbol 12-myristate 13-acetate (PMA). This activation-induced expression was abrogated when the cells were pretreated with 300 nM cyclosporin A.

Conclusion

These results demonstrate that the SINIL-2pr retrovector leads to activation-inducible transgene expression in Jurkat T-cell lines. We propose that this design can be potentially exploited in several cellular immunotherapy applications.

Background

T-cells have several features that render them attractive target cells for cell and gene therapy applications. In addition to their diverse and critical role in the immune system, T-cells are readily available from human peripheral blood, they can be easily isolated, activated, expanded to large numbers in vitro, and reinfused with a potential to circulate for a long time in vivo. Indeed, T-cells have been successfully utilized for adoptive immunotherapy of cancer, viral infections, and autoimmune diseases.

Since adoptive transfer of allogeneic T-cells has been associated with acute graft-versus-host disease (GVHD), a life-threatening complication initiated by the donor alloreactive T-cells against the normal host tissues, alternative approaches have been attempted to reduce the risk of GVHD while maintaining the beneficial aspects of this treatment modality. One scenario was to use allogeneic T-cells genetically-engineered to express the herpes simplex virus thymidine kinase (HS-VTK) suicide gene that would lead to the depletion of the gene-modified T-cells following treatment with the drug ganciclovir if GVHD arises [1]. This strategy gave promising results in a number of clinical studies [2,3]. In a second approach, in vitro enriched antigen-specific T-cells were used instead of total allogeneic T-cell populations to mount antiviral or antitumour responses. Donor-derived cytomegalovirus (CMV)-specific cytotoxic T lymphocyte (CTL) clones were successfully used to treat patients suffering from CMV viral infection following marrow transplants [4]. Similarly, infusion of Epstein-Barr virus (EBV)-specific CTLs was used to prevent and to treat EBV lymphoma in immuno-compromised patients post BMT [5]. Interestingly, more recent approaches in cancer immunotherapy investigated the use of T-cells primed in vitro against minor histocompatibility antigens [6], or against tumor associated antigens [7,8].

Moreover, autologous T-cells genetically engineered to express antiviral transgenes have been used for the treatment of HIV [9,10]. Similar studies investigated the use of autologous T-cells engineered to express anti-inflammatory cytokines for the treatment of autoimmune diseases [11,12]. The success of these approaches is dependent on the stable transfer of the therapeutic gene into all infused T-cells and its subsequent expression upon T-cell activation. On the other hand, successful use of antigen-specific T-cells is dependent upon their rapid in vitro selection and enrichment. Therefore, stable and selective expression of a gene marker in activated antigen-specific T-cells would improve their therapeutic utility.

Integrating retroviral vectors have been successfully used for stable expression of transgenes in T-cells [2,10,13]. Recently, the development of new retroviral pseudotypes and improved transduction protocols have made it possible to gene modify T-cells with high efficiency [14]. Furthermore, it was documented that the level of transgene expression can vary with the activation status of T-cells [15]. In fact, it has long been known that the enhancer machinery within the retroviral long terminal repeat (LTR) such as that of Moloney murine leukemia virus (MoMLV) incorporates elements that confer transcriptional preference to activated T-cells [16,17]. However, LTR-driven transgene expression is neither exclusive to the activated subset of gene-modified T-cells nor does it guarantee that all activated transduced T-cells express the transgene if the vector integrates in a transcriptionaly silent chromosomal background [18-20].

Activation of T-cells, physiologically by triggering of the T-cell receptor (TCR) complex or pharmacologically by using reagents such as ionomycin and PMA results in the production of IL-2 which plays a key role in T-cell proliferation and differentiation. Therefore, IL-2 gene transcription by induction of the IL-2 promoter is specific to activated T-cells [21]. Plasmid vectors in which reporter gene expression is under the control of the full-length IL-2 promoter or synthetic promoters composed of either single or multiple binding motifs within the IL-2 promoter have been used in many investigative studies of T-cell activation [22-24]. In these studies, reporter gene expression was induced following activation of the transfected T-cell lines. Hence, we propose that incorporation of the IL-2 promoter or its synthetic derivatives into a transcriptionaly silent retroviral vector would lead to stable and activation-induced transgene expression in T-cells. Our objective is to develop a retroviral vector for stable activation-induced transgene expression in T-lymphocytes.

In this study, we first compared the transcriptional potency and stringency of the full-length IL-2 promoter with that of a synthetic promoter composed of 3 NFAT elements following stimulation of T-cells. Transient transfections of Jurkat Tag cells with plasmids harboring either one of the two promoters upstream of a luciferase reporter showed that although the NFAT3 synthetic promoter resulted in a stronger induction of reporter expression following T-cell activation, the IL-2 promoter had a more stringent activation-dependent expression. Consequently, we tested if a self-inactivating retroviral vector incorporating the full-length IL-2 promoter namely SINIL-2 pr would serve as an integrated platform for stable and specific activation-induced transgene expression in T-cell lines.

Results

Activation-Induced Luciferase Reporter Expression in Jurkat Tag Cells

We compared the transcriptional capacity of the full-length IL-2 promoter with that of a synthetic promoter composed of 3 NFAT elements in transient transfection assays using plasmids harboring either one of the two promoters upstream of a luciferase reporter to determine which would allow for more stringent T-cell activation-dependent reporter expression. Jurkat T-cells expressing the large T Ag were transfected with IL-2 promoter-luciferase plasmid (Figure 1A) or NFAT3 promoter-luciferase plasmid (Figure 1B). Co-stimulation of transfected cells with 1 μM ionomycin and 10 ng/ml PMA for ~6 hr resulted in 6.4 ± 0.4 – fold increase in IL-2 promoter-driven luciferase reporter expression (P = 0.002) and 14.6 ± 1.3 – fold increase in NFAT3 promoter-driven luciferase expression

(P = 0.008) which were both significant relative to transfected untreated cells or cells stimulated with either drug alone (n = 3). Although T-cell activation resulted in an average 2.3 ± 0.2 – fold stronger induction capacity of luciferase reporter expression from the NFAT3 promoter compared to the IL-2 promoter (P < 0.001), the basal transcriptional activity of the NFAT3 promoter in the absence of T-cell activation was 4.6 ± 0.2 – fold higher than that of the IL-2 promoter (P < 0.001). These results indicated that the IL-2 promoter serves as more stringent T-cell activation-dependent promoter than the synthetic NFAT3 promoter.

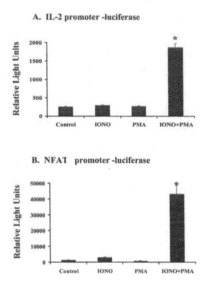

Figure 1. Activation-induced luciferase reporter expression in Jurkat TAg cells. A. Jurkat cells that express the large T Ag were transiently transfected by electroporation with an IL-2 promoter-luciferase construct. Co-stimulation of these T-cells with 1 μM ionomycin and 10 ng/ml PMA for ~6 hr resulted in 6.4 ± 0.4 – fold increase in IL-2 promoter-driven luciferase reporter expression relative to control non-stimulated cells (P = 0.002). B. Similar co-stimulation of Jurkat T Ag cells that were transfected with NFAT3 promoter-luciferase construct reulsted 14.6 ± 1.3 – fold increase in NFAT3 promoter-driven luciferase expression relative to control (P = 0.008).

Synthesis of the SINIL-2pr Retrovector

To develop a retroviral vector for activation-induced stable transgene expression in T-cells, we generated a self-inactivating retroviral vector (SIN) by creating extensive deletions to the U3 region of the Moloney Murine Leukemia Virus 3' long terminal repeat (3'LTR) in pGFP/TKfus (Figure 2A). Retroviral promoter and enhancer machinery were replaced by the full-length human IL-2 promoter as described in the methods section. The resulting construct SINIL-2pr incorporated the EGFP fused to HSVTK as a reporter/suicide gene (Figure 2B). Subsequently, the SINIL-2pr plasmid was stably transfected into the 293GPG packaging cell line to generate polyclonal as well as single clone retroviral producer populations. Vesicular Stomatitis Virus-G Protein (VSV-G) pseudotyped retroparticles were produced and used to transduce the Jurkat T-cell line and the ZAP-70-deficient Jurkat P116 T-cell line. Since recombinant retroviral vectors can be be susceptible to rearrangements prior to their final integration as a DNA proviral genome, we characterized the integrated conformation in transduced target cells by Southern Blot analysis (Figure 2C). Genomic DNA was extracted from null & from transduced Jurkat cells and P116 cells, digested with KpnI and probed with [P32]-labeled DNA sequences complementary to the GFP reporter cDNA. We detected DNA bands consistent with the predicted sized fragment expected from KpnI digest of integrated unrearranged proviral DNA.

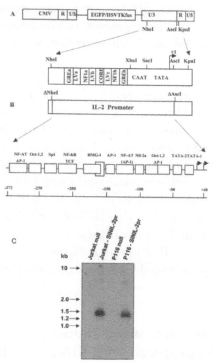

Figure 2. Design of the SINIL-2pr retrovector. A. pGFP/TKfus plasmid bears a full-length 3'LTR that incorporates all the promoter and enhancer machinery intrinsic to the wild-type MoMLV retrovirus. The CMV promoter element which substitutes the U3 region in the 5' LTR drives the expression of the retroviral genome in transfected packaging cells for the production of replication-defective retrovirus. B. SINIL-2 pr retrovector is designed by creating an NheI-AscI deletion to the 3'LTR of pGFP/TKfus and by replacing the U3 with the full-length human IL-2 promoter which results in a self-inactivating retrovector whereby expression of the EGFP and HSVTK fusion protein is dependent on the IL-2 promoter in cells transduced with SINIL-2pr retroparticles. C. Southern Blot analysis of the integrated proviral DNA in Jurkat and p116 cells transduced with SINIL-2pr retrovector at similar MOI. Genomic DNA was extracted from transduced and control null cells, digested with KpnI and probed with [P32]-labeled probe complementary to the GFP reporter cDNA. The detected DNA bands are consistent with the predicted sized fragment indicating no rearrangement in the integrated proviral DNA.

Cyclosporin-A-Sensitive Activation-Induced EGFP Expression in Jurkat Cells Transduced With SINIL-2pr Retroparticles

In order to examine whether the SINIL-2pr configuration leads to activation-induced expression in Jurkat T-cells, we compared the level of EGFP reporter expression in Jurkat T-cells transduced with SINIL-2pr retroparticles at a multiplicity of infection (MOI) of 5, (three consecutive transduction rounds) following drug stimulation relative to that of untreated cells (Figure 3). Mean EGFP expression (MnX) was quantified at ~48 hr post -activation using flow cytometry. Figure 3C shows the flow cytometry profile of one representative experiment whereby a mixed population of SINIL-2pr-transduced Jurkat T-cells had detectable EGFP fluorescence (MnX = 3.22) relative to control non gene-modified cells (MnX = 1.33, Figure 3A). Since the SINIL-2pr design is expected to display a "self-inactivation" phenotype with no expression driven from the incorporated IL-2 promoter in the absence of any T-cell activation, we speculated that this low level of EGFP expression resulted from the basal level of activation in Jurkat T-cells since these cells are dividing continuously in culture. Co-stimulation of the mixed SINIL-2pr-transduced population with 1 µM ionomycin and 10 ng/ml PMA for ~6 hr resulted in enhanced EGFP reporter expression (MnX = 8.37, Figure 3D) as compared to stimulated null cells (MnX = 1.62, Figure 3B). Results obtained from three independent experiments (Figure 3E) showed an average 2.0 ± 0.1 – fold increase (mean ± SEM) in relative mean EGFP expression in SINIL-2pr – transduced Jurkat cells following ionomycin and PMA treatment which was significantly higher than that in untreated cells (P = 0.011). This activation-induced expression was abrogated when the cells were pretreated for 30 minutes with 300 nM CsA, a potent inhibitor of calcineurin and therefore of the signal transduction cascade that leads to IL-2 promoter transcriptional activation, indicating that the enhanced expression of EGFP was under the regulatory control of the IL-2 promoter.

Cyclosporin A-Sensitive Activation-Induced EGFP Expression in Jurkat P116 Cells Transduced with SINIL-2pr Retroparticles

To determine if the low levels of EGFP reporter expression obtained in Jurkat cells transduced with SINIL-2pr prior to drug stimulation were indeed a consequence of basal state of activation in this tumor T-cell line, we tested the SINIL-2pr design in Jurkat P116 T-cells that are deficient in ZAP-70, a kinase implicated in early steps in T-cell activation. Though expandable in culture, Jurkat P116 cells exhibit a longer doubling time compared to that of the parent cell line reflecting a lower basal activation state (data not shown). Figure 4 shows the flow cytometry profile of one representative experiment whereby a mixed polyclonal population of SINIL-2pr-transduced Jurkat P116 T-cells showed level of EGFP fluorescence (MnX = 1.86, Figure 4C) that was similar to background fluorescence of control null cells (MnX = 1.12, Figure 4A). Co-stimulation of the SINIL-2 pr-modified cells with 1 µM ionomycin and 10 ng/ml PMA for ~6 hr resulted in a marked increase in EGFP reporter expression (MnX = 8.28, Figure 4D) as compared to stimulated null cells (MnX = 1.16, Figure 4B). Moreover, following drug stimulation, there was a clear-cut segregation between the EGFP-expressing population and the EGFP-negative population. Results obtained from three independent experiments (Figure 4E) showed an average 3.4 ± 0.4 – fold increase in relative mean EGFP expression in SINIL-2pr – transduced Jurkat P116 cells following ionomycin and PMA treatment which was significantly higher than that in untreated transduced cells (P = 0.029). This activation-induced expression was inhibited when the Jurkat P116 cells were pretreated for 30 minutes with 300 nM CsA.

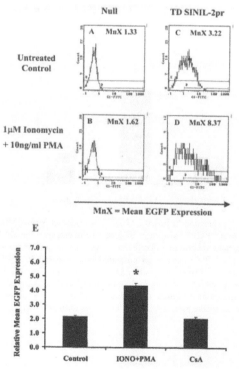

Figure 3. Activation-induced EGFP expression in Jurkat cells transduced with SINIL-2pr retroparticles. A-D. Flow cytometry analysis of mean EGFP expression in Jurkat T-cells transduced with SINIL-2pr retroparticles relative to Jurkat null cells. Expression was measured at ~48 hr post-activation with 1 µM ionomycin and 10 ng/ml PMA relative to untreated cells. E. Cyclosporin A-sensitive induction of EGFP expression in Jurkat cells transduced with SINIL-2pr. Co-stimulation of the SINIL-2pr-gene modified Jurkat T-cells with 1 µM ionomycin and 10 ng/ml PMA resulted in a 2.0 ± 0.1 – fold increase in relative mean EGFP expression measured at ~48 hrs (P = 0.011). This activation-induced EGFP expression was abrogated when the cells were pretreated for ~30 minutes with 300 nM CsA.

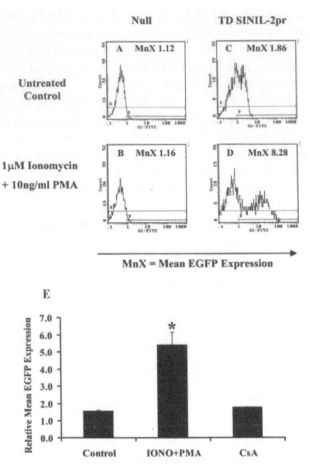

Figure 4. Activation-induced EGFP expression in Jurkat P116 cells transduced with SINIL-2pr retroparticles. A-D. Flow cytometry analysis of mean EGFP expression in Jurkat cells deficient for ZAP-70 (Jurkat P116) transduced with SINIL-2pr retroparticles relative to null cells. Expression was measured at ~48 hr post-activation with 1 μM ionomycin and 10 ng/ml PMA relative to untreated counterparts. E. Cyclosporin A-sensitive induction of EGFP expression in Jurkat P116 cells transduced with SINIL-2pr. Co-stimulation of the SINIL-2pr- gene modified Jurkat P116 T-cells with 1 μM ionomycin and 10 ng/ml PMA resulted in a 3.4 ± 0.4 – fold increase in relative mean EGFP expression measured at ~48 hrs (P = 0.029). This activation-induced EGFP expression was abrogated when the cells were pretreated for ~30 minutes with 300 nM CsA.

Discussion

In this study, we first compared the transcriptional potency of two T-cell activation-dependent promoters, the IL-2 promoter and a synthetic promoter composed of 3 repeats of the NFAT element following pharmacological stimulation of transiently transfected T-cells. Plasmid vectors incorporating one of the two promoters were used to transfer luciferase reporter expression into Jurkat T-cells expressing the large T-antigen. Subsequent stimulation of the cells with the drugs ionomycin (a calcium ionophore) and PMA induces polyclonal T-cell activation by causing direct and rapid increase in intracellular $[Ca^{2+}]$ [25]. Although the composite NFAT promoter resulted in stronger induction of luciferase reporter expression following T-cell activation, the full-length IL-2 promoter had much lower basal levels of luciferase reporter expression in the absence of any stimulation (Figure 1). These results indicate that unlike the synthetic NFAT promoter, the IL-2 promoter in the plasmid background was strictly controled as there was no considerable constitutive reporter expression in un-stimulated cells. Furthermore, costimulation with ionomycin and PMA resulted in a significant increase in IL-2 promoter-driven luciferase expression compared to unstimulated cells.

It was shown that reporter gene expression driven by the full-length IL-2 promoter in transfected human peripheral blood T lymphocytes and the Jurkat T- cell line paralleled that of the endogenous IL-2 gene following T-cell activation [26]. Their results also indicated that the NFAT sites were less implicated in transcriptional regulation in normal T cells than they were in the tumor T-cell line. A later study showed that activation of NFAT element is insufficient for activation of the IL-2 promoter, and that the NFAT and IL-2 promoters differ in their requirements for co-stimulation [27]. Furthermore, Hooijberg E et al., used a self-inactivating retroviral vector in which GFP reporter expression was controlled by multiple (3 or 6) repeats of NFAT elements followed by a minimal IL-2 promoter to visualize and isolate antigen-stimulated Jurkat cells, primary T-cell blasts and antigen-specific T-cell clones [28]. Although the design resulted in activation-induced GFP expression, the authors reported that a percentage of the gene-modified T-cells expressed GFP constitutively independently of activation. In addition, not all T-cells in an antigen-specific clonal population had upregulated GFP expression following activation. The sum of these studies and our results suggest that the full-length IL-2 promoter may serve as a more specific and stringent activation-dependent promoter in T-cells.

Subsequently, we examined the potential of the full-length IL-2 promoter within the context of a retroviral vector to drive stable and activation-induced reporter expression in transduced Jurkat T-cells. The cDNA for human IL-2 promoter was incorporated into a self-inactivating retroviral vector replacing the enhancer and promoter machinery in the 3'LTR of MoMLV. The resulting SINIL-2pr construct incorporated EGFP fused to HSVTK as a reporter/suicide gene (Figure 2). VSV-G pseudotyped retroparticles were generated and were used to transduce the Jurkat T-cell line. Flow cytometry analysis showed that expression of the EGFP reporter was markedly enhanced post co-stimulation of the gene-modified Jurkat T-cells with ionomycin and PMA (Figure 3). Furthermore, activation-induced EGFP expression from the SINIL-2 design was specific to induction of the IL-2 promoter as it was sensitive to pre-treatment of the cells with cyclosporin A (Figure 3E), an immunosuppressant drug that blocks IL-2 promoter transcriptional activation by interfering with the enzymatic activity of the phosphatase calcineurin [29].

However, although there was no significant reporter expression from the IL-2 promoter in the plasmid backbone, the SINIL-2 pr design did result in measurable EGFP expression in unstimulated transduced Jurkat cells. We speculated that this basal level of EGFP expression reflects a basal status of activation in the Jurkat tumor T-cell line since the cells are constantly dividing in culture that is not the case for unactivated primary T-cells that are quiescent in the absence of stimulation. To investigate this we transduced the Jurkat P116 T-cells. Due to their defectiveness in ZAP-70, a kinase involved in early steps of TCR signaling, the cells divide at a much slower rate compared to the parent cell line reflecting a lower basal status of activation. However, since T-cell activation by co-stimulation with ionomycin and PMA bypasses these early steps, Jurkat P116 cells can be activated with these drugs. Interestingly, the Jurkat P116 cells transduced with SINIL-2 pr had lower basal levels of EGFP expression compared to gene-modified Jurkat cells (Figure 4). Furthermore, as speculated, the mutant cells resulted in a stronger induction of reporter expression following drug stimulation whereby the segregation between the GFP +ve and the GFP -ve cells was more evident.

Zhang P-X and Fuleihan RL used a recombinant adeno-associated virus (AAV) vector incorporating the IL-2 promoter to drive activation-dependent luciferase reporter expression in the Jurkat T-cell line [30]. However, there was low efficiency of gene transfer with very few clonal AAV integration events. Despite in vitro selection in G418, some of the transferred material was lost or silenced over time. Furthermore, to induce luciferase reporter expression following T-cell activation, the cells had to be subjected to additional stimulation such as heat shock and irradiation due to the nature of the vector used. Hence, taking all this into consideration, the AAV system would be greatly limited if applied to primary T-cells.

Since cytotoxic CD8+ T-cells do not significantly induce IL-2 promoter as compared to CD4+ T-cells, we propose that the SINIL-2pr system would be especially useful in applications such as negative selection of alloreactive CD4+ T-cells when the disparity between donor and recipient is at the MHC Class II locus and for preferential enrichment of antigen-specific CD4+ T-cells which have been shown to confer an advantage in several immunotherapy applications. In one study, unselected donor T-lymphocytes with a predominance of CD4+ population resulted in a stronger expansion of virus-specific T-cells [31]. Similarly, another study showed that there was a strikingly higher proportion of CD4+ EBV-specific T-cells transduced after early sensitization with EBV-presenting B-cells [19]. In adoptive immunotherapy of cancer, CD4+ T-cells were also shown to play an essential role by providing early stimuli for proliferation of tumor-reactive T-cells [32]. In addition, co-expression of the EGFP reporter and the HSVTK suicide gene as a fusion protein in the SINIL-2pr design allows not only for selection of the engineered T-cells but also conditional destruction following GCV treatment. The utility of the EGFP-HSVTK fusion as a "bifunctional" reporter/suicide transgene and its potential use in primary human T-cells was already demonstrated in previously published work by our group [33].

Conclusion

We conclude that the SINIL-2 pr design can serve as platform for stable, specific, and activation-induced expression in T-cell lines. Further experiments need to be done to investigate the function of this system in primary T-cells and to exploit its use in several T-cell immunotherapy applications such as negative selection of alloreactive T-cells and their elimination from the donor T-cell population prior to infusion or for ex vivo sorting of antigen-specific T-cells and their subsequent in vivo tracking.

Methods

Cell Lines and Plasmids

The IL-2 promoter-luciferase plasmid (IL-2Luc) and the NFAT3 promoter-luciferase plasmid (NFATLuc) were previously described [34]. The Jurkat human acute lymphoblastic leukemia T-cell line was obtained from the American Type Culture Collection (ATCC, Rockville, MD). Jurkat cells expressing the SV40 large T antigen (Jurkat TAg) were previously described [34]. ZAP-70-deficient Jurkat P116 cells were generously provided by Dr. R. T. Abraham (Department of Immunology, Mayo Clinic, Minnesota). All T-cell lines were maintained in RPMI media (Gibco-BRL, Gaithesburg, MD) supplemented with 10% heat-inactivated FBS (Gibco-BRL) and 1% penicillin-streptomycin. pJ6Ω Bleo plasmid and 293GPG retroviral packaging cell line were generous gifts from Dr. Richard. C. Mulligan (Children's Hospital, Boston, MA). 293GPG cells were maintained in media composed of DMEM (Gibco-BRL), 10% heat-inactivated FBS supplemented with 0.3 mg/ml G418 (Mediatech, Herndon, VA), 2 µg/ml puromycin (Sigma, Oakville, ONT), and 1 µg/ml tetracycline (Fisher Scientific, Nepean, ONT).

Transfection of Jurkat TAg Cells and Drug Stimulation

20×10^6 Jurkat T-cells expressing the large T antigen were co-transfected with a total of ~10 μg DNA of either the IL-2 promoter-luciferase or the NFAT3 promoter-luciferase plasmid and PEF-GFP (empty vector) at a ratio of 10:1. A control sample was transfected with ~10 μg PEF-GFP DNA only to normalize for the efficiency of transfection. Transient transfections were performed using standard electroporation at 960 μF and 240 V (Bio-Rad, Hercules, CA). The samples were then cultured in 75-cm2 tissue culture flasks in complete RPMI media at 37°C and 5%CO2 for two days. Afterwards, samples were analyzed by flow cytometry (FACStar sorter, Becton Dickinson, Mountain View, CA) to determine transfection efficiency based on EGFP fluorescence. Each sample was then divided into 4 subsets that were either stimulated with 1 μM ionomycin (Sigma, St. Louis, MO) or/and 10 ng/ml PMA (Sigma) or left untreated for ~6 hr.

Luciferase Assays

Transfected Jurkat TAg cells were harvested ~6 hr post stimulation and washed twice with phosphate-buffered saline (PBS). Afterwards, cells in each sample were lysed using 250 μl lysis buffer (Luciferase Reporter Gene Assay, Boehringer Mannheim). Then, a 20 μl aliquot of each lysate supernatant was mixed with 50 μl luciferase reagent buffer containing the luciferin substrate (Luciferase Reporter Gene Assay, Boehringer Mannheim). Within 30 sec of starting the reaction per sample, luciferase activity was measured as light units based on light emission at 562 nm using the Lumat LB-9507 luminometer (Perkin Elmer Instruments, Rodgau-Juegesheim, Germany) according to manufacturer's specifications. All measurements were done in triplicates.

SINIL-2pr Retrovector Design and Synthesis

We used a derivative of pGFP/TKfus [33] to generate the SINIL-2pr design. pGFP/TKfus contains the cDNA for the enhanced green fluorescent protein (EGFP) reporter and the herpes simplex virus thymidine kinase (HSVTK) suicide gene as a fusion and incorporates the CMV promoter in the 5'LTR which drives expression in transfected viral packaging cells. We derived a self-inactivating (SIN) vector from this plasmid by creating a 341-bp NheI-AscI/Klenow deletion in the 3'LTR to remove all the retroviral promoter and enhancer machinery. An insert encoding for the human IL-2 promoter (-372 to +48) was generated by PCR (PTC-100™ Programmable Thermal Controller, MJ Research Inc.) using CTA GCT AGC (introducing NheI site) GAA AGG AGG AAA AAC TGT TTC AT sense primer and C GAG CTC (introducing Ecl136II site) TAG GGA

AC T CTT GAA CAA GAG antisense primer (Sheldon Biotechnology Centre, McGill University, Montreal). The amplified insert was then digested with the corresponding enzymes, purified, and ligated into the SIN-derivative in order to generate the SINIL-2pr plasmid (Figure 2). Nucleotide sequences of the mutated 3'LTR and the inserted IL-2 promoter were confirmed by DNA sequencing (GenAlyTic Inc., University of Guelph, Ontario).

Generation of The SINIL-2pr Retroviral Producers and Production of VSV-G Pseudotyped Retroviral Particles

The SINIL-2pr retroviral producers were generated by stable transfection of the 293GPG packaging cell line as previously described [35]. In brief, stable producer cells were generated by co-transfection of 5 µg FspI-linearized SINIL-2pr retrovector and pJ6Ω Bleo plasmid at a 10:1 ratio. Transfected packaging cells were subsequently selected in 293GPG media supplemented with 100 µg/ml Zeocin (Invitrogen, San Diego, CA) for 3–4 weeks. Resulting stable polyclonal as well as isolated single clone producer populations were utilized to generate VSV-G pseudotyped retroviral particles. All viral supernatants were filtered through 0.45-µm pore size syringe-mounted filters (Gelman Sciences, Ann Arbor, MI) and stored at -20°C. Subsequent 100 × concentration of the retroviral preparations was performed using ultracentrifugation according to a previously described procedure [35] after which samples were stored at -80°C.

SINIL-2pr Viral Titer Determination and RCR Assay

Jurkat T-cells were suspension-cultured in 12-well plates at ~2×10^5 cells per well per 1 ml of the concentrated retroviral sample serially diluted in complete RPMI media supplemented with 6 µg/ml Polybrene (Sigma). Cells transduced with the various viral dilutions were then incubated overnight at 37°C and 5 % CO_2. Afterwards, the samples were spun at 1000 g for 5 min to pellet the cells and the virus-containing supernatent was discarded. The cells for each sample were re-suspended in 2 ml fresh complete RPMI media and expanded in culture. FACS analysis was performed on these samples on 5th day post transduction to ascertain retrovector expression and gene transfer efficiency as measured by EGFP fluorescence. Based on the number of target cells per well per dilution sample at the time of viral addition and assuming that each cell takes 1 retroparticle when the gene transfer is not saturated (less than 40%), the viral titer was estimated as ~1×10^7 infectious particles per ml. Viral preparations were devoid of replication competent retrovirus (RCR) as determined by standard EGFP marker rescue assay performed on null untransduced Jurkat cells exposed to conditioned supernatant collected from transduced Jurkat cells.

Transduction Of Jurkat and P116 T-Cell Lines, Southern Blot, and Drug Stimulation

Each of Jurkat and P116 T-cell lines was suspension-cultured at ~2 × 10^5 cells per well per 1 ml of complete RPMI media. The cells were transduced with SINIL-2-pr concentrated retrovirus at an MOI of 5 in whole media supplemented with 6 µg/ml Polybrene. This procedure was repeated daily for ~6 hr exposure for three consecutive days after which stably transduced cells were expanded. No clonal selection was performed, and mixed populations of transduced cells were used for all subsequent experiments. Southern blot analysis was performed on 15 µg of overnight KpnI digested genomic DNA extracted from stably transduced cells as well as untransduced control cells. Blots were hybridized with P32 labeled cDNA probes, washed and exposed on photographic film. T-cell stimulation was done using 1 µM ionomycin and/or 10 ng/ml PMA in whole media for ~6 hr and in some experiments following 30 min exposure to 300 nM cyclosporin A (CsA) (Novartis Pharma Canada Inc.). FACS analysis was performed to ascertain retrovector expression as measured by mean EGFP fluorescence of positively gated cells.

Statistics

Student T test was applied using Microsoft Excel software.

Competing Interests

The author(s) declare that they have no competing interests.

Authors' Contributions

DJ carried out the cloning and generation of the various constructs, retroviral production and titer assays, transduction and subsequent analysis of target cells, activation assays, luciferase assays, data acquisition and analysis, and drafted the manuscript. LL did the Southern Blot, and revised the manuscript. CC contributed to the conception of the designs in the study, acquisition of funding, and revised the manuscript. JG had substantial contribution to the conception of the study, the experimental designs, general supervision of the research, acquisition of funding, and critically revised the manuscript. All authors approved of the final manuscript version.

Acknowledgements

We thank Franca Sicilia (Jewish General Hospital, Montreal) for flow cytometry analysis. We also thank Dr. R. T. Abraham (Department of Immunology, Mayo Clinic, Minnesota) for generously providing the ZAP-70 deficient Jurkat P116 cell line as well as Dr. J. Hiscott and Dr. MA. Alaoui -Jamali (Lady Davis Institute for Medical Research, Montreal) for providing access to needed equipment. This work was supported in part by the Leukemia Research Fund of Canada and by the Cancer Research Society Inc.

References

1. Qasim W, Gaspar HB, Thrasher AJ: T cell suicide gene therapy to aid hae-matopoietic stem cell transplantation. Current gene therapy 2005, 5(1):121–132.

2. Bonini C, Ferrari G, Verzeletti S, Servida P, Zappone E, Ruggieri L, Ponzoni M, Rossini S, Mavilio F, Traversari C, Bordignon C: HSV-TK gene transfer into donor lymphocytes for control of allogeneic graft-versus-leukemia. Science 1997, 276(5319):1719–1724.

3. Verzeletti S, Bonini C, Marktel S, Nobili N, Ciceri F, Traversari C, Bordignon C: Herpes simplex virus thymidine kinase gene transfer for controlled graft-versus-host disease and graft-versus-leukemia: clinical follow-up and improved new vectors. Human gene therapy 1998, 9(15):2243–2251.

4. Riddell SR, Watanabe KS, Goodrich JM, Li CR, Agha ME, Greenberg PD: Res-toration of viral immunity in immunodeficient humans by the adoptive transfer of T cell clones. Science 1992, 257(5067):238–241.

5. Rooney CM, Smith CA, Ng CY, Loftin SK, Sixbey JW, Gan Y, Srivastava DK, Bowman LC, Krance RA, Brenner MK, Heslop HE: Infusion of cytotoxic T cells for the prevention and treatment of Epstein-Barr virus-induced lymphoma in allogeneic transplant recipients. Blood 1998, 92(5):1549–1555.

6. Fontaine P, Roy-Proulx G, Knafo L, Baron C, Roy DC, Perreault C: Adoptive transfer of minor histocompatibility antigen-specific T lymphocytes eradicates leukemia cells without causing graft-versus-host disease. Nature medicine 2001, 7(7):789–794.

7. Yee C, Thompson JA, Byrd D, Riddell SR, Roche P, Celis E, Greenberg PD: Adoptive T cell therapy using antigen-specific CD8+ T cell clones for the treat-ment of patients with metastatic melanoma: in vivo persistence, migration, and

antitumor effect of transferred T cells. Proceedings of the National Academy of Sciences of the United States of America 2002, 99(25):16168–16173.

8. Zhao Y, Zheng Z, Robbins PF, Khong HT, Rosenberg SA, Morgan RA: Primary human lymphocytes transduced with NY-ESO-1 antigen-specific TCR genes recognize and kill diverse human tumor cell lines. Journal of immunology (Baltimore, Md : 1950) 2005, 174(7):4415–4423.

9. Cooper D, Penny R, Symonds G, Carr A, Gerlach W, Sun LQ, Ely J: A marker study of therapeutically transduced CD4+ peripheral blood lymphocytes in HIV discordant identical twins. Human gene therapy 1999, 10(8):1401–1421.

10. Ranga U, Woffendin C, Verma S, Xu L, June CH, Bishop DK, Nabel GJ: Enhanced T cell engraftment after retroviral delivery of an antiviral gene in HIV-infected individuals. Proceedings of the National Academy of Sciences of the United States of America 1998, 95(3):1201–1206.

11. Hansen G, McIntire JJ, Yeung VP, Berry G, Thorbecke GJ, Chen L, DeKruyff RH, Umetsu DT: CD4(+) T helper cells engineered to produce latent TGF-beta1 reverse allergen-induced airway hyperreactivity and inflammation. The Journal of clinical investigation 2000, 105(1):61–70.

12. Tarner IH, Slavin AJ, McBride J, Levicnik A, Smith R, Nolan GP, Contag CH, Fathman CG: Treatment of autoimmune disease by adoptive cellular gene therapy. Annals of the New York Academy of Sciences 2003, 998:512–519.

13. Mavilio F, Ferrari G, Rossini S, Nobili N, Bonini C, Casorati G, Traversari C, Bordignon C: Peripheral blood lymphocytes as target cells of retroviral vector-mediated gene transfer. Blood 1994, 83(7):1988–1997.

14. Movassagh M, Boyer O, Burland MC, Leclercq V, Klatzmann D, Lemoine FM: Retrovirus-mediated gene transfer into T cells: 95% transduction efficiency without further in vitro selection. Human gene therapy 2000, 11(8):1189–1200.

15. Agarwal M, Austin TW, Morel F, Chen J, Bèohnlein E, Plavec I: Scaffold attachment region-mediated enhancement of retroviral vector expression in primary T cells. Journal of virology 1998, 72(5):3720–3728.

16. Elsholtz HP, Mangalam HJ, Potter E, Albert VR, Supowit S, Evans RM, Rosenfeld MG: Two different cis-active elements transfer the transcriptional effects of both EGF and phorbol esters. Science 1986, 234(4783):1552–1557.

17. Cooper LJ, Topp MS, Pinzon C, Plavec I, Jensen MC, Riddell SR, Greenberg PD: Enhanced transgene expression in quiescent and activated human CD8+ T cells. Human gene therapy 2004, 15(7):648–658.

18. Dobie KW, Lee M, Fantes JA, Graham E, Clark AJ, Springbett A, Lathe R, McClenaghan M: Variegated transgene expression in mouse mammary gland

is determined by the transgene integration locus. Proceedings of the National Academy of Sciences of the United States of America 1996, 93(13):6659–6664.

19. Koehne G, Gallardo HF, Sadelain M, O'Reilly RJ: Rapid selection of antigen-specific T lymphocytes by retroviral transduction. Blood 2000, 96(1):109–117.

20. Tsukahara T, Agawa H, Matsumoto S, Matsuda M, Ueno S, Yamashita Y, Yamada K, Tanaka N, Kojima K, Takeshita T: Murine leukemia virus vector integration favors promoter regions and regional hot spots in a human T-cell line. Biochemical and biophysical research communications 2006, 345(3):1099–1107.

21. Farrar JD, Asnagli H, Murphy KM: T helper subset development: roles of instruction, selection, and transcription. The Journal of clinical investigation 2002, 109(4):431–435.

22. Parra E, Varga M, Hedlund G, Kalland T, Dohlsten M: Costimulation by B7-1 and LFA-3 targets distinct nuclear factors that bind to the interleukin-2 promoter: B7-1 negatively regulates LFA-3-induced NF-AT DNA binding. Molecular and cellular biology 1997, 17(3):1314–1323.

23. Gringhuis SI, de Leij LF, Verschuren EW, Borger P, Vellenga E: Interleukin-7 upregulates the interleukin-2-gene expression in activated human T lymphocytes at the transcriptional level by enhancing the DNA binding activities of both nuclear factor of activated T cells and activator protein-1. Blood 1997, 90(7):2690–2700.

24. Williams S, Couture C, Gilman J, Jascur T, Deckert M, Altman A, Mustelin T: Reconstitution of T cell antigen receptor-induced Erk2 kinase activation in Lck-negative JCaM1 cells by Syk. European journal of biochemistry / FEBS 1997, 245(1):84–90.

25. Morgan AJ, Jacob R: Ionomycin enhances Ca2+ influx by stimulating store-regulated cation entry and not by a direct action at the plasma membrane. The Biochemical journal 1994, 300:665–672.

26. Hughes CC, Pober JS: Transcriptional regulation of the interleukin-2 gene in normal human peripheral blood T cells. Convergence of costimulatory signals and differences from transformed T cells. The Journal of biological chemistry 1996, 271(10):5369–5377.

27. Shapiro VS, Mollenauer MN, Weiss A: Nuclear factor of activated T cells and AP-1 are insufficient for IL-2 promoter activation: requirement for CD28 upregulation of RE/AP. Journal of immunology (Baltimore, Md : 1950) 1998, 161(12):6455–6458.

28. Hooijberg E, Bakker AQ, Ruizendaal JJ, Spits H: NFAT-controlled expression of GFP permits visualization and isolation of antigen-stimulated primary human T cells. Blood 2000, 96(2):459–466.

29. Schreiber SL, Crabtree GR: The mechanism of action of cyclosporin A and FK506. Immunology today 1992, 13(4):136–142.

30. Zhang PX, Fuleihan RL: Transfer of activation-dependent gene expression into T cell lines by recombinant adeno-associated virus. Gene therapy 1999, 6(2):182–189.

31. Lucas KG, Small TN, Heller G, Dupont B, O'Reilly RJ: The development of cellular immunity to Epstein-Barr virus after allogeneic bone marrow transplantation. Blood 1996, 87(6):2594–2603.

32. Toes RE, Ossendorp F, Offringa R, Melief CJ: CD4 T cells and their role in antitumor immune responses. The Journal of experimental medicine 1999, 189(5):753–756.

33. Paquin A, Jaalouk DE, Galipeau J: Retrovector encoding a green fluorescent protein-herpes simplex virus thymidine kinase fusion protein serves as a versatile suicide/reporter for cell and gene therapy applications. Human gene therapy 2001, 12(1):13–23.

34. Northrop JP, Ullman KS, Crabtree GR: Characterization of the nuclear and cytoplasmic components of the lymphoid-specific nuclear factor of activated T cells (NF-AT) complex. The Journal of biological chemistry 1993, 268(4):2917–2923.

35. Jaalouk DE, Eliopoulos N, Couture C, Mader S, Galipeau J: Glucocorticoid-inducible retrovector for regulated transgene expression in genetically engineered bone marrow stromal cells. Human gene therapy 2000, 11(13):1837–1849.

An Inducible and Reversible Mouse Genetic Rescue System

Hongkui Zeng, Kyoji Horie, Linda Madisen, Maria N. Pavlova,
Galina Gragerova, Alex D. Rohde, Brian A. Schimpf,
Yuqiong Liang, Ethan Ojala, Farah Kramer, Patricia Roth,
Olga Slobodskaya, Io Dolka, Eileen A. Southon,
Lino Tessarollo, Karin E. Bornfeldt, Alexander Gragerov,
George N. Pavlakis and George A. Gaitanaris

ABSTRACT

Inducible and reversible regulation of gene expression is a powerful approach for uncovering gene function. We have established a general method to efficiently produce reversible and inducible gene knockout and rescue in mice. In this system, which we named iKO, the target gene can be turned on and off at will by treating the mice with doxycycline. This method combines two genetically modified mouse lines: a) a KO line with a tetracycline-dependent transactivator replacing the endogenous target gene, and b) a line with a tetracycline-inducible cDNA of the target gene inserted into a tightly regulated (TIGRE) genomic locus, which provides for low basal expression and high inducibility. Such a locus occurs infrequently in the genome and we have developed a method to easily introduce genes into the TIGRE site of mouse embryonic stem (ES) cells by recombinase-mediated insertion. Both KO and TIGRE lines have been engineered for high-throughput, large-scale and cost-effective production of iKO mice. As a proof of concept, we have created iKO mice in the apolipoprotein E (ApoE) gene, which allows for sensitive and quantitative phenotypic analyses. The results demonstrated reversible switching of ApoE transcription, plasma cholesterol levels, and atherosclerosis progression and regression. The iKO system shows stringent regulation and is a versatile genetic system that can easily incorporate other techniques and adapt to a wide range of applications.

Introduction

In the post-genome era, a major challenge is deciphering the function of thousands of newly identified genes. One of the main approaches for studying gene function involves inactivation of genes in cells or animals using random (chemical or insertional) mutagenesis or gene targeting. A common problem with these methods stems from the fact that the gene of interest is usually mutated throughout the animal's life. As a result, 1) in many cases the mutation leads to embryonic or neonatal lethality, precluding the assessment of the gene's function in later life; 2) in viable mutants interpretation of observed phenotypes is often complicated by the inability to distinguish the direct effects of the gene loss at the time of observation from the results of developmental abnormalities caused by the gene loss earlier in life; 3) in still other cases, life-long absence of a gene product causes compensatory adjustments of activities of other genes precluding the elucidation of the function of the gene of interest. Conditional knockout and gene expression technologies, such as the Cre/lox-mediated tissue-specific knockout [1] and the tetracycline (Tet) regulated transcriptional activation system [2], can regulate gene expression in a more spatially and temporally controlled fashion. However, these technologies are often laborious to establish and the results are frequently variable.

Here we report the development of a system that provides for the inducible and reversible gene inactivation in the mouse and can also be readily scaled up for high-throughput applications. The iKO system is a binary approach based on the Tet-dependent regulatory technology. It involves the combination of two mouse lines – a KO line that expresses the Tet-transactivator (tTA or rtTA) in place of the gene of interest, and a TIGRE (for tightly regulated) line that contains the gene of interest under the control of the Tet-responsive element (TRE) at a predetermined genomic locus. It has the advantage of, 1) ability to turn genes on or off at will by adding or removing doxycycline (Dox) at any time during the animal's life, thus minimizing embryonic lethality, developmental effects, and compensatory effects; 2) high degree of regulation to any gene inserted at the TIGRE locus, which has been selected to confer minimal basal expression and high inducibility, and to insert any gene of interest in a single step by Cre/loxP recombination; 3) efficiency; the design allowing streamlined production of both KO and TIGRE mice makes it possible to generate iKOs for a large number of genes in a cost-effective manner; 4) flexibility; KO and TIGRE lines can be engineered independently and combined in numerous ways, making a wide range of applications possible.

As a proof of concept, we report the characterization of an iKO of the apolipoprotein E gene (ApoE iKO). ApoE plays a key role in regulating cholesterol metabolism and atherosclerosis progression. ApoE KO mice develop hypercholesterolemia and atherosclerosis that closely resemble the human conditions and are rapidly reversed when APOE protein is supplied [3],[4],[5],[6],[7],[8]. Thus, inducible and reversible regulation of ApoE expression could result in rapid physiological changes, which in turn can help assess the iKO technology. Furthermore, the phenotype of ApoE deficiency is quantifiable and very sensitive to leaky expression, allowing for the evaluation of the stringency of gene regulation by iKO technology [9]. Here we demonstrate that in the ApoE iKO mice, ApoE gene expression, as well as blood cholesterol levels, is tightly controlled by Dox. In the presence of Dox, ApoE is expressed and the cholesterol levels are low; in its absence, the reverse is observed. Furthermore, on examination of aortic atherosclerosis in the ApoE iKO mice we found that Dox treatment before the onset of atherosclerotic lesions completely prevented lesion formation and Dox treatment after extensive lesions had already formed resulted in regression of the lesions. These results demonstrate the reversibility of the iKO, leading to phenotype switching within the same animal. ApoE iKO is also useful in its own right as a novel model system for the study of molecular mechanisms underlying atherosclerosis progression and regression.

Results

Principle of the iKO System

As illustrated in Figure 1A, two genetically modified mouse strains are created. The first is a KO line in which a Tet-dependent transactivator (rtTA in this example) is inserted into the target gene (Gene X). The insertion inactivates Gene X, and places rtTA under the control of the endogenous promoter of Gene X. The KO line can be generated via either homologous recombination or insertional mutagenesis. The second line (TIGRE) contains an additional copy of Gene X cDNA (or genomic fragments) driven by TRE-promoter inserted in a specific locus in the genome (the TIGRE locus), which has been pre-selected for low basal transcriptional activity and high inducibility. When these two lines are crossed, rtTA protein produced from the KO allele can activate the TRE-Gene X in the TIGRE locus only in the presence of Dox. (Alternatively, tTA can be used, which works in the opposite way – Tet-off instead of Tet-on.)

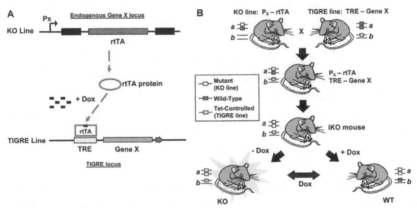

Figure 1. Schematic diagram of the principle of the iKO system. (A) Two components of the iKO system, a KO line and a TIGRE line. (B) Generation of iKO mouse by crossing the KO and TIGRE lines. The iKO is a composite mouse in which: 1) both copies of the endogenous Gene X are disrupted by the insertion; 2) rtTA is driven by the promoter of Gene X; and 3) the cDNA of Gene X is in the TIGRE locus under the control of TRE. Functional Gene X is only expressed from the TIGRE locus when rtTA is bound to Dox, and therefore is inducibly and reversibly regulated by Dox.

Figure 1B illustrates the breeding of KO and TIGRE lines to produce the iKO mouse, which is homozygous for the KO locus and carries one copy of the TIGRE allele. The status of the iKO mouse is regulated by Dox. In the absence of Dox, rtTA protein is produced, but is inactive. As a result, TRE-Gene X is silent, Gene X protein is not produced and the KO phenotype is manifested. In the presence of Dox, rtTA stimulates the synthesis of Gene X protein from the TIGRE locus in the same cells in which the endogenous gene would normally be expressed. Expression complements the missing endogenous Gene X activity and leads to phenotypically normal animals. Thus, one can switch between wild type and KO state of animals by simply adding or removing Dox (e.g. with food).

Selection of TIGRE Loci

To screen for TIGRE loci, we constructed a Moloney murine leukemia virus (MoMLV)-based retroviral vector, pRTonZ (Figure 2A), in which the TRE-controlled lacZ gene was used as a reporter for gene regulation. Retroviral transduction at low multiplicity of infection ensures integration of a single copy of the TRE-lacZ unit into the genome. pRTonZ contains a modified neomycin phosphotransferase gene, loxneo, in which initiating AUG has been placed upstream of a loxP site in frame with the neo coding region. Once optimal locus is selected, it is utilized as a target site for transgene-integration by the scheme shown in Figures 2B and 2C. The TRE-lacZ unit is removed by Cre-mediated recombination of flanking loxP sites, leaving one loxP site and the neo gene in the genome (Figure 2B). Since the promoter of the loxneo gene as well as the initiating AUG are also removed, ES cells become G418-sensitive. In this configuration, any gene of interest can be introduced into the same locus by Cre/loxP recombination (Figure 2C). Recombinant ES clones can be selected by G418-resistance because the neo expression unit is reconstituted. PCR screening showed that >90% of these G418-resistant clones had correct insertion of the new gene.

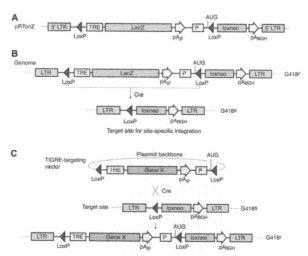

Figure 2. Retroviral vector used to search for tightly regulated loci and strategy to introduce a new gene into these predetermined loci. (A) Structure of the pRTonZ retroviral vector. To prevent effects of viral enhancer on the TRE promoter, the enhancer sequence was deleted in the 3′ long terminal repeat (LTR). Subsequent transduction into target cells is expected to lead to enhancer deletion in both LTRs. The insert was cloned in the opposite orientation of the LTRs, so that the polyA addition signals would not decrease the viral titer. (B) Excision of the lacZ reporter gene from TIGRE loci. LoxP-flanked sequences within the retroviral vector are removed by transient expression of Cre recombinase in ES cells, leaving a single copy of loxP site in the genome. ES cells become G418-sensitive since the loxneo gene loses its promoter and initiating AUG. (C) Introduction of a new gene into the TIGRE locus. The G418-sensitive ES cells selected in (B) are cotransfected with the TIGRE-targeting vector carrying a new gene (gene X) and the Cre expression vector. Proper Cre-mediated recombination between the TIGRE-targeting vector, containing the new gene, and the TIGRE site introduces the new gene into the TIGRE locus and converts the G418 sensitive ES cells into G418 resistant (the expected recombinant leads to neo expression by placing the PGK promoter and initiating AUG upstream of the loxneo gene). Symbols are: pAgl, rabbit β-globin gene polyA addition signal; P, mouse phosphoglycerate kinase-1 gene promoter; AUG, initiating AUG of neomycin phosphotransferase gene; loxneo, neomycin phosphotransferase gene having loxP sequence in-frame to initiating AUG; pABGH, bovine growth hormone gene polyA addition signal; G418r, G418 resistant; G418s, G418 sensitive.

Optimal loci were initially screened in ES cells. The ES cell line used was derived from CJ7 [10], of 129/Sv background. Following infection with the pR-TonZ retroviral vector, G418-resistant (G418r) clones were stained with X-gal (Figure 3A). Most clones showed mosaic staining pattern. By the percentage of the X-gal-stained cell population, ES clones were classified into 4 categories (Figure 3A): class I (<1% of X-gal-stained cells) to class IV (>50% of X-gal-stained cells). From 242 ES clones analyzed, 55 clones were classified into class I. Inducibility was examined by β-galactosidase (β-gal) activity after transfecting 43 class I clones with a tTA expression vector (Figure 3B). We set the cut-off value for high induction level at 1500 μunits/mg of protein and nine clones belonged to this category.

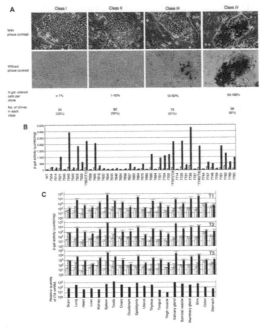

Figure 3. Characterization of gene regulation at TIGRE loci. (A) Screening for optimal integration sites in ES cells and classification of ES clones by X-gal staining. A representative ES clone is shown for each class with and without phase contrast for better imaging of ES cell morphology and X-gal staining, respectively. (B) Screening of class I clones for high inducibility. Forty three class I ES clones were transfected with a tTA expression vector and β-gal activity was quantified 48 hours post-transfection. Bars: open, without tTA; filled, with tTA. A luciferase expression vector was cotransfected to normalize β-gal activities. Three ES clones were designated as T1, T2 and T3 as shown in the figure, and were used to generate mice. (C) Gene regulation in mice generated from tightly regulated ES clones. Three mouse strains were established from ES clones T1, T2 and T3, and crossed to MMTV-tTA mouse. Top three panels: β-gal activity was measured in the following three genotypes: nontransgenic mice (lacZ(–)tTA(–), open bars); mice with lacZ gene but without tTA (lacZ(+)tTA(–), lightly shaded bars); mice with both lacZ gene and tTA gene (lacZ(+)tTA(+), filled bars). Values are shown as means with error bars of standard deviations from five male and five female animals. Values of nontransgenic mice (open bars) are common in all three panels. Bottom panel: tTA mRNA expression level quantified by real-time PCR from one male and one female mouse. Mean values are presented by filled bars. Note that values are shown in logarithmic scale.

Those loci were further examined in mice. Three independent TRE-lacZ mouse lines were generated from class I ES clones T1, T2 and T3 (Figure 3C). Heterozygous TRE-lacZ mice were crossed to MMTV-tTA mice, which has been reported to be transcriptionally active in a wide variety of cell types [11], allowing for the examination of lacZ induction in various tissues. Three genotypes of mice (lacZ(–)tTA(–), lacZ(+)tTA(–), and lacZ(+)tTA(+)) were analyzed for β-gal activity (Figure 3C). Activity of lacZ(–)tTA(–) represents endogenous eukaryotic β-gal activity. Difference between lacZ(–)tTA(–) and lacZ(+)tTA(–) indicates basal activity of lacZ gene in the absence of tTA, and comparison of lacZ(+)tTA(–) and lacZ(+)tTA(+) reveals induction levels in the presence of tTA. Overall, the three mouse lines showed similar pattern of β-gal activity, although some differences were also seen. Basal activities were low but detectable in many tissues, and they were comparable to the values measured in the parental ES clones. β-gal activity was inducible in almost every tissue, and overall induction levels correlated with the expression levels of tTA (Figure 3C, bottom panel).

Improvement of Gene Regulation Control at the TIGRE Loci

Although the three loci T1, T2 and T3 showed tight regulation of gene expression, basal activity was still detectable in many tissues. This could result from enhancers in the vicinity of the integration sites. To solve this problem, we flanked the TRE-lacZ reporter by the insulator sequence derived from the chicken β-globin locus [12],[13]. The insulator sequences were introduced into all three loci (T1, T2, T3) by Cre-mediated recombination according to the scheme shown in Figures 2B and 2C and regulation of the lacZ gene was examined by transient expression of tTA in ES cells (Figure 4A). The insulators reduced basal β-gal activity to levels indistinguishable from wild type ES cells. In contrast, inducibility was not impaired by the insulator sequence, indicating its effectiveness for increasing stringency of gene regulation. The insulators were also introduced into class II, III and IV ES clones, which showed high basal activity (Figure 3A). Although the insulator sequences were effective, basal activity was still clearly detectable in every clone (Figure 4B). To test whether tight regulation is achieved by using other genes, we replaced the lacZ gene with a luciferase gene. With this reporter, insulators reduced basal activity by 14, 20 and 11 fold in T1, T2 and T3 loci, respectively, bringing it close to the instrument detection limit (Figure 4C). Calculation of the number of luciferase molecule per ES cell (Supporting Methods) revealed that on average only 1, 0.7 and 0.3 luciferase molecule were expressed per cell in T1, T2 and T3 loci, respectively (Figure 4D). Importantly, induced levels were not impaired by the insulator (Figure 4C), leading to high induction ratio of luciferase activity. Similar basal activity levels could also be achieved by expressing transrepressor. However, our system is simpler because no additional protein expression is required.

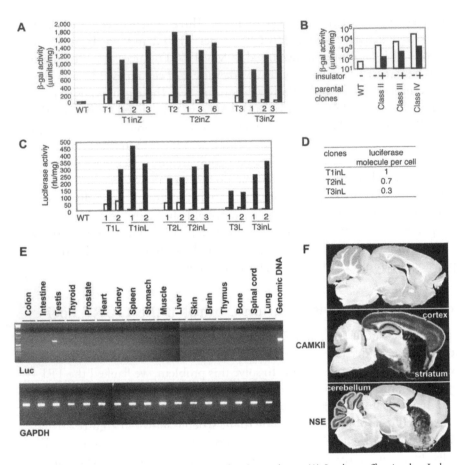

Figure 4. Tightening of gene regulation at TIGRE loci by insulators. (A) Insulator effect in class I clones. Parental ES clones without the insulator (T1, T2, T3) and clones with the insulator (T1inZ, T2inZ, T3inZ) were transfected by tTA expression vector, and β-gal activity was measured 48 hours post-transfection. Three independent integrant clones with insulator were analyzed in each integration site. A luciferase expression vector was cotransfected to normalize the β-gal activities. WT, wild type ES cells. Bars: open, without tTA; filled, with tTA. The same symbol was used in (B) and C). (B) Insulator effect on basal β-gal activity in class II, III and IV clones. Insulator sequence was introduced into three ES clones categorized in class II, III and IV of Figure 3A. Note that values are shown in logarithmic scale. (C) Insulator effect on the regulation of luciferase gene. The lacZ gene of the three class I clones (T1, T2, T3) were replaced by luciferase gene without (T1L, T2L, T3L) or with (T1inL, T2inL, T3inL) the insulator. Two independent integrants were established in each case. A lacZ expression vector was cotransfected to normalize the luciferase activities. (D) Number of luciferase molecule per cell in class I clones with the insulator sequence. (E) RT-PCR of the luciferase (Luc, upper panel) and positive control GAPDH (lower panel) transcripts in a TIGRE line (T1) with TRE-Luc and insulators. Each tissue has an RT-PCR reaction (RT+, left lane) and an RT- control (right lane) run simultaneously to exclude any possibility of genomic DNA contamination. The last lane is a positive control of genomic DNA PCR just for Luc. (F) β-gal staining of brain sagital sections (50 μm) of mice carrying a TRE-LacZ (+insulators) TIGRE (T1) line alone (top panel) or combined with either αCaMKII-tTA (middle panel) or NSE-tTA (bottom panel) transgene. There is no detectable β-gal staining in the absence of inducer (top panel). When TRE-LacZ TIGRE line is combined with αCaMKII-tTA, β-gal staining is seen in the same regions αCaMKII-tTA is expressed – mainly cortex, hippocampus and striatum (middle panel). When TRE-LacZ TIGRE line is combined with NSE-tTA, β-gal staining is seen in the same regions NSE-tTA is expressed – mainly striatum, dentate gyrus and cerebellum (bottom panel).

To evaluate the insulator effect in vivo, we also generated mice containing TRE-lacZ gene and insulators, or TRE-luciferase (Luc) gene and insulators at the TIGRE locus (T1), and bred them with mice containing tTA under the control of various promoters. Basal expression was examined in multiple tissues of TRE-Luc mice by RT-PCR (Figure 4E). Luc mRNA was undetectable in all tissues examined except testis, indicating very low basal levels of expression throughout the body. To examine if the Luc transcript detected in testis can produce functional Luc proteins, we conducted luciferase activity assays from protein extracts of all these tissues. The luciferase activity in testis was at the similar low basal level as all other tissues examined (data not shown) as well as the original TIGRE ES clone containing TRE-Luc with insulators that had been shown to express approximately one Luc molecule per cell (Figure 4D), suggesting that the RT-PCR band detected in testis resulted from aberrant transcription that did not generate functional protein. To examine induced expression, we used lines of mice with tTA under the control of two brain-specific promoters: α-CaMKII (calcium/calmodulin-dependent protein kinase II) [14] or NSE (neuron-specific enolase) [15]. Sagital sections of 50 μm thickness across the entire brain from TRE-lacZ, PCAMKII-tTA/TRE-lacZ or PNSE-tTA/TRE-lacZ mice were stained for β-galactosidase (gal) activity and representative sections are shown in Figure 4F. In TRE-lacZ mice, no β-gal staining was seen in any parts of the brain. In PCAMKII-tTA/TRE-lacZ or PNSE-tTA/TRE-lacZ mice, intense β-gal staining was observed in specific regions of the brain defined by the two promoters respectively. These results demonstrate tight control of the TIGRE locus in animals.

Characterization of the TIGRE Locus

Using the splinkerette PCR method [16], we obtained genomic fragments covering either 5′ or 3′ junctions of the TIGRE vector insertion site in the T1 ES cell line. After sequencing the fragments, we determined the precise integration site of the viral TIGRE vector in the T1 TIGRE locus, which is located on chromosome 9. Genomic sequences surrounding the T1 TIGRE locus are shown in Figure 5A. Characteristic of retrovirus insertions, four nucleotides immediately adjacent to the insertion were duplicated and the viral TIGRE vector was inserted exactly in between the duplication. BLAT search of the UCSC Mouse Genome Browser with genomic sequences surrounding T1 revealed the localization of T1 locus to chr9 qA3 (Figure 5B). The insertion site is flanked by two genes: AB124611 and Carm1. The insertion site is located 3′ to the hypothetical gene AB124611 with unknown function and undetermined polyA site. The insertion site is also ~1.5 kb upstream of the transcriptional start of Carm1. Carm1 is ubiquitously expressed and Carm1−/− mice are embryonic lethal [17]. However, we have not observed overt developmental or other abnormalities in heterozygous or homozygous T1

TIGRE reporter lines TRE-lacZ or TRE-Luc (both with insulators), indicating that the viral insertion did not disrupt the nearby Carm1 gene.

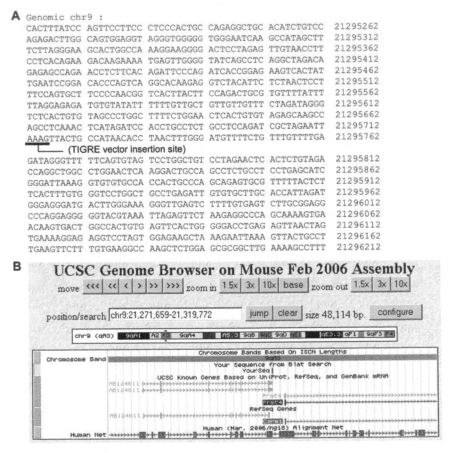

Figure 5. Chromosomal location of the T1 TIGRE locus. (A) Genomic sequences surrounding the T1 TIGRE locus. The underlined AAAG sequence was duplicated upon viral integration and the viral TIGRE vector was inserted exactly in between the duplication. (B) BLAT search of the UCSC Mouse Genome Browser (http://genome.ucsc.edu/cgi-bin/hgBlat) with genomic sequences (as in (A)) surrounding T1 revealed the localization of T1 locus to chr9 qA3. This panel is a screen shot of the BLAT search result. The location of the sequences used for the search is indicated by a vertical bar next to "YourSeq". The insertion site is in between two genes: AB124611 and Carm1 (alternative name Prmt4), and does not seem to disrupt either gene.

KO Lines Produced from an ES Cell Library Mutagenized by a Retroviral Vector

KO lines with target genes replaced by rtTA or tTA can be produced by any gene targeting or insertional mutagenic methods. To implement high-throughput production of iKO mice, we have utilized a large-scale insertional mutagenesis ES cell library developed in house for the KO production [18]. Figure 6A upper panel illustrates the structure of the retroviral vector used. In the particular case of ApoE, the virus is inserted in the third intron. The vector contains splice acceptor, stop codons, polyA signal and transcriptional terminator to ensure gene inactivation, which we confirmed to be the case by showing that ~90% of isolated gene-specific ES clones were null alleles [18]. (The remaining were mostly knockdowns, and in nearly all these cases the retrovirus was inserted into 5′ UTR (exon or intron), suggesting that retroviral insertions upstream of the coding regions of genes should be avoided.) The vector also includes the rtTA gene immediately downstream of the splice acceptor, stop codons and internal ribosome entry site (IRES), so that rtTA protein can be synthesized from the ApoE-IRES-rtTA hybrid transcript.

To examine if random insertion of the retroviral vector containing rtTA into a gene results in rtTA expression reflecting the expression patterns of the inactivated gene, we compared rtTA transcription with the transcription of the endogenous gene by RT-PCR in different tissues for 26 different G protein coupled receptor (GPCR) KO lines generated from the library. Heterozygous mice carrying one allele of the intact endogenous gene and one allele interrupted by the rtTA-bearing retroviral vector were used to prepare total RNA samples from different tissues and amplify gene-specific transcript and rtTA transcript from the same RNA preps for side-by-side comparison. Figure 6B shows three examples of such comparison for genes P2Y6, RE2 and LGR6 respectively. The retroviral vector was inserted into a different part of each of the three genes – a 5′UTR intron of P2Y6, an intron within the coding region of RE2 and the 15th coding exon of LGR6. In all three cases, rtTA expression profiles closely resemble those of the endogenous genes to be inactivated. Real-time qPCR for 16 tissues also showed high degree of correlation between rtTA and endogenous gene from tissue to tissue. Overall, out of the 26 lines examined, 19 lines showed good correlation between rtTA and the endogenous gene's expression , i.e. rtTA is expressed in the tissues where the corresponding endogenous gene is expressed and the relative ratio across different tissues for each transcript also appears similar between rtTA and endogenous gene. In the remaining 7 lines in which rtTA was not expressed well, 5 lines had the retroviral vector inserted upstream of the first (and usually fairly large) intron of each gene. It has been recognized that the first intron (especially if it is large) could contain essential transcriptional regulatory elements, and thus insertions in this region might disrupt the transcriptional regulation and so should be avoided. Otherwise, our data showed that in the majority of insertional sites, rtTA could be expressed through the endogenous promoter upon integration via the retroviral vector.

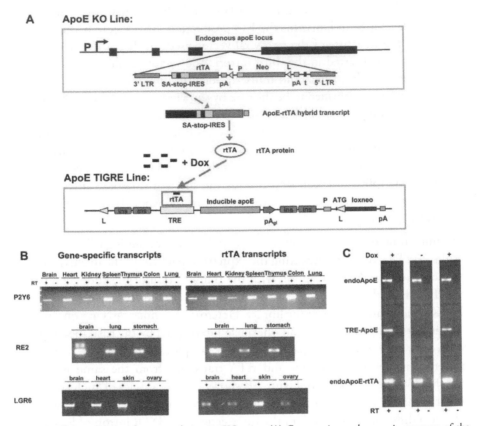

Figure 6. Characterization of gene regulation in iKO mice. (A) Construction and genomic structure of the ApoE iKO mice. Endogenous ApoE gene comprises of 4 exons. In the ApoE KO line, retroviral vector is inserted into the third intron of the ApoE gene, 205 bp upstream of the fourth exon (the largest coding exon). The retroviral vector contains the virus backbone (including 5′LTR and 3′LTR), a splice acceptor (SA) – stop codon (stop) – IRES cassette immediately followed by rtTA, a PGK promoter (P) driven neo selection marker flanked by two loxP sites (L), and a transcriptional terminator sequence (t). pA, polyadenylation sequence. From this locus, transcription initiated from the endogenous ApoE gene continues through rtTA to form an ApoE-SA-Stops-IRES-rtTA-polyA hybrid transcript in place of the full-length endogenous ApoE transcript. The rtTA protein is produced from this hybrid transcript through IRES-mediated translation, and in turn turns on the expression of the TRE-ApoE from the TIGRE locus only when Dox is present. The ApoE TIGRE locus contains an exogenous copy of ApoE cDNA driven by TRE and flanked by four copies of chicken β-globin insulator (ins) sequences, two on each side, and a PGK promoter (P) driven loxneo selection marker. (B) Side-by-side comparison of the expression of rtTA and each individual endogenous gene in various tissues by semi-quantitative RT-PCR. Three GPCR genes are shown here: P2Y6, RE2 and LGR6. Heterozygous mice are used so that the endogenous transcripts from the WT allele and the rtTA-containing hybrid transcripts from the KO allele can be amplified from the same RNA preps. Each RT-PCR reaction (RT+) has an RT- control run simultaneously to exclude any possibility of genomic DNA contamination. (C) Comparison of three transcripts by RT-PCR from the same RNA prep of the liver, the major site of normal ApoE production, of the ApoE+/−;TRE-ApoE mice – endogenous ApoE transcript (endoApoE), TRE-ApoE transcript from the TIGRE locus (TRE-ApoE) and ApoE-rtTA hybrid transcript (endoApoE-rtTA). Mice were fed either with or without Dox. As expected, expression of endoApoE and rtTA were independent of Dox. However, expression of TRE-ApoE was strictly dependent on Dox – it was undetectable in its absence and significantly expressed in its presence, indicating high degree of ApoE regulation achieved in the mice.

Tightly Regulated ApoE Expression and Blood Cholesterol Levels by Dox in the ApoE iKO Mice

To prove the concept of iKO system, we generated ApoE iKO mice by creating ApoE KO line and ApoE TIGRE line separately and breeding them together (Figure 6A), and attempted to model human conditions such as hypercholesterolemia and atherosclerosis. ApoE KO line was created by screening the mutant ES cell library as mentioned above. ApoE TIGRE line was created by inserting a TRE-ApoE transgene, flanked with 4 copies of chicken β-globin insulators, into the T1 TIGRE locus. RT-PCR analysis of heterozygous mice (ApoE+/−; TRE-ApoE) showed that expression of TRE-ApoE was strictly dependent on Dox (Figure 6C). Real-time qPCR using primers specific for the endogenous ApoE or TRE-ApoE (2 sets of primers for each gene) showed that endogenous ApoE mRNA level is very high (compared to 18s rRNA), and the TRE-ApoE mRNA level in the presence of Dox is ~6.7 fold lower than the endogenous ApoE (ΔCt = 2.7 between TRE-ApoE and endogenous ApoE), while in the absence of Dox TRE-ApoE is >55,000 fold lower than the endogenous ApoE (ΔCt>15.8). This shows the TRE-ApoE induction by Dox is >8,000 fold.

Next, we carried out phenotypic analysis of homozygous ApoE iKO mice (ApoE−/−; TRE-ApoE), i.e. mice having both endogenous ApoE alleles inactivated and carrying TRE-ApoE in the TIGRE locus. We analyzed blood cholesterol levels in these mice in the absence and presence of Dox. Constitutive KO (i.e. ApoE−/− without TRE-ApoE in the TIGRE locus) and WT group mice that were littermates of iKO were used as controls. As shown in Figure 7A, in the absence of Dox the iKO mice had high cholesterol levels similar to that of the KO mice; in the presence of Dox, the iKO showed normal cholesterol levels, demonstrating that inducible expression of ApoE can lead to the reversion of the KO phenotype of hypercholesterolemia. When Dox was withdrawn, the cholesterol levels in the iKO mice rose again. These on/off switches occurred rapidly, within a few days after Dox administration or withdrawal.

Dox-Regulated Atherosclerosis Progression and Regression in the ApoE iKO Mice

We examined the atherosclerotic lesion formation in the aortas of the ApoE iKO mice. As controls we used KO and WT group mice that were littermates of iKO. In the absence of ApoE protein, aortic atherosclerotic lesions start to form around 3–4 months of age and progress continually with time. One set of iKO, KO and WT group mice were treated with Dox-containing food throughout their life, and were compared with mice fed with normal food. Figure 7B shows the atherosclerotic lesions formed around the arch region of the aorta, as visualized by Sudan IV staining. By 7 months of age, extensive lesions had formed in KO mice, regardless of whether they were treated with Dox or not, whereas WT mice did not have any lesions in the absence (Figure 7B) or presence (data not shown) of Dox. The iKO mice developed extensive lesions in the absence of Dox (similar to KO mice), whereas in the presence of Dox no lesions had formed.

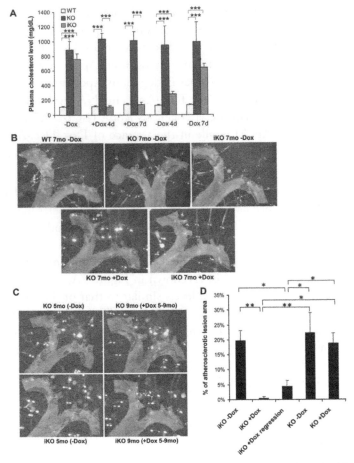

Figure 7. Plasma cholesterol levels and atherosclerotic lesion progression/regression regulated by Dox in ApoE iKO mice. (A) Plasma cholesterol levels in the ApoE iKO, KO and WT group mice in the absence and presence of Dox. (Littermate mice of various genotypes other than homozygous KO or iKO, i.e. ApoE+/+, ApoE+/−, ApoE+/+;TRE-ApoE and ApoE+/−;TRE-ApoE, all displayed normal and indistinguishable blood cholesterol levels, and never developed lesions under any treatment regime used in our study. Consequently they were lumped together as the "WT group".) Plasma cholesterol levels were first measured in mice fed with normal food without Dox (−Dox) and both ApoE iKO and KO mice showed significantly higher cholesterol levels compared to WT group mice (p = 0.35 between iKO and KO, p<0.0001 between iKO and WT or between KO and WT, Student's t-test). The mice were then switched to Dox-containing food, and plasma cholesterol levels were measured again 4 days (+Dox 4d) and 7 days (+Dox 7d) later. Cholesterol level of ApoE iKO mice dropped

to WT levels in less than 4 days while that of ApoE KO remained high (p<0.0001 between iKO and KO or between KO and WT, p = 0.86 between iKO and WT). Sometime later, some Dox-treated mice were switched back to normal food, and plasma cholesterol levels were measured again 4 days (–Dox 4d) and 7 days (–Dox 7d) after Dox withdrawal. Cholesterol level of ApoE iKO mice significantly elevated by day 4 (p<0.001 between iKO and WT, p<0.01 between iKO and KO) and approached pre-Dox treatment level by day 7 (p<0.0001 between iKO and WT, p = 0.13 between iKO and KO). ***P<0.001. (B) Atherosclerotic lesion progression. Aortas were stained with Sudan IV to visualize the lesions in red. The arch region of the aorta contains the most extensive areas of lesions and is shown here. A group of iKO, KO and WT mice were treated with Dox-containing food starting before the onset of lesions, and were compared with mice fed with normal food. At 7 months of age, aortas were dissected from these mice and lesions were examined. ApoE iKO mice showed extensive aortic lesions as the KO mice in the absence of Dox, and yet no lesions at all as the WT mice in the presence of Dox. (C) Atherosclerotic lesion regression. ApoE iKO and KO mice of 5 months of age were switched from normal food to Dox-containing food. Aortic lesions were examined before (at 5 months) and after (at 9 months) the Dox treatment. After 4 months of Dox treatment, the lesions in KO mice continued to grow, whereas in the iKO mice the lesions had regressed. (D) Quantification of the aortic atherosclerotic lesion areas in the arch region above the first intercostal artery, as expressed by the percentage of lesion areas versus the whole aortic area in this segment. All genotypes are matched with ages for different Dox treatment. Dox-treated groups are: iKO+Dox: Dox food started at 2–4 months of age (before the onset of atherosclerosis); iKO+Dox regression: Dox food started at 5–6 months of age (after the onset of atherosclerosis); KO+Dox: Dox food started at 2–4 months of age (before the onset of atherosclerosis). The results are compared using one-way ANOVA followed by Neuman-Keul's post hoc test. Both iKO+Dox and iKO+Dox regression groups had significantly reduced atherosclerotic areas compared to the remaining groups. *P<0.05, **P<0.01. The iKO mice without Dox, as well as KO mice either with or without Dox, all developed comparable areas of lesions (p>0.05 in all pair-wise comparisons). The iKO mice treated with Dox before the onset of atherosclerosis had nearly no lesions and were significantly different from the above groups (p<0.01 iKO+Dox versus iKO–Dox; p<0.01 iKO+Dox versus KO–Dox; p<0.05 iKO+Dox versus KO+Dox). The iKO mice treated with Dox after the onset of atherosclerosis had significantly reduced atherosclerotic areas compared to iKO mice without Dox or KO mice either with or without Dox (p<0.05 in all comparisons between iKO+Dox regression versus iKO–Dox, KO–Dox or KO+Dox).

We further investigated what happens if ApoE protein expression is turned on after the atherosclerotic lesions have already formed. ApoE iKO and KO mice of 5 months of age were switched from normal food to Dox-containing food for the next 4 months. Aortic atherosclerotic lesions were examined before (at 5 months) and after (at 9 months) the Dox treatment. As shown in Figure 7C, at 5 months of age, both iKO and KO had developed similar levels of lesions. After 4 months of Dox treatment, the lesions in KO mice continued to grow, whereas in the iKO mice, the lesions had regressed nearly completely with only scar-like tissues remaining, suggesting that the lipid-containing foam cells have disappeared from the lesions. The results were verified by quantification and statistical analysis of the lesions by one-way ANOVA followed by Neuman-Keul's post hoc test (Figure 7D).

Discussion

We have developed a reversible and inducible rescue system for gene KO in mice and have applied this method into the ApoE gene. Our system complements existing inducible gene expression approaches and provides certain advantages. The tamoxifen-dependent Cre-ERT2 [19],[20] recombination can drive inducible knockout of the endogenous gene, however, it is irreversible and the efficiency of tamoxifen inducibility throughout the body is yet to be demonstrated. The usefulness of the Tet-inducible system [21],[22] is critically dependent on the tightness of transcription induction and suppression. When either the Tet-transactivator or the target gene is randomly integrated into the genome, they are subjects to positional effects [23]. It is often necessary to screen through multiple transgenic lines every time a new transgene is introduced. Attempts to target both tTA and TRE together into an endogenous gene locus to achieve inducible activation and inactivation were successful in a few cases [24],[25],[26], but this method is not applicable to most other genes as the TRE is easily subject to activation from nearby enhancers independent from tTA. Our iKO technique utilizes each gene's own promoter to direct the expression of transcriptional activators, e.g. rtTA and tTA. Therefore it is not limited to certain tissues and available tissue-specific promoter driven transgenic lines. It could be applicable to any gene in any tissue. Dox-regulated expression can be turned on and off rapidly, i.e. within a few days, and at any time in development or adult. It also allows analysis of the effects of gene inactivation in the same animal. The TIGRE locus has been selected to confer little or no basal transgene expression throughout the body while maintaining high inducibility, enabling stringent control of the on/off switching of the target gene. Our system also has the flexibility to allow for further improvements. For example, in some cases it may be more desirable to put the genomic copy of the gene under TRE promoter into the TIGRE locus instead of the cDNA to more precisely mimic the expression of the endogenous gene. Also, the IRES we have used to drive rtTA translation may not work uniformly well in all tissues, and can be replaced by other approaches such as using the viral 2A-like sequences for bicistronic translation [27] or direct targeting of tTA/rtTA into the ATG start codon of the endogenous gene.

Unique chromosomal loci for predicted gene expression provide fundamental tools for genetic studies. The most widely used is the ROSA26 locus [28] in which ubiquitous gene expression is achieved by endogenous promoter activity of this locus. The TIGRE loci identified in this study allow a different mode of gene regulation – they were selected from hundreds of insertion sites for tight gene regulation by an exogenous promoter. Therefore, the TIGRE loci offer a platform for easy insertion of any gene in a tightly regulated locus, applicable to not only the tetracycline system but also other gene expression systems utilizing exogenous promoters such as constitutively active promoters, tissue specific promoters, or other inducible promoters regulated by reagents such as ecdysone [29], mifepristone [30] and streptogramin [31]. Recently a similar approach was also applied to a human fibrosarcoma cell line to pre-screen optimal integration sites for transgenes [32]. In addition, here we demonstrated that the stringency of the regulation at TIGRE loci is further enhanced by the incorporation of insulators, i.e. basal expression level was reduced to less than one luciferase molecule per cell without impairing inducibility (Figure 4C, D). Insulators were also shown previously to improve inducibility of randomly integrated TRE-reporters [33]. It should be noted that in our study the insulator effect was limited in many other loci (Figure 4A versus 4B), demonstrating the uniqueness of the TIGRE loci. The genomic location of the T1 TIGRE locus, which was most extensively used in our study including the ApoE iKO mice, was determined (Figure 5). Further manipulation of this locus would be possible to expand its application.

The stringent gene regulation is demonstrated in the ApoE iKO mice. It is known that regulation of blood cholesterol levels is very sensitive to the plasma ApoE protein levels. Even with the production of 3% of wild-type level of ApoE protein, the hypercholesterolemia and atherosclerosis phenotypes of the ApoE KO mice can be reversed [9]. Given this high sensitivity, the fact that our ApoE iKO mice in the uninduced state (i.e. in the absence of Dox) exhibit similarly high levels of cholesterol compared to ApoE KO mice indicates that there is hardly any expression of functional APOE. Reversing the KO at will is a particularly powerful approach in atherosclerotic regression studies. Dox-treatment of ApoE iKO mice results in expression of ApoE, marked reduction of plasma cholesterol levels, and regression of aortic atherosclerotic lesions. These findings are consistent with previous studies showing that aggressive lipid lowering or expression of ApoE can induce regression of pre-existing atherosclerotic lesion [7],[8],[34],[35]. It is becoming increasingly clear that lesion regression is regulated by a complex interplay between lipids, inflammation and the immune system [35]. The ApoE iKO mice will allow detailed studies on the roles of specific genes in these complex interactions.

The binary nature of the iKO system is inherently simple, with the KO line serving dual roles: it could be used as a constitutive KO or combined with the TIGRE line to produce an inducible and reversible iKO. The 10-million clone ES cell library we utilized [18] has been estimated to contain insertional mutations for >90% of genes, and individual ES clones with retroviral insertions in a specific target gene can be rapidly identified through a PCR pooling strategy and subsequently isolated in a streamlined process. The identification and modification of the TIGRE locus allows rapid insertion of any gene of interest via co-transfection with Cre. Therefore, each component of the binary system, the KO or the TIGRE line, is amenable for high-throughput production to generate inducible and reversible KOs for a large number of genes.

It should be noted that the iKO system may not be applicable in certain situations where highly stringent gene regulation is required. For example, even though the system had enough stringency in low basal activity and high induction, the induced gene expression level is usually not exactly the same as the endogenous gene's level and this could be a problem for genes that are highly sensitive to gene dosage effect or show haploid deficiency. The kinetics of the system (days) may be also too slow for some developmental problems where transient expression of developmental genes are critical, although it should be noted that it could still work well in carefully thought-out developmental studies (as in [24]). In addition, it has been reported that rtTA often can not induce sufficient gene expression in the brain, at least partially due to developmental inactivation of the TRE promoter in neurons [36]. It appears that the tTA (Tet-off) system is better suited for the use in brain [14], as substantial β-gal induction was observed in brain with tTA (Figure 4F).

Our results of the ApoE iKO mice and the quantitative data using lacZ and luciferase reporters suggest that the iKO system could be a useful tool in addressing a variety of biological questions. The two components of the iKO system can be independently modified, and pairing of their different forms can generate numerous combinations. In KO lines, genes with unique expression patterns are tagged with a transcription transactivator, which can control, in an inducible fashion, the expression of a variety of genes derived from TIGRE lines, enabling a number of additional applications in specific types of tissues or cells. Those include: 1) introducing mutant forms of the target gene into the TIGRE locus for better functional probing of different domains, splice variants, post-transcriptional modifications (e.g. phosphorylation), or modeling human diseases; 2) humanizing target genes by placing a human ortholog of the mouse gene under TRE control, which can facilitate drug efficacy studies; 3) placing a cytotoxic gene under TRE control to allow inducible cell-type specific ablation; 4) introducing a marker gene such as GFP, protein interacting probe, or transneuronal tracer, into the TIGRE locus for cell-type specific tagging, functional analysis, isolation of specific cell population, or mapping neuronal networks; 5) combining with recently developed RNAi techniques [37],[38],[39] to down-regulate any genes of interest in a tissue-specific and inducible manner; 6) re-engineering the TIGRE locus to place a TRE-driven Cre, and combining it with a tissue-specific rtTA or tTA line and floxed target genes to achieve inducible region-specific gene knockout [40].

Materials and Methods

Screening and Isolation of ES Clones with Insertions in Target Genes

Construction of ES cell library infected with the retrovirus and screening and isolation of ES clones with genes of interest inactivated by the viral insertion is described previously [18]. Specifically for ApoE, insertions were found by nested

PCR analysis of the ES DNA using two vector-specific and gene-specific primer pairs. The ApoE-specific primers are antisense and located in the fourth exon. Several independent retroviral insertions in the ApoE gene were identified. PCR fragments were sequenced to confirm insertion into the gene. A library tube with a clone of interest identified by the PCR also contains a few hundred other ES cell clones. The sole desired clone was isolated from the mixture by three rounds of cell sorting and growing followed by PCR using the same pair of primers to identify positive clones. Additional PCRs using ApoE primers located in the third intron flanking the viral insertion site were conducted to confirm the precision of the insertion and integrity of the genomic sequence of ApoE.

Generation of TIGRE ES Clones Containing Target Genes

Full length cDNAs for the coding sequences of target genes were cloned into the TIGRE-targeting vector containing TRE, insulators, a PGK promoter and a pair of loxP sites (Figure 5A lower panel). The TIGRE-targeting vectors were co-transfected with a Cre-expressing plasmid into the neo-sensitive ES cells carrying the minimal TIGRE locus with a single loxP site and the promoterless, ATGless loxneo marker. When the TIGRE-targeting vector is integrated into the TIGRE locus through Cre/lox-mediated recombination, neo-resistance is restored to the ES cells by the addition of PGK promoter and in-frame fusion of ATG to the loxneo marker. Correctly integrated ES clones were identified and confirmed by PCR screening and southern blot analysis.

Animal Production and Maintenance

ES cell clones were injected into blastocysts of C57BL/6J mice following standard techniques. Chimeric mice were bred with C57BL/6J mice to test germline transmission and generate heterozygous mice. Mice from the corresponding KO lines and TIGRE lines were crossed with each other to produce inducible KO mice according to the scheme shown in Figure 1B. All mice used for studies in this paper were in a mixed genetic background of 50% 129S1/SvImJ and 50% C57BL/6. For the ApoE KO line, southern blot using rtTA coding sequence as probe confirmed the correct insertion of the retroviral vector into the endogenous ApoE gene. It also revealed an additional retroviral insertion somewhere else in the genome. The additional insertion was selectively bred out, and the ApoE iKO colony was maintained with a single retroviral insertion at the ApoE locus. MMTV-tTA, PCAMKII-tTA and PNSE-tTA mice were purchased from The Jackson Laboratory (Bar Harbor, ME). All experimental procedures were approved by the Institutional Animal Care and Use Committee of NCI and Nura, Inc. in accordance with NIH guidelines.

Dox was administered to the mice through Dox-containing food, which was custom made by Bio-Serv (Frenchtown, NJ) to contain 2 g Dox per kilogram of food. The nutrition content of the Dox food (e.g. 19% protein and 8.6% fat) was very similar to the regular diet (20% protein, 9% fat) used in our colony. Therefore the switching of food types did not result in change of cholesterol levels in mice.

Acknowledgements

We thank Gary Felsenfeld for providing pJC13-1, Hermann Bujard for pUHG10-3 and pUHD15-1neo, Wolfgang Hillen for pCMV-TetR(B/E)-KRAB, Douglas Powell for statistical analysis, and Vincent Keng for comments on the manuscript. We thank Susan Reid and Janet Blair-Flynn for excellent technical assistance. Requests for the TIGRE ES clones should be addressed to GNP.

Author Contributions

Conceived and designed the experiments: HZ KH LM LT KB GP GAG. Performed the experiments: HZ KH LM MP GG AR BS YL EO FK PR OS ID ES AG. Analyzed the data: HZ KH KB AG GAG. Contributed reagents/materials/analysis tools: FK ES LT GP. Wrote the paper: HZ KH GAG.

References

1. Lewandoski M (2001) Conditional control of gene expression in the mouse. Nat Rev Genet 2: 743–755.

2. Gossen M, Bujard H (2002) Studying gene function in eukaryotes by conditional gene inactivation. Annu Rev Genet 36: 153–173.

3. Plump AS, Smith JD, Hayek T, Aalto-Setala K, Walsh A, et al. (1992) Severe hypercholesterolemia and atherosclerosis in apolipoprotein E-deficient mice created by homologous recombination in ES cells. Cell 71: 343–353.

4. Zhang SH, Reddick RL, Piedrahita JA, Maeda N (1992) Spontaneous hypercholesterolemia and arterial lesions in mice lacking apolipoprotein E. Science 258: 468–471.

5. Boisvert WA, Spangenberg J, Curtiss LK (1995) Treatment of severe hypercholesterolemia in apolipoprotein E-deficient mice by bone marrow transplantation. J Clin Invest 96: 1118–1124.

6. Linton MF, Atkinson JB, Fazio S (1995) Prevention of atherosclerosis in apo-lipoprotein E-deficient mice by bone marrow transplantation. Science 267: 1034–1037.

7. Tsukamoto K, Tangirala R, Chun SH, Pure E, Rader DJ (1999) Rapid regression of atherosclerosis induced by liver-directed gene transfer of ApoE in ApoE-deficient mice. Arterioscler Thromb Vasc Biol 19: 2162–2170.

8. Raffai RL, Loeb SM, Weisgraber KH (2005) Apolipoprotein E promotes the regression of atherosclerosis independently of lowering plasma cholesterol levels. Arterioscler Thromb Vasc Biol 25: 436–441.

9. Thorngate FE, Rudel LL, Walzem RL, Williams DL (2000) Low levels of extrahepatic nonmacrophage ApoE inhibit atherosclerosis without correcting hypercholesterolemia in ApoE-deficient mice. Arterioscler Thromb Vasc Biol 20: 1939–1945.

10. Swiatek PJ, Gridley T (1993) Perinatal lethality and defects in hindbrain development in mice homozygous for a targeted mutation of the zinc finger gene Krox20. Genes Dev 7: 2071–2084.

11. Hennighausen L, Wall RJ, Tillmann U, Li M, Furth PA (1995) Conditional gene expression in secretory tissues and skin of transgenic mice using the MMTV-LTR and the tetracycline responsive system. J Cell Biochem 59: 463–472.

12. Chung JH, Whiteley M, Felsenfeld G (1993) A 5′ element of the chicken beta-globin domain serves as an insulator in human erythroid cells and protects against position effect in Drosophila. Cell 74: 505–514.

13. Gaszner M, Felsenfeld G (2006) Insulators: exploiting transcriptional and epigenetic mechanisms. Nat Rev Genet 7: 703–713.

14. Mayford M, Bach ME, Huang YY, Wang L, Hawkins RD, et al. (1996) Control of memory formation through regulated expression of a CaMKII transgene. Science 274: 1678–1683.

15. Chen J, Kelz MB, Zeng G, Sakai N, Steffen C, et al. (1998) Transgenic animals with inducible, targeted gene expression in brain. Mol Pharmacol 54: 495–503.

16. Devon RS, Porteous DJ, Brookes AJ (1995) Splinkerettes–improved vectorettes for greater efficiency in PCR walking. Nucleic Acids Res 23: 1644–1645.

17. Yadav N, Lee J, Kim J, Shen J, Hu MC, et al. (2003) Specific protein methylation defects and gene expression perturbations in coactivator-associated arginine methyltransferase 1-deficient mice. Proc Natl Acad Sci U S A 100: 6464–6468.

18. Gragerov A, Horie K, Pavlova M, Madisen L, Zeng H, et al. (2007) Large-scale, saturating insertional mutagenesis of the mouse genome. Proc Natl Acad Sci U S A 104: 14406–14411.

19. Feil R, Brocard J, Mascrez B, LeMeur M, Metzger D, et al. (1996) Ligand-activated site-specific recombination in mice. Proc Natl Acad Sci USA 93: 10887–10890.

20. Vallier L, Mancip J, Markossian S, Lukaszewicz A, Dehay C, et al. (2001) An efficient system for conditional gene expression in embryonic stem cells and in their in vitro and in vivo differentiated derivatives. Proc Natl Acad Sci USA 98: 2467–2472.

21. Gossen M, Bujard H (1992) Tight control of gene expression in mammalian cells by tetracycline-responsive promoters. Proc Natl Acad Sci USA 89: 5547–5551.

22. Gossen M, Freundlieb S, Bender G, Muller G, Hillen W, et al. (1995) Transcriptional activation by tetracyclines in mammalian cells. Science 268: 1766–1769.

23. Kistner A, Gossen M, Zimmermann F, Jerecic J, Ullmer C, et al. (1996) Doxycycline-mediated quantitative and tissue-specific control of gene expression in transgenic mice. Proc Natl Acad Sci USA 93: 10933–10938.

24. Shin MK, Levorse JM, Ingram RS, Tilghman SM (1999) The temporal requirement for endothelin receptor-B signalling during neural crest development. Nature 402: 496–501.

25. Bond CT, Sprengel R, Bissonnette JM, Kaufmann WA, Pribnow D, et al. (2000) Respiration and parturition affected by conditional overexpression of the Ca2+-activated K+ channel subunit, SK3. Science 289: 1942–1946.

26. Gross C, Zhuang X, Stark K, Ramboz S, Oosting R, et al. (2002) Serotonin1A receptor acts during development to establish normal anxiety-like behaviour in the adult. Nature 416: 396–400.

27. Szymczak AL, Workman CJ, Wang Y, Vignali KM, Dilioglou S, et al. (2004) Correction of multi-gene deficiency in vivo using a single 'self-cleaving' 2A peptide-based retroviral vector. Nat Biotechnol 22: 589–594.

28. Zambrowicz BP, Imamoto A, Fiering S, Herzenberg LA, Kerr WG, et al. (1997) Disruption of overlapping transcripts in the ROSA beta geo 26 gene trap strain leads to widespread expression of beta-galactosidase in mouse embryos and hematopoietic cells. Proc Natl Acad Sci USA 94: 3789–3794.

29. No D, Yao TP, Evans RM (1996) Ecdysone-inducible gene expression in mammalian cells and transgenic mice. Proc Natl Acad Sci USA 93: 3346–3351.

30. Wang Y, DeMayo FJ, Tsai SY, O'Malley BW (1997) Ligand-inducible and liver-specific target gene expression in transgenic mice. Nat Biotechnol 15: 239–243.

31. Fussenegger M, Morris RP, Fux C, Rimann M, von Stockar B, et al. (2000) Streptogramin-based gene regulation systems for mammalian cells. Nat Biotechnol 18: 1203–1208.

32. Brough R, Papanastasiou AM, Porter AC (2007) Stringent and reproducible tetracycline-regulated transgene expression by site-specific insertion at chromosomal loci with pre-characterised induction characteristics. BMC Mol Biol 8: 30.

33. Anastassiadis K, Kim J, Daigle N, Sprengel R, Scholer HR, et al. (2002) A predictable ligand regulated expression strategy for stably integrated transgenes in mammalian cells in culture. Gene 298: 159–172.

34. MacDougall ED, Kramer F, Polinsky P, Barnhart S, Askari B, et al. (2006) Aggressive very low-density lipoprotein (VLDL) and LDL lowering by gene transfer of the VLDL receptor combined with a low-fat diet regimen induces regression and reduces macrophage content in advanced atherosclerotic lesions in LDL receptor-deficient mice. Am J Pathol 168: 2064–2073.

35. Trogan E, Feig JE, Dogan S, Rothblat GH, Angeli V, et al. (2006) Gene expression changes in foam cells and the role of chemokine receptor CCR7 during atherosclerosis regression in ApoE-deficient mice. Proc Natl Acad Sci USA 103: 3781–3786.

36. Zhu P, Aller MI, Baron U, Cambridge S, Bausen M, et al. (2007) Silencing and un-silencing of tetracycline-controlled genes in neurons. PLoS ONE 2: e533.

37. Ventura A, Meissner A, Dillon CP, McManus M, Sharp PA, et al. (2004) Cre-lox-regulated conditional RNA interference from transgenes. Proc Natl Acad Sci USA 101: 10380–10385.

38. Dickins RA, Hemann MT, Zilfou JT, Simpson DR, Ibarra I, et al. (2005) Probing tumor phenotypes using stable and regulated synthetic microRNA precursors. Nat Genet 37: 1289–1295.

39. Stegmeier F, Hu G, Rickles RJ, Hannon GJ, Elledge SJ (2005) A lentiviral microRNA-based system for single-copy polymerase II-regulated RNA interference in mammalian cells. Proc Natl Acad Sci USA 102: 13212–13217.

40. Monteggia LM, Barrot M, Powell CM, Berton O, Galanis V, et al. (2004) Essential role of brain-derived neurotrophic factor in adult hippocampal function. Proc Natl Acad Sci USA 101: 10827–10832.

Hormone-Induced Protection of Mammary Tumorigenesisin Genetically Engineered Mouse Models

Lakshmanaswamy Rajkumar, Frances S. Kittrell,
Raphael C. Guzman, Powel H. Brown, Satyabrata Nandi
and Daniel Medina

ABSTRACT

Introduction

The experiments reported here address the question of whether a short-term hormone treatment can prevent mammary tumorigenesis in two different genetically engineered mouse models.

Methods

Two mouse models, the p53-null mammary epithelial transplant and the c-neu mouse, were exposed to estrogen and progesterone for 2 and 3 weeks, respectively, and followed for development of mammary tumors.

Results

In the p53-null mammary transplant model, a 2-week exposure to estrogen and progesterone during the immediate post-pubertal stage (2 to 4 weeks after transplantation) of mammary development decreased mammary tumorigenesis by 70 to 88%. At 45 weeks after transplantation, analysis of whole mounts of the mammary outgrowths demonstrated the presence of premalignant hyperplasias in both control and hormone-treated glands, indicating that the hormone treatment strongly affects the rate of premalignant progression. One possible mechanism for the decrease in mammary tumorigenesis may be an altered proliferation activity as the bromodeoxyuridine labeling index was decreased by 85% in the mammary glands of hormone-treated mice. The same short-term exposure administered to mature mice at a time of premalignant development also decreased mammary tumorigenesis by 60%. A role for stroma and/or systemic mediated changes induced by the short-term hormone (estrogen/progesterone) treatment was demonstrated by an experiment in which the p53-null mammary epithelial cells were transplanted into the cleared mammary fat pads of previously treated mice. In such mice, the tumor-producing capabilities of the mammary cells were also decreased by 60% compared with the same cells transplanted into unexposed mice. In the second set of experiments using the activated Her-2/neu transgenic mouse model, short-term estradiol or estradiol plus progesterone treatment decreased mammary tumor incidence by 67% and 63%, and tumor multiplicity by 91% and 88%, respectively. The growth rate of tumors arising in the hormone-treated activated Her-2/neu mice was significantly lower than tumors arising in non-hormone treated mice.

Conclusion

Because these experiments were performed in model systems that mimic many essential elements of human breast cancer, the results strengthen the rationale for translating this prevention strategy to humans at high risk for developing breast cancer.

Introduction

It has long been recognized that hormones have an intimate and decisive role in the development and progression of mammary tumorigenesis. The earliest treatment for prevention of breast tumor recurrence in humans (published in 1896) was removal of the ovaries, the source of the reproductive hormones estrogen and progesterone [1]. The most efficacious treatment for the prevention of recurrence in modern therapy for breast cancer is treatment with drugs that target the estrogen pathway, either the estrogen receptor or estrogen metabolism [2]. Many risk factors for breast cancer development-nult involve the hormonal history of the patient, including early age of menarche, late age of menopause, late age of first pregnancy, nulliparity, and estrogen/progesterone hormone replacement exposure. Indirectly, postmenopausal obesity and alcohol consumption are thought to enhance breast cancer risk through altering estrogen levels [3].

Paradoxically, one of the strongest and extensively documented protective factors for breast cancer in both humans and rodent models is short-term stimulation with hormone, either by means of an early full-term pregnancy or with the hormones estrogen and progesterone. The protective properties of estrogen and progesterone were initially shown in 1962 by Charles Huggins [4] and the extensive studies in this area have been reviewed [5,6]. Studies over the past 30 years to document and understand the mechanistic basis for this phenomenon have used the traditional rat and mouse chemical-carcinogen-induced mammary tumor models. These models use etiological agents that are thought not to be risk factors in human breast cancer. One of the known risk factors in human breast cancer is exposure to radiation [7]. Ironically, in the only study done in the rat where radiation was the initiating agent, hormones were not protective [8]. These considerations raise the question of the relevance of the traditional models for understanding the mechanism of the protective effects of hormones.

Over the past 10 years, there have been many new mouse models of mammary cancer based on the enhanced or deleted expression of specific genes known to have a role in human breast cancer [9]. One of the best-studied models is deletion of the p53 gene in BALB/c mammary epithelium [10,11]. In this model, the deletion of p53 gene expression does not perturb the normal growth and differentiation of the mammary epithelium nor its normal dependence on hormones. However, risk for spontaneous mammary cancer is increased over a 14-month period, and this risk is enhanced by prolonged stimulation with estrogen and progesterone, by irradiation, or by chemical carcinogens [10-12]. The cancers that arise in these mice are locally invasive, exhibit metastases to the lung, and about 20% retain hormone dependence. In addition, the premalignant phenotype mimics that of human ductal carcinoma in situ both morphologically and with respect to retaining estrogen receptor expression [11]. A study of gene expression profiles of tumors arising in the p53-null mammary epithelium with a randomly selected group of stage 1 and 2 human breast cancers found a large number of genes that were commonly expressed in both sets of tumors [13].

A second and more tumorigenic model is the activated Her-2/neu (murine-mammary-tumor virus (MMTV)-c-neu) transgenic mouse model [14,15]. HER-2/neu, an abbreviation for human epidermal growth factor receptor-2, is a proto-oncogene, which when activated by mutation or overexpressed has a role in uncontrolled cancer cell growth. The protein product of the Her-2/neu gene is overexpressed in 25 to 30% of human breast cancers. A significant proportion of human intraductal breast carcinomas (ductal carcinoma in situ) demonstrate Her-2/neu amplification/overexpression, suggesting that the functional activity of this oncogene is enhanced early in the progression of malignant breast disease. Virgin activated Her-2/neu transgenic mice have a 100% incidence of mammary cancer and a high multiplicity of mammary cancers [14,15]. A full-term pregnancy does not increase the incidence or multiplicity of mammary adenocarcinomas in the activated neu transgenic mice but does decrease the size and metastatic potential of the mammary tumors in the activated neu transgenic mice [16]. As these two models mimic many features of major subsets of human breast cancer, we tested whether a short-term exposure to estrogen and progesterone (p53-null and MMTV-neu models) or just estrogen (MMTV-neu model) would induce a protective effect on spontaneous tumorigenesis.

Materials and Methods

Mice

BALB/c mice were bred and maintained at Baylor College of Medicine. The donor mice were BALB/c p53 homozygous null, and the recipient mice were p53 wild type. FVB (activated neu) mice were bred and maintained at the Cancer Research Laboratory, University of California, Berkeley, California, and at Texas Tech University of Health Sciences Center, El Paso, Texas. All mice were maintained in conventional mouse facilities with food and water provided ad libitum, and the room temperature was set at 70°F (21°C). The animal facilities at Baylor College of Medicine and University of California are accredited by the American Association of Laboratory Animal Care. All experiments followed NIH guidelines for the care and use of mice.

p53-Null Model

The basic transplantation protocol for the experiments using the p53-null model was as described [10]. In brief, fragments of mammary ducts from 8 to 10-week-old female p53-null mice were transplanted into the cleared mammary fat pads of 3-week-old or 11-week-old female mice. Although the p53 deletion is the same in all donor mice, the array of secondary alterations important for neoplastic development include both common and unique events. The consequence is that the tumorigenic capabilities of mammary gland fragments vary over a small range between donor mice in the same host environment. Thus, the time to palpable tumor occurrence in 50% of animals in any two untreated groups in two different experiments may be different (for example 60 weeks versus 50 weeks). Thus, each experiment always has an untreated control group for assessment of the effect of a particular treatment. In all the transplantation experiments described here, three different donors were used for each experiment with equal representation in the different groups.

In experiment 1 there were two groups of mice. Group 1 comprised untreated control mice and group 2 comprised mice that received Silastic tubing containing 50 μg of estradiol-17β and 20 mg of progesterone for the period of weeks 2 to 4 after transplantation. The Silastic tubings were implanted subcutaneously dorsally and removed after the 2-week period. Mice were palpated weekly until 45 weeks after transplantation. At that time, the mammary fat pads free of palpable tumors were either collected for the preparation of epithelial cell pellets and frozen or prepared as whole mounts. This experiment was repeated identically and designated experiment 1A. Additionally, transplants were collected between weeks 8 and 16 after transplantation to assess proliferation activity by bromodeoxyuridine (BrdU) immunohistochemistry. The mice were injected intraperitoneally with BrdU 2 hours before tissue collection and the samples were processed as described previously [17]. A total of 500 cells were counted from each fat pad under ×10 magnification. There were four fat pads in each treatment group. In experiments 1 and 1A, a total of 66 transplants per group were assessed for tumorigenic potential.

Experiment 2 was started 12 months after experiment 1A and addressed the question of whether the developmental stage of the mammary epithelium influenced the response to the short-term hormone exposure. In this experiment there were four groups of mice. Groups 1 and 2 were identical to the groups in experiment 1. Thus, epithelial cells were actively proliferating as the ducts were filling the fat pad. The proliferation index at this stage was about 8% [11]. Group 3 comprised mice that received the hormone treatment at 23 to 25 weeks after transplantation at a period when the mammary epithelium had filled the fat pad and proliferation was in a steady state but tumors were starting to appear. We included a positive control as group 4, in which the mice were exposed to a 5 mg pellet of tamoxifen between weeks 11 and 24. Previous experiments had shown that lifetime exposure to tamoxifen prevented the development of mammary tumors in this model system [18]. The mice were palpated weekly until 50 weeks after transplantation.

Experiment 3 tested whether pretreatment of the recipient mice with hormones could provide a protective effect on mammary epithelium that had not been directly exposed to the added hormone exposure. In this experiment there were four groups of mice. Groups 1 and 2 were as in experiment 1, in which the duct fragments were transplanted into 3-week-old mice and estrogen/progesterone was provided at 2 to 4 weeks after transplantation. Groups 3 and 4 contained mice that received the hormones for 2 weeks in one group (the other group was untreated age-matched controls), followed by a rest period of 4 weeks, followed by transplantation of the duct fragments into the cleared fat pads of 11-week-old mice. The mice were palpated weekly until 60 weeks after transplantation.

Whole-mount preparations were made from four mammary glands from each of the groups in experiments 1 to 3 at 4 weeks after the removal of the Silastic tubing implants. In all experiments there was no significant difference between the treated and untreated outgrowths in the percentage of mammary fat pad filled by the implants. All palpable tumors were fixed in 4% paraformaldehyde for 2 to 4 hours, embedded in paraffin and processed for staining with hematoxylin/ eosin. As the transplanted mammary epithelium develops primarily mammary adenocarcinomas but also hemangiosarcomas (at 10% incidence, presumably because of endothelium that is also transplanted as part of the duct fragment), we assessed all tumors histologically to confirm the cell type of origin. In the text throughout, tumor refers to mammary adenocarcinoma. Mammary adenocarcinoma incidences were evaluated statistically with Fisher's exact test. The BrdU immunohistochemistry was performed as described [11].

Activated MMTV-Neu Model

The design of the experiments using the MMTV-neu model was slightly different. FVB transgenic mice were treated for 3 weeks, starting at 7 weeks of age, with 100 µg of estradiol in Silastic capsules. Another set of transgenic animals were treated with 100 µg of estradiol and 15 mg of progesterone, also in Silastic capsules. The doses of estradiol used result in pregnancy levels of estradiol in the circulation [19]. The control animals received empty Silastic capsules for the same duration. Mice were palpated once every week for 8 months to monitor for mammary cancer development. Histopathological examination was performed to confirm the carcinomatous nature of the palpable tumors. Mammary cancer incidence was evaluated statistically with the $\chi 2$ test.

Hormone Preparations

All mice received either empty Silastic tubing or the hormones in Silastic tubing. For both the p53-null model and the neu model, the tubings were prepared with the same protocol. The hormones were packed in individual Silastic capsules (Dow Corning; size 0.078 inch (2 mm) internal diameter, 0.125 inch (3.2 mm) outside diameter, 2 cm in length). Estradiol-17β (Sigma, St Louis, MO, USA) was packed in the Silastic capsules in a cellulose matrix. Progesterone (15 mg;

Sigma) was packed into the Silastic capsules also in a cellulose matrix. All these capsules were primed by soaking overnight in medium 199 (Gibco, Grand Island, NY, USA) at 37°C. Silastic capsules were implanted subcutaneously dorsally.

RNA Analysis

A group (n = 3) of control animals and animals treated with estradiol or estradiol plus progesterone were killed immediately after the 3 weeks of hormone treatments. Mammary glands were removed from all the groups for the transgene expression analysis. Another set of animals (n = 15) received the above-mentioned hormone treatments and were killed at 9 months of age. Mammary tumor and mammary tissue adjacent to the mammary tumors were excised, snap-frozen in liquid nitrogen, and stored at -80°C for molecular analysis. Total RNA isolated from the normal mammary gland, mammary tumor, and mammary tissue adjacent to the mammary tumor was subjected to real-time RT-PCR analyses to study the expression of the transgene [20]. RNA was isolated with Trizol reagent, and real-time reverse transcription was performed with the one-step QuantiTect SYBR Green kit (Qiagen Inc., Valencia, CA, USA) in accordance with the manufacturer's specifications. 18S RNA was used as standard and positive control in these assays. Fold changes between samples for relative Her-2/neu transgene mRNA expression were calculated from the differences in ΔCt values between the two samples ($\Delta\Delta Ct$) and the equation Fold change = $2-\Delta\Delta Ct$. Real-time RT-PCR was performed with the oligonucleotide primers 5'-GGAACCTTACTTCTGTGGTGT-3' and 5'-GGAAAGTCCTTGGGGTCTTCT-3' targeting the SV40 poly(A) region of the transgene. Cyclin D1 expression was performed on the mammary tumors from the control and hormone-treated groups with the oligonucleotide primers 5'-CATCAAGTGTGACCCGGACTG-3' and 5'-CCTCCTC-CTCAGTGGCCTTG-3'.

Results

Effect of Short-Term Hormone Stimulation on P53-Null Mammary Tumorigenesis

The tumorigenic response of the p53-null mammary epithelium exposed to estrogen and progesterone combination at 2 to 4 weeks after transplantation is shown in Figure 1 and Table 1. In the control transplants, 16/66 (24%) produced tumors by 45 weeks after transplantation, with an initial tumor latency of 20 weeks. This incidence is consistent with previous studies [10,18]. In the hormone-treated mice, only 2/66 (3%) produced tumors (p < 0.05). This experiment demonstrates conclusively that a short-term hormone treatment can delay tumor development in a non-carcinogen model of mammary cancer.

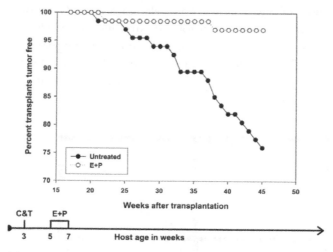

Figure 1. Effect of short-term hormone treatment on tumorigenesis in p53-null mammary epithelial transplants. The data are the sum of two identical experiments (experiments 1 and 1A). Mice treated with estradiol and progesterone had a significant decrease in tumor incidence (p < 0.05). Filled circles, untreated; open circles, hormone-treated. The flow diagram illustrates the experimental treatment plan. C, clear mammary gland; T, transplant; E, estrogen; P, progesterone.

Table 1. Tumorigenesis in short-term hormone-treated mice

Tumorigenesis in short-term hormone-treated mice

Experiment	Group	Tumors/transplants (percentage)	Percentage inhibition	Experiment duration (weeks)
1/1A	Untreated	16/66 (24)	-	45
	E/P (5–7)	2/66 (03)[a]	87.5	45
2	Untreated	10/20 (50)	-	48
	E/P (5–7)	3/20 (15)[a]	70	48
	E/P (23–25)	4/20 (20)[a]	60	48
	Tamoxifen (11–24)	0/20 (0)[a]	100	48
3	Untreated, trans. (3)	9/20 (45)	-	58
	E/P (5–7), trans. (3)	3/15 (20)[a]	56	58
	Untreated, trans. (11)	9/20 (45)	-	58
	E+P (5–7), trans. (11)	4/20 (20)[a]	56	58
4	Untreated	15/15[b] (100)		24
	E/P	6/15[b] (37) [a]	63	32
	E	5/15[b] (33) [a]	67	32

E, estrogen; P, progesterone; trans., transplantation. [a]p < 0.05; [b]in this experiment, the column refers to the number of tumor-bearing mice/total number of mice.

Whole-mount analysis of the glands in experiments 1 and 1A at 4 weeks and at 8 to 12 weeks after transplantation did not indicate a significant difference in outgrowth morphology. The outgrowths exhibited normal alveolar differentiation to 2 weeks of estrogen/progesterone exposure at 4 weeks after transplantation. At 8 to 12 weeks after transplantation, the outgrowths appeared as normal mammary duct arborization with no evidence of ductal hyperplasia (Figure 2). However, whole-mount analysis of the glands at 45 weeks after transplantation revealed extensive ductal hyperplasia in both control and hormone-treated glands. Premalignant lesions were present in 13/31 glands in the estrogen/progesterone-treated group (42%) and 6/13 glands in the untreated controls (46%) (Figure 2; $p > 0.05$). These data suggest that hormones act to block premalignant progression and not the onset of hyperplastic growth.

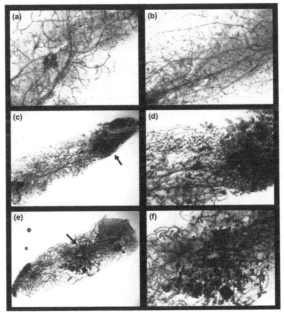

Figure 2. Effect of short-term hormone treatment on gland morphology in p53-null transplants. (a,b) Mammary whole mounts were prepared 4 weeks after the removal of hormones in mice from experiment 1A. The ductal organization at the histological and microscopic (hematoxylin/eosin-stained sections, not shown) was similar in the untreated (a) and treated (b) mice (original magnification ×2). (c–f) Whole mounts were prepared at 45 weeks after transplantation in transplants that had not yet developed palpable tumors. A large number of transplants in the untreated group (46%) (original magnifications ×1 (c) and ×2 (d)) and in the treated group (42%) (original magnifications ×1 (e) and ×2 (f)) contained areas of ductal hyperplasias (arrows).

Effect of Developmental State on Tumorigenesis in P53-Null Mammary Epithelium

The tumorigenic response of the mammary epithelium exposed to estrogen and progesterone combination at 23 to 25 weeks after transplantation is shown in

Figure 3 and Table 1. In the control transplants, 10/20 (50%) produced tumors by 48 weeks after transplantation, with an initial tumor latency of 24 weeks. In the transplants exposed to hormone combination when they were actively filling the fat pad, the tumor incidence was 3/20 (15%), with an initial tumor latency of 44 weeks. In contrast, the transplants exposed to the hormone combination much later after transplantation had an initial tumor latency equivalent to the controls, but the final tumor incidence (4/20; 20%) was not significantly different (p > 0.05) from that in the group exposed to hormones early after transplantation. The transplants exposed to tamoxifen for only 3 months did not develop any tumors after 50 weeks. This experiment demonstrates that a short-term treatment administered at either the actively proliferating or the steady-state stage of mammary development can delay tumorigenesis.

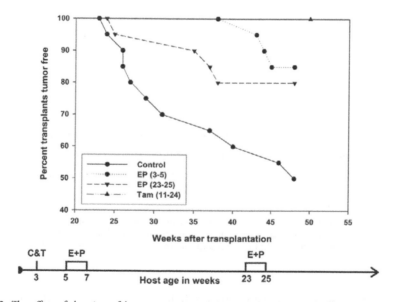

Figure 3. The effect of duration of hormone exposure on tumorigenesis in p53-null mammary epithelial transplants. The effect of hormone exposure was compared with that of a hormone receptor antagonist, tamoxifen. Mice treated with estradiol and progesterone, either early or late, or treated with tamoxifen, had a significant decrease in tumor incidence (p < 0.05). Filled circles and solid line, control; filled circles and dotted line, early hormone exposure; inverted triangles and broken line, late hormone exposure; upright triangle, tamoxifen. The flow diagram illustrates the experimental treatment plan.

Effect of Hormone Pretreatment of the Host on Tumorigenesis in P53-Null Mammary Epithelium

The tumorigenic capability of the p53-null mammary epithelial cells either directly exposed to estrogen/progesterone or just to the estrogen/progesterone-treated host is shown in Figure 4 and Table 1. The tumorigenic capability was the same whether the target epithelial cells were directly exposed to estrogen/progesterone or transplanted into an estrogen/progesterone-treated host environment (3/15 versus 4/20, respectively; p > 0.05). The tumorigenic response was decreased from 45% (18/40 combined controls) to 20% (7/35 combined treatment; p < 0.05).

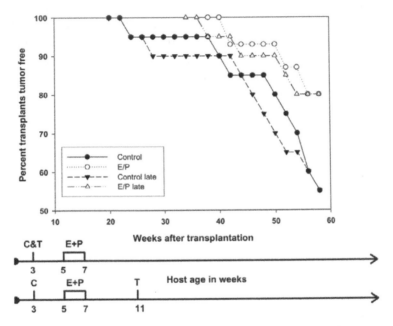

Figure 4. The effect of hormone-induced systemic changes on tumorigenesis in p53-null mammary epithelial transplants. Mammary epithelial transplants in a host exposed to estradiol and progesterone before transplantation (open triangles and dot-dashed line) also had a significant decrease in tumor incidence (p < 0.05) compared with mice not exposed to the hormones (filled triangles and broken line). The flow diagrams illustrate the experimental treatment plans.

The proliferative state of the mammary epithelial cells was assessed at 4 to 12 weeks after removal of the hormones in all three experiments (Figures 5 and 6). In experiment 1, at 4 and 8 weeks after hormone removal, the control transplants showed a BrdU-labeling index of 9.8 and 8.2, respectively, in comparison with 1.5 and 1.8, respectively, in the hormone-treated transplants (p < 0.05). In experiment 2, at 4 weeks after hormone removal (that is, 26 weeks after transplantation), the control transplants showed a BrdU-labeling index of 19.8 in comparison with 7.8 in the hormone-treated transplants (p < 0.05). In experiment 3, at 12 weeks after hormone removal (that is, 8 weeks after transplantation into 11-week-old mice), the control transplants had a BrdU-labeling index of 9.5 in comparison with 4.5 in the transplants in the hormone-treated mice (p < 0.05).

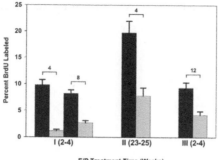

Figure 5. Bromodeoxyuridine-labeling index (number labeled per 500 cells) in hormone-treated p53-null mammary epithelial transplants. Transplants from each experiment (1, 2 and 3) were assayed for bromodeoxyuridine. The black bars represent untreated transplants, the gray bars hormone-treated transplants. The number above each pair of bars indicates the number of weeks after the removal of hormones. In experiment 3, 12 weeks after the removal of hormone represents 8 weeks after transplantation. There were four transplants per treatment group. Five hundred cells were counted in each transplant. All four comparisons were significantly different ($p < 0.05$).

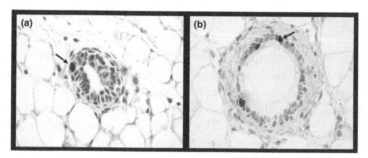

Figure 6. Immunohistochemistry of bromodeoxyuridine-labeled mammary epithelial cells. The nuclei (arrows) represent the uptake of bromodeoxyuridine into cells undergoing DNA synthesis. (a) Untreated; (b) treated with estrogen plus progesterone.

Effect of Hormone Treatment on MMTV-Neu Mammary Tumorigenesis

The marked response of the p53-null mammary epithelium to a short-term exposure to hormone raised the question of whether a similar response would occur in a more tumorigenic model of mammary tumorigenesis. The MMTV-activated neu model provided such a model, because tumors develop rapidly and with high multiplicity. The effect of either short-term treatment with estradiol or estradiol plus progesterone on mammary carcinogenesis was tested. The data in Figure 7 and Table 1 indicate that both treatments were equally effective in providing protection. Mammary cancer incidence after treatment with estradiol alone (33%) or with estradiol plus progesterone (37%) was significantly decreased in comparison with the controls (100%).

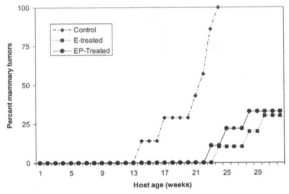

Figure 7. Effect of short-term hormone treatment on mammary carcinogenesis in the activated Her-2/neu transgenic mice. Mice treated with estradiol alone (7/21) or estradiol plus progesterone (9/24) had a significant decrease in the incidence of mammary tumors compared with the controls (20/20) at 8 to 9 months of age (p < 0.01).

All the control mice developed mammary cancers by 5 months of host age with a multiplicity of 6.8 cancers per mouse. Treatment with estradiol (0.6 cancers per mouse) or estradiol plus progesterone (0.8 cancers per mouse) drastically reduced mammary cancer multiplicity and also approximately doubled the mammary cancer latency (Figures 8 and 9). This experiment was repeated once with identical results. A second repeat experiment was not informative because the control untreated mice did not develop tumors with their usual early latency of 13 to 15 weeks of host age but started to develop tumors only at 30 weeks.

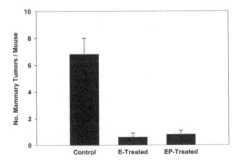

Figure 8. Effect of short-term hormone treatment on mammary cancer multiplicity in the activated Her-2/neu transgenic mice. The group of mice (n = 21) treated with estradiol alone developed a total of 13 mammary tumors and the group of mice (n = 24) treated with estradiol plus progesterone developed a total of 20 mammary tumors by 8 to 9 months of age; in comparison, the group of control mice (n = 20) developed a total of 136 mammary tumors (p < 0.005). Results are expressed as means ± SEM.

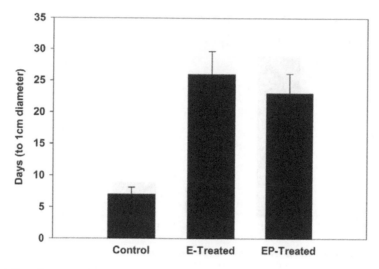

Figure 9. Effect of short-term hormone treatment on tumor growth rate in the activated Her-2/neu transgenic mice. Treatment with estradiol alone or estradiol plus progesterone significantly increased the time taken for the mammary tumors to grow from initial palpation to a diameter of 1 cm in comparison with the controls. Results are expressed as means ± SEM.

Mice were examined twice weekly beginning after the hormone treatments, and the earliest tumors were detectable by palpation (Figure 9). The sizes of tumors were recorded after each examination. In control mice the average age at which palpable tumors were first detected was 95 days. Tumors were initially palpated 142 days in estradiol-treated mice and at 138 days in mice treated with estradiol plus progesterone. Tumor growth was monitored until the tumor in each mouse had reached a diameter of 1 cm. The difference in tumor growth rate between the control and hormone-treated mice was significant. There was almost a threefold difference in the time required for the mammary tumors to grow to 1 cm in diameter (after initial palpation) in mice treated with estradiol or estradiol plus progesterone compared with the controls.

Real-time PCR analyses demonstrated that expression of the transgene was not significantly altered by the transgene in the normal mammary tissue, mammary tumors and the mammary tissue adjacent to the mammary tumor. There was a significant difference in the expression levels of cyclin D1 (Figure 10). Mammary tumors from the control group had a significantly higher (about 5.6-fold) expression of cyclin D1 in comparison with animals treated with estradiol alone or estradiol plus progesterone.

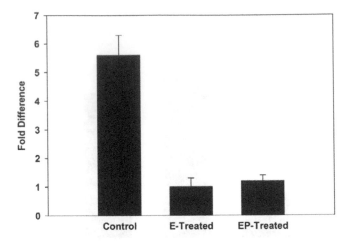

Figure 10. Cyclin D1 mRNA expression levels in mammary tumors arising in activated Her-2/neu mice. There was an approximately sixfold increase in the level of cyclin D1 in mammary tumors arising in control mice in comparison with tumors arising in hormone-treated mice. Results are expressed as means ± SEM.

Discussion

The experiments reported here are the first to address the question of whether short-term hormone treatment can delay tumorigenesis in genetically engineered models of mammary cancer. All previous experiments, with one exception in which radiation was used, were performed in rodent models treated with chemical carcinogen. The results show clearly that a short-term hormone treatment of estrogen with or without progesterone can significantly delay tumorigenesis in two different genetically engineered mouse models. The models differ in fundamental mechanisms of mammary tumorigenesis. The p53-null epithelium is a model in which a major tumor suppressor gene is deleted and aneuploidy is a major feature of the mammary tumors. The tumors arise over a 14-month period and the incidence reaches only 50 to 60% during this period. In contrast, the activated-neu model represents the overexpression of an oncogene, and tumors arise very rapidly and with high multiplicity. These results need to be repeated with other genetically engineered mouse models, such as the BRCA1 and c-myc models, to determine the wider applicability of this effect of hormones. Conceptually, this result is important because the genetically engineered models replicate more faithfully basic features of human breast cancer than do the chemical carcinogen models.

Several results are of general interest. In the p53-null model, the mature gland as well as the developing (that is, immediately post-pubescent) gland was responsive to the protective state induced by the hormone treatment. This result implies that there is no unique developmental state of susceptibility. In the MMTV-activated neu model, the effect of hormones was tested on the developing mammary gland at 7 to 10 weeks of age. These results suggest that the translation of this approach to humans is not limited to the young post-pubescent female but can be applied to the young to middle-aged adult female. Although the emphasis from the human epidemiology studies has always been on early first pregnancy as a critical determinant for protection, the use of a specific hormone combination for short durations might be applicable to a wider age range than previously thought. This conclusion is in line with experiments in the rat model that show a protective effect of hormones even after initiation by a chemical carcinogen. Huggins and colleagues originally reported that estradiol and progesterone given for 30 days, beginning 15 days after carcinogen administration, inhibited the appearance of mammary cancers in rats treated with chemical carcinogens [4]. We also demonstrated that protection against mammary carcinogenesis could be achieved by treatment with physiological levels of estradiol and progesterone for 21 days or less [21,22]. In the MMTV-activated neu model, 100 µg of estradiol in the Silastic tubing yielded a circulating level of serum estradiol of 98.56 ± 8.37 pg/ml (SEM) (n = 6) at 21 days after implantation of the tubing. The groups treated with estradiol plus progesterone yielded similar results (L Rajkumar, unpublished data).

In support of the idea that the mature gland is responsive to the protective effects of a short-term exposure to estrogen and progesterone is the observation that this hormone combination seems to be acting to delay premalignant progression. The presence of frequent hyperplasia in the hormone-treated gland but the absence of invasive cancers supports this conclusion. There is some limited information in the literature that supports this idea. Reddi and colleagues have presented data that show the presence of microtumors in the glands of hormone-treated rats under conditions in which the controls had a high incidence of invasive cancers [23]. The data presented in the present study would be the first demonstration of this result in mice. Experiments that test the growth potential of these microtumors or hyperplasias by transplantation into control animals have yet to be reported. It is evident that there is a point in premalignant progression at which the cells are no longer susceptible to the preventive effects of this hormone combination. Examination of the tumor incidence curves of mice that were exposed to hormones at 23 to 25 weeks of age clearly show that the first tumors appeared with a latency similar to that in control mice. Thereafter, this group behaved similarly to the mice that received the early exposure to hormone.

One of the cellular mechanisms underlying the protective effect involves a diminution of the proliferative potential of the mammary cells. All experiments in the p53-null model demonstrated that the proliferative index of the hormone-treated cells was reduced by 53% to 85% of that of untreated control mammary cells at the steady-state level observed in the mature gland. Interestingly, the ability of the short-term hormone treatment to stimulate proliferation and differentiation during the expansion period at 2 to 6 weeks after transplantation was not compromised, because the extent of filling of the fat pad was the same in the two groups at 6 weeks after transplantation. This suggests that the mechanism for controlling proliferation in the two states (that is, expanding versus a steady-state cell population) might be different either at the level of the cell type that is proliferating in the two states or at the molecular level. We have not yet evaluated the proliferative indices of the hyperplasias in the control and hormone-treated mice because the original observations that determined the presence of these hyperplasias in the hormone-treated mice were performed on whole mounts of the glands, and, indeed, the result was surprising to us. However, in the MMTV-activated neu model, the decrease in proliferative activity was also observed in the tumors arising in the hormone-treated mice. This decrease was manifested at the cellular level in a decrease in cyclin D1 expression.

Perhaps the most surprising result is the apparent systemic effect of the hormones. This experiment demonstrated conclusively that the effect of the hormones can be mediated, in part, by changes induced at the systemic level and/or the mammary stroma. This idea was presented by Thordarson and colleagues in studies on the carcinogen-induced rat mammary system [24]. Attempts to test this hypothesis were only partly successful [25]. The results presented here demonstrate conclusively that hormone-induced effects at the systemic level and/or at the mammary stroma can affect tumorigenesis in the p53-null mammary cells. Such an effect is not without precedent. Barcellos-Hoff and Ravani demonstrated that irradiated stroma can alter premalignant progression of a mouse mammary outgrowth line, COMMA-D [26]. Maffini and colleagues showed that stroma treated with a chemical carcinogen (N-methyl-N-nitrosourea) can enhance the progression of rat mammary cells to mammary tumors [27]. Schedin and colleagues [28] demonstrated that the extracellular matrix of mammary stroma from different reproductive states alters mammary epithelial morphogenesis as well as mammary epithelial growth rate. Specifically, they showed that matrix isolated from parous stroma delayed glandular morphogenesis. The cellular and molecular mechanisms underlying the changes systemically or at the mammary stroma are beginning to be identified. Alterations in both transforming growth factor-β signaling [26] and growth hormone signaling [23,24] have been implicated. Interestingly, we could not demonstrate altered systemic insulin-like growth factor-1 levels in our hormone-treated mice at 4 weeks after hormone removal (D Medina and A Lee, unpublished data).

Finally, it is apparent that the preventive activity can be induced by a modest dose of estrogen alone (100 μg) or a combined dose of estrogen and progesterone.

Previously we had shown that short-term sustained exposure to 100 µg of estradiol resulted in pregnancy levels of estradiol in circulation [19,29]. We further showed that a pregnancy level of estradiol alone or a combination of estradiol and progesterone was highly effective in inducing refractoriness to chemical carcinogen-induced rat mammary carcinogenesis. Similarly to the protection observed in rats by a pregnancy level of estradiol, activated Her-2/neu mice are also rendered refractory to mammary cancer development by short-term hormone treatments.

These studies illustrate two of the major paradoxes of the role of hormones in mammary tumorigenesis. On the one hand, a short duration of estrogen and progesterone or estrogen alone imparts a protective effect on tumor development. On the other hand, continuing the same dose of hormones for a prolonged period strongly stimulates the development of tumors in this same system as well as in other mouse models [5,18,30]. This same result has been reported in the rat mammary tumor system [31]. Different mechanisms underlying the protective effects have been proposed by several investigators. Sivaraman and colleagues and Ginger and colleagues [21,32,33] emphasize the induction of a different developmental fate as a consequence of hormone exposure. Thordarson and colleagues have argued for a systemic effect involving the downregulation of pituitary hormones [24,34,35]. In either event, one would have to conclude that continuing exposure to hormones overrides these mechanisms. Mechanistically, the basis for this override is not clear.

The second paradox is that by either exposing the mammary gland to a short duration of hormones or blocking the same hormone pathway (for example by exposure to tamoxifen) a similar result is generated, namely a decrease in tumorigenic potential. However, the cellular pathways perturbed by the two treatments might be entirely different because tamoxifen-treated outgrowths do not show the presence of hyperplasias that occur in the hormone-treated outgrowths.

In summary, these studies provide a further rationale for considering the use of short-term hormone exposure as a preventive modality, particularly in high-risk individuals. Despite the extensive documentation of the preventive potential of early full-term pregnancy and its mimicry by estrogen and progesterone, there is great resistance to the use of these hormones as a preventive modality. In part, the resistance is due to the overwhelming data showing that prolonged exposure to estradiol and progesterone increases the risk for breast cancer [36]. This resistance might be mitigated by recent data indicating that hormone replacement therapy with estrogen alone does not increase the risk for breast cancer [37]. Perhaps this resistance will be overcome once the mechanisms underlying the preventive effects of specific hormone combinations and duration of exposure are understood.

Conclusion

These studies demonstrate that short doses of the hormones estrogen and progesterone induce a long-lasting protective effect on mammary tumorigenesis in two genetically engineered mouse models. At least part of the effects of the hormone treatment is mediated through systemic and/or mammary stroma alterations.

Abbreviations

MMTV = murine-mammary-tumor virus; PCR = PCR; RT = reverse transcriptase.

Competing Interests

The authors declare they have no competing interests.

Authors' Contributions

Each author contributed equally to the results reported in this manuscript. All authors read and approved the final manuscript.

Acknowledgements

We gratefully acknowledge the technical assistance of Dave Edwards, Himo Garricks, Jamal Hill and Edward Blank. This research was supported by NCI research grants PO1-CA64255 (DM), RO1-CA101211 (PB), and CBCRP-8PB-0132 (SN).

References

1. Beaston GL: On the treatment of inoperable cases of carcinoma of the mamma: suggestions for a new method of treatment with illustrative cases. Lancet 1896, ii:104–107.

2. Howell A, Cuzick J, Baum M, Buzdar A, Dowsett M, Forbes JF, Hoctin-Boes G, Houghton J, Locker GY, Tobias JS, ATAC Trialists' Group: Results of the ATAC (Arimidex, Tamoxifen, Alone or in Combination) trial after completion of 5 years' adjuvant treatment for breast cancer. Lancet 2005, 365:60–62.

3. Willett WC, Rockhill B, Hawkinson SE, Hunter D, Colditz GA: Nongenetic factors in the causation of breast cancer. In Diseases of the Breast. 3rd edition. Edited by: Harris JA, Lippman ME, Morrow M, Osborne CK. Philadelphia: Lippincott, Williams & Wilkins; 2004:223–276.

4. Huggins C, Moon RC, Morii S: Extinction of experimental mammary cancer. I. Estradiol-17β and progesterone. Proc Natl Acad Sci USA 1962, 48:379–386.

5. Sivaraman L, Medina D: Hormone-induced protection against breast cancer. J Mammary Gland Biol Neoplasia 2002, 7:77–92.

6. Medina D: Mammary developmental fate and breast cancer risk. Endocr Relat Cancer 2005, 12:483–495.

7. Tokunaga M, Land CE, Tokuoka S, Nishimori I, Soda M, Akiba S: Incidence of female breast cancer among atomic bomb survivors 1950–1985. Radiation Res 1994, 138:209–223.

8. Holtzman S, Stone JP, Shellabarger CJ: Radiation-induced mammary carcinogenesis in virgin, pregnant, lactating, and postlactating rats. Cancer Res 1982, 42:50–53.

9. Cardiff RD: Mouse models of human breast cancer. Comp Med 2003, 53:250–253.

10. Jerry DJ, Kittrell FS, Kuperwasser C, Laucirica R, Dickinson ES, Bonilla PJ, Butel JS, Medina D: A mammary-specific model demonstrates the role of the p53 tumor suppressor gene in tumor development. Oncogene 2000, 19:1052–1058.

11. Medina D, Kittrell FS, Shepard A, Stephens LC, Jiang C, Lu J, Allred DC, McCarthy M, Ullrich RL: Biological and genetic properties of the p53 null preneoplastic mammary epithelium. FASEB J 2002, 16:881–883.

12. Medina D, Ullrich R, Meyn R, Wiseman R, Donehower L: Environmental carcinogens and p53 tumor suppressor gene interactions in a transgenic mouse model for mammary carcinogenesis. Environ Mol Mutagen 2002, 39:178–183.

13. Hu Y, Sun H, Drake J, Kittrell F, Abba MC, Deng L, Gaddis S, Sahin A, Baggerly K, Medina D, Aldaz CM: From mice to humans: Identification of commonly deregulated genes in mammary cancer via comparative SAGE studies. Cancer Res 2004, 64:7748–7755.

14. Muller WJ, Sinn E, Pattengale PK, Walace R, Leder P: Single-step induction of mammary adenocarcinoma in transgenic mice bearing the activated c-neu oncogene. Cell 1988, 54:105–115.

15. Guy CT, Webster MA, Schaller M, Parsons TJ, Cardiff RD, Muller WJ: Expression of the neu protooncogene in the mammary epithelium of transgenic mice induces metastatic disease. Proc Natl Acad Sci USA 1992, 89:10578–10582.

16. Anisimov VN, Popovich IG, Alimova IN, Zabezhinski MA, Semenchenko AV, Yashin AI: Number of pregnancies and ovariectomy modify mammary carcinoma development in transgenic HER-2/neu female mice. Cancer Lett 2003, 193:49–55.

17. Said TK, Conneely O, Medina D, O'Malley BW, Lydon JP: Progesteronem, in addition to estrogen, induces cyclin D1 expression in mammary epithelial cells in vivo. J Endocrinology 1997, 138:3933–3939.

18. Medina D, Kittrell FS, Hill J, Shepard A, Thordarson G, Brown P: Tamoxifen inhibition of estrogen receptor-α-negative mouse mammary tumorigenesis. Cancer Res 2005, 65:3493–3496.

19. Rajkumar L, Guzman RC, Yang J, Thordarson G, Talamantes F, Nandi S: Short-term exposure to pregnancy levels of estrogen prevents mammary carcinogenesis. Proc Natl Acad Sci USA 2001, 98:11755–11759.

20. Bouchard L, Lamarre L, Tremblay PJ, Jolicoeur P: Stochastic appearance of mammary tumors in transgenic mice carrying the MMTV/c-neu oncogene. Cell 1989, 57:931–936.

21. Sivaraman L, Stephens LC, Markaverich BM, Clark JA, Krnacik S, Conneely OM, O'Malley BW, Medina D: Hormone-induced refractoriness to mammary carcinogenesis in Wistar-Furth rats Carcinogenesis 1998, 19:1573–1581.

22. Guzman RC, Yang J, Rajkumar L, Thordarson G, Chen X, Nandi S: Hormonal prevention of breast cancer: mimicking the protective effect of pregnancy. Proc Natl Acad Sci USA 1999, 96:2520–2525.

23. Reddy M, Nguyen S, Farjamrad F, Laxminarayan S, Rajkumar J, Guzman RC, Yang J, Nandi S: Short-term hormone treatment with pregnancy levels of estradiol prevents mammary carcinogenesis by preventing promotion of carcinogen-initiated cells. Proceedings of the 93rd American Association for Cancer Research Annual Meeting: 6–10 April, 2002; San Francisco, CA 43:824.

24. Thordarson G, Jin E, Guzman RC, Swanson SM, Nandi S, Talamantes F: Refractoriness to mammary tumorigenesis in parous rats: is it caused by persistent changes in the hormonal environment or permanent biochemical alterations in the mammary epithelia? Carcinogenesis 1995, 16:2847–2853.

25. Abrams TJ, Guzman RC, Swanson SM, Thordarson G, Talamantes F, Nandi S: Changes in the parous rat mammary gland environment are involved in parity-associated protection against mammary carcinogenesis. Anticancer Res 1998, 18:4115–4121.

26. Barcellos-Hoff MH, Ravani SA: Irradiated mammary gland stroma promotes the expression of tumorigenic potential by unirradiated epithelial cells. Cancer Res 2000, 60:1254–1260.

27. Maffini MV, Soto AM, Calabro JM, Ucci AA, Sonnenschein C: The stroma as a crucial target in rat mammary gland carcinogenesis. J Cell Sci 2004, 117:1495–1502.

28. Schedin P, Mitrenga T, McDaniel S, Kaeck KM: Mammary ECM composition and function are altered by reproductive state. Mol Carcinog 2004, 41:207–220.

29. Rajkumar L, Guzman RC, Yang J, Thordarson G, Talamantes F, Nandi S: Prevention of mammary carcinogenesis by short-term estrogen and progestin treatments. Breast Cancer Res 2004, 6:R31-R37.

30. Swanson SM, Guzman RC, Collins G, Tafoya P, Thordarson G, Talamantes F, Nandi S: Refractoriness to mammary carcinogenesis in the parous mouse is reversible by hormonal stimulation induced by pituitary isografts. Cancer Lett 1995, 90:171–181.

31. Thordarson G, Van Horn K, Guzman RC, Nandi S, Talamantes F: Parous rats regain high susceptibility to chemically induced mammary cancer after treatment with various mammotropic hormones. Carcinogenesis 2001, 22:1027–1033.

32. Ginger MR, Gonzalez-Rimbau MF, Gay JP, Rosen JM: Persistent changes in gene expression induced by estrogen and progesterone in the rat mammary gland. Mol Endocrinol 2001, 15:1993–2009.

33. Ginger MR, Rosen JM: Pregnancy-induced changes in cell-fate in the mammary gland. Breast Cancer Res 2003, 5:192–197.

34. Thordarson G, Semaan S, Low C, Ochoa D, Leong H, Rajkumar L, Guzman RC, Nandi S, Talamantes F: Mammary tumorigenesis in growth hormone deficient spontaneous dwarf rats; effects of hormonal treatments. Breast Cancer Res Treat 2004, 87:277–290.

35. Thordarson G, Slusher N, Leong H, Ochoa D, Rajkumar L, Guzman R, Nandi S, Talamantes F: Insulin-like growth factor (IGF)-I obliterates the pregnancy-associated protection against mammary carcinogenesis in rats: evidence that IGF-I enhances cancer progression through estrogen receptor-α activation via the mitogen-activated protein kinase pathway. Breast Cancer Res 2004, 6:R423-R436. 36. Chlebowski RT, Hendrix SL, Langer RD, Stefanick ML, Gass M, Lane D, Rodabough RJ, Gilligan MA, Cyr MG, Thomson CA, et al.: Influence of estrogen plus progestin on breast cancer and mammography in healthy postmenopausal women: the Women's Health Initiative Randomized Trial. JAMA 2003, 289:3243–3253.

37. Stefanick ML, Anderson GL, Margolis KL, Hendrix SL, Rodabough RJ, Paskett ED, Lane DS, Hubbell FA, Assaf AR, Sarto GE, et al.: Effects of conjugated equine estrogens on breast cancer and mammography screening in postmenopausal women with hysterectomy. JAMA 2006, 295:1647–1657.

Human Neural Stem Cells Genetically Modified to Overexpress Akt1 Provide Neuroprotection and Functional Improvement in Mouse Stroke Model

Hong J. Lee, Mi K. Kim, Hee J. Kim and Seung U. Kim

ABSTRACT

In a previous study, we have shown that human neural stem cells (hNSCs) transplanted in brain of mouse intracerebral hemorrhage (ICH) stroke model selectively migrate to the ICH lesion and induce behavioral recovery. However, low survival rate of grafted hNSCs in the brain precludes long-term therapeutic effect. We hypothesized that hNSCs overexpressing Akt1 transplanted into the lesion site could provide long-term improved survival of hNSCs, and behavioral recovery in mouse ICH model. F3 hNSC was genetically modified with a mouse Akt1 gene using a retroviral vector. F3 hNSCs expressing Akt1 were found to be highly resistant to H2O2-induced cytotoxicity in vitro. Following transplantation in ICH mouse brain, F3.Akt1 hNSCs induced behavioral improvement and significantly increased cell survival (50–100% increase) at 2 and 8 weeks post-transplantation as compared to parental F3 hNSCs. Brain transplantation of hNSCs overexpressing Akt1 in ICH animals provided functional recovery, and survival and differentiation of grafted hNSCs. These results indicate that the F3.Akt1 human NSCs should be a great value as a cellular source for the cellular therapy in animal models of human neurological disorders including ICH.

Introduction

Two major types of stroke are cerebral infarction (ischemia) and intracerebral hemorrhage (ICH). ICH causes severe neurological deficits and extensive death rate in patients. Since medical therapy against ICH such as mechanical removal of hematoma, prevention of edema formation by drugs and reduction of intracranial pressure shows only limited effectiveness, alternative approach is required such as stem cell-based cell therapy [1], [2].

Recent progress in stem cell biology has opened up a new way to therapeutic strategies to replace lost neural cells by transplantation of neural stem cells (NSCs) in CNS injury and disease [3]–[8]. Previous studies have indicated that NSCs or neural progenitor cells engrafted in animal models of stroke survive and ameliorate neurological deficits in the animals [9]–[17]. Among these studies, human neural progenitor cells isolated from fetal brain have been transplanted into the brain of stroke animal models and found to restore brain function [13], [14]. This approach, however, is not widely acceptable for stroke patients because of moral, religious and logistic problems associated with the use of human fetal tissues. In addition, primary human NSCs derived from fetal tissues can be provided for only a limited time before they undergo senescence, and it is difficult to secure sufficient numbers and homogeneous populations of human NSCs from fetal brain. These problems can be circumvented by the use of stable, permanent cell lines of human NSCs. We have previously reported that human NSC line ameliorate neurological deficits in animal models of Parkinson disease [18], Huntington disease [19], [20], amyotrophic lateral sclerosis [21] and lysosomal storage disease [22] following their transplantation into the brain or spinal cord. In stroke animal models, intravenously transplanted human NSCs migrated selectively to the damaged brain sites caused by ischemia and ICH, differentiated into neurons and astrocytes, and promoted functional recovery in these animals [9]–[12], [15]–[17]. However, low survival rate of grafted F3 NSCs in ischemia and ICH rats in the previous studies is a grave concern; less than 50% of grafted NSCs survived in ICH mice at 2-weeks post transplantation and 30% at 8-weeks [15], [16].

One possible way to promote extended survival of transplanted NSCs in animal brain is to modulate properties of the NSCs, and this might be accomplished by over-expressing Akt1 protein which is known as a general mediator of cell survival signals in the NSCs. Akt, a serine/threonine kinase, plays a critical role in the modulation of cell proliferation, growth, and survival. The PI3K-Akt signal pathway is well-known for the cell survival and it exhibits anti-apoptotic effects against a variety of apoptotic paradigms including withdrawal of extracellular signaling factors, oxidative and osmotic stress, irradiation and ischemic shock [23]–[26]. Previous studies have demonstrated that overexpression of Akt prevents cerebellar granule cells from apoptotic cell death during growth factor withdrawal [23], and promotes cell survival during free radical exposure to free radical or hypoxia in hippocampal neurons [27]–[29].

Considering evidence of functional recovery in stroke animals following brain transplantation of human NSCs and Akt1 protein as a general mediator of survival signals, the present study is designed to investigate whether human NSCs overexpressing Akt1 can lead to the prolonged cell survival of grafted human NSCs and functional recovery in the mouse ICH stroke model.

Materials and Methods

Cell Culture

Primary dissociated cell cultures of fetal human brain tissues of 14 weeks gestation were prepared as described previously [30], [31]. The cells were grown in T25 flasks in Dulbecco's modified Eagle medium with high glucose (DMEM; HyClone, Logan, UT), 5% fetal bovine serum (FBS) and 20 µg/ml gentamicin (Sigma, St Louis, MO) (Sigma). The medium was changed twice a week. The permission to use the fetal tissues was granted by the Clinical Research Screening Committee involving Human Subjects of the University of British Columbia, and the fetal tissues were obtained from the Anatomical Pathology Department of Vancouver General Hospital.

Stable, immortal human neural stem cell line, HB1.F3 (F3), was generated from the primary fetal human brain cell culture as described previously [15], [32]–[34]. PA317 amphotropic packaging cell line was infected with the recombinant replication-incompetent retroviral vector pLNX.v-myc, and the supernatants from the packaging cells were used to infect NSCs in human telencephalon cultures. Stably transfected colonies were selected by neomycine resistance. Several stable clones of human NSCs (hNSCs) were isolated, and one of them, HB1.F3 (F3 hereafter), was expanded for the present study. F3 hNSCs express ABCG2, nestin and Musashi1, which are cell type specific markers for NSCs [15], [34].

To generate Akt1 overexpressing hNSC line (F3.Akt1), Akt1 cDNA (Upstate, Charlottesville, VA) was ligated into multiple cloning sites of the retroviral vector pLHCX (Clontech, Mountain View, CA). Before the ligation, mouse Akt1 cDNA was PCR-amplified by using forward primer 5'-CTAGTTAAGCTTATGGGGAGCAGC-3'; reverse primer 5'- GATATGATC-GATTGATC AGAGGGTTTA -3'. PA317 amphotropic packaging cell line was infected with the recombinant retroviral vector, and the supernatants from the packaging cells were added to the F3 hNSCs. Stably transfected colonies were selected by hygromycin resistance.

RT-PCR Analysis

Reverse transcription was performed with M-MLV reverse transcriptase (Promega, Madison, WI) for 1 hr at 42°C, inactivated for 15 min at 95°C and cooled at 4°C. PCR reaction solution consisted of DNA polymerase buffer, containing cDNA 1 μL, 5 mM $MgCl_2$, 1 mM dNTPs, 10 pM primers and 2.5 units Taq polymerase (Promega). The cDNA was amplified using 30 PCR cycles and RT-PCR products were separated electrophoretically on 1.2% agarose gel containing ethidium bromide and visualized under UV light. The primers used for the RT-PCR for nestin, neurofilament (NF)-L, NF-M, NF-H, glial fibrillary acidic protein (GFAP), GAPDH and Akt1 are listed in Table 1.

Table 1. PCR primer sequences for cell type-specific markers (all human).

Gene	Sequence
Akt1	Sense: 5'-ACCTCTGAGACTGACACCATG-3'
	Antisense: 5'-CACTGGCTGAGTAGGAGAAC-3'
Nestin	Sense: 5'-CTCTGACCTGTCAGAAGAAT-3'
	Antisense: 5'-GACGCTGACACTTACAGAAT-3'
NF-L	Sense: 5'-TCCTACTACACCAGCCATGT-3'
	Antisense: 5'-TCCCCAGCACCTTCAACTTT-3'
NF-M	Sense: 5'-TGGGAAATGGCTCGTCATTT-3'
	Antisense: 5'-CTTCATGGAAGCGGCCAATT-3'
NF-H	Sense: 5'-CTGGACGCTGAGCTGAGGAA-3'
	Antisense: 5'-CAGTCACTTCTTCAGTCACT-3'
GFAP	Sense: 5'-GCAGAGATGATGGGAGCTCAATGACC-3'
	Antisense: 5'-GTTTCATCCTGGAGCTTCTGCCTCA-3'
GADPH	Sense: 5'-CATGACCACAGTCCATGCCATCACT-3'
	Antisense: 5'-TGAGGTCCACCACCCTGTTGCTGTA-3'

Hydrogen Peroxide Treatment

F3 and F3.Akt1 cells were plated 1×10^4 cells per well in 96 well-plates (Falcon, Becton Dickinson, Franklin lakes, NJ) with DMEM containing 5% FBS and incubated for overnight. Hydrogen peroxide (Sigma) was added to each well containing F3 and F3.Akt1 cells to give a final H2O2 concentration in the range from 0 to 0.5 mM. Control wells contained normal medium. Cells were left for 6 hr and 24 hr for viability assay and for 6 hr for western blot analysis.

Cell Viability Assay

Cell viability was determined by the conversion of MTT [3-(4,5-dimethylthiaxol-2-yl)-2,5 diphenyl tetrazolium bromide] to formazan utilizing NADH and NADPH pyridine nucleotide cofactors. After stimulation with H2O2 at different dose for 6 or 24 hr, MTT [2 mg/ml in phosphate buffered saline (PBS)] was added to each well and incubated for 2 hr. The media containing MTT was removed, 100 μL of dimethyl sulfoxide (DMSO, Sigma) added to each well and absorbance read at 570 nm.

Western Blot Analysis

Western blot analysis was performed with 50 μg total protein extract separated on 10% SDS-PAGE gels that were subsequently transferred to polyvinylidene difluoride (PVDF) membrane (Millipore, Billerica, MA). Blocking of membranes with 5% skim milk in TBST and washed in TBST, membranes were incubated with anti-phospho-Akt1-Thr 308 (1:500, Upstate), anti-caspase-3 (1:1000, Chemicon) and anti-beta-actin antibody (1:10,000, Santa Cruz) at 4°C overnight, membrane washed in Tris-buffered saline containing 0.05% tween 20 for 1 hr at RT, then incubated in secondary antibody [horseradish peroxidase-conjugated anti-rabbit or anti-mouse IgG] (Sigma) for 2 hr at RT. Immunoreactive bands were detected by chemiluminescence using ECL system. Western blots analyses were performed on samples from twice separate experiments.

Mouse Intracerebral Hemorrhage Stroke Model

All experimental procedures were approved by the Animal Care Committee of the University of British Columbia. ICH was induced by stereotaxic, intrastriatal administration of bacterial collagenase by previously described methods [12], [16], [17]. In brief, after an intraperitoneal injection of 1% ketamine (30 mg/kg) and xylazine hydrochloride (4 mg/kg), the mice were placed in a stereotaxic frame (Kopf Instruments, Tujunga, CA). A burr hole was made, and a 30-gauge needle was inserted through the burr hole into the striatum (0.1 mm posterior, 4.0 mm ventral, and 2.0 mm lateral to the bregma). ICH was induced by the administration of collagenase type IV (0.5 μL saline containing 0.078 U, Sigma) over a period of 5 min. After remaining in place for another 3 min, the needle was gently removed.

Brain Transplantation

F3 and F3.Akt1 hNSCs were dissociated into single cells by a brief trypsin treatment and suspended in PBS at 4×10^7 cells/0.1 ml and kept on ice until transplanted. Randomly selected ICH mice of one week after ICH surgery received 2 μL (2×10^5 cells) of F3.Akt1 cell suspension (n = 9), F3 cell suspension (n = 9) and killed F3 cell suspension (n = 10). F3 cells in glass tube were killed by placing the tube in boiling water for 1 min, injected slowly for 5 minutes into overlying cortex of the hemorrhage lesion (0.1 mm posterior, 2.0 mm ventral, and 2.0 mm lateral to the bregma). In another control group, 2 μL of PBS was injected into the ICH mice (n = 10). Immunosuppressant was not used in any of the animals. In order to identify the migration potential to the lesion site of grafted cells, F3.Akt1 hNSCs were infected with an adenovirus vector encoding LacZ gene (pAV.LacZ) in vitro at 100 MOI (PU/cell) for 24 hr before transplantation (n = 2).

Behavioral Testing

Behavioral testing was performed weekly with the rotarod (Harvard Instrument) by 2 individuals blinded to mice treatment status [16], [17]. In the rotarod test, mice were placed on the rotarod cylinder, and the time the animals remained on the rotarod was recorded. The speed was slowly increased from 10 to 40 rpm within a period of 2 min. The trial was ended if the animal fell off the rungs or gripped the device and spun around for 2 consecutive revolutions. The animals were trained for 3 days before ICH operation. The maximum duration (in seconds) on the device was recorded with 3 rotarod measurements 1 day before ICH induction. Motor test data are presented as percentages of the maximal duration compared with the internal baseline control (before ICH). The modified limb-placing test is a version of a test previously described [16], [17]. The test consists of 2 limb-placing tasks that assess the sensorimotor integration of the forelimb and the hind limb by checking responses to tactile and proprioceptive stimulation. First, the mouse is suspended 10 cm over a table, and the stretch of the forelimbs toward the table is observed and evaluated: normal stretch, 0 points; flexion with a delay (2 sec) and/or incomplete, 1 point; abnormal flexion, 2 point. Next, the mouse is positioned along the edge of the table, with its forelimbs suspended over the edge and allowed to move freely. Each forelimb (forelimb, second task; hind limb, third task) is gently pulled down, and retrieval and placement are checked. Finally, the mouse is placed toward the table edge to check for lateral placement of the forelimb. The 3 tasks are scored in the following manner: normal performance, 0 points; performance with a delay (2 sec) and/or incomplete, 1 point; no performance, 2 points. A total of 9 points means maximal neurological deficit, and 0 points means normal performance. Additionally, the body weights of all animals were checked weekly for 8 weeks.

Histology and Immunohistochemistry

Histology and immunohistochemistry of brain sections were performed as described previously [16], [17]. At the end of behavioral testing, each animal was anesthetized and perfused through the heart with cold saline followed by 4% paraformaldehyde in 0.1 M phosphate buffer. The brains were post-fixed in same fixative for 24 hr, followed with followed with cryoprotection in 30% sucrose for 24 hr and then 30 μm sections were prepared on a cryostat (Leica CM 3000). The sections through the needle entry site which was identifiable on the brain surface, and sites 1.0 mm anterior and 1.0 mm posterior to plane were processed for X-gal staining to analyze the hemisphere area. These sections are representative of the core of the ICH lesion. The morphometric analyses involved computer-assisted hand delineation of the area of the striatum, cerebral cortex, and ventricles, as well as the whole hemisphere. Adjacent serial coronal sections were processed for double immunofluorescence staining of human nuclear matrix antigen (hNuMA, 1:100, mouse monoclonal, Oncogene) and antibodies specific for cell type specific markers. Antibodies specific for neurofilament low molecular weight protein (NF-L, 1:1000, rabbit, Chemicon), neurofilament high molecular weight protein (NF-H, 1:1000, rabbit, Chemicon), microtubule associated protein-2 (MAP2, 1:500, rabbit, Chemicon), glial fibrillary acidic protein (GFAP, 1:1000, rabbit, DAKO, Carpinteria, CA), and phospho-Akt1 (1:100, rabbit, Upstate) were used for cell type identification of neurons and astrocytes. Brain sections were incubated in mixed solution of primary antibodies overnight at 4°C as free floating sections, followed by mixed secondary antibodies of Alexa Fluor 488-conjugated anti-mouse IgG (1:1000, Molecular Probes, Eugene, OR) and Alexa Fluor 594-conjugated anti-rabbit IgG (1:1000, Molecular Probes) for 1 hr at RT. Negative control sections from each animal were prepared for immunohistochemical staining in an identical manner except the primary antibodies were omitted. Stained sections were then examined under an Olympus confocal laser scanning biological microscope (Olympus, Tokyo, Japan).

Stereological Cell Counts

Total number of human NuMA-positive F3 (n = 3) and F3.Akt1 (n = 3) hNSCs in the brain sections from ICH animals was determined by stereological estimation as described previously [16], [17]. The sections used for cell count covered the entire striatum with hemorrhage lesion and overlying cortex. This generally yielded six or seven sections in a series. Sampling was done using the Computer assisted stereological toolbox system, version 2.1.4 (Olympus), using an Olympus BX51 microscope, a motorized microscope stage (Prior Scientific, Rockland, NY) run by an IBM compatible computer, and a microcator (Heidenhain ND 281B, Schaumberg, IL) connected to the stage and feeding the computer with the distance information in the z-axis. The counting areas were delineated at a 1.25× objective and generated counting areas of 150×150 μm. A counting frame (1612 μm2) was placed randomly on the first counting area and systemically moved through all counting areas until the entire delineated area was sampled. Actual counting was performed using a 100× oil objective. Guard volumes (4 μm from the top and 4–6 μm from the bottom of the section) were excluded from both surfaces to avoid the problem of lost caps, and only the profiles that came into focus within the counting volume (with a depth of 10 μm) were counted. The estimate of the total number of HuNuMA-positive F3 and F3.Akt1 calculated according to the optical fractionator's formula [35].

Statistical Analysis

The statistical significance between group comparisons for behavioral data was determined by one-way ANOVA and two-way ANOVA. P values<0.001 were considered to be statistically significant (version 12.0, SPSS, Chicago, IL).

Results

Stable Human Neural Stem Cell Line Overexpressing Akt1

F3 hNSC line was infected with a retroviral vector encoding mouse Akt1 gene (Figure 1A), and clones resistant to hygromycin were selected and expanded. One of the clones was chosen and used in the present study. The morphology of the selected hNSC line, F3.Akt1 does not differ from the parental F3 hNSCs with bipolar- or multipolar-morphology (Figures 1B, C). Results of RT-PCR analysis of mRNAs isolated from F3 and F3.Akt1 cells are shown in Figure 1D. Transcripts for nestin (an NSC specific marker), neurofilament triplet proteins (NF-L, NF-M and NF-H, cell type-specific markers for neurons), glial fibrillary acidic protein

(GFAP, a specific marker for astrocytes) and Akt1 are all expressed by both F3 and F3.Akt1 cells. However, GFAP band at F3.Akt1 lane detected was lighter than F3 parental cells. In addition transcriptional level of Akt1 gene is higher in F3.Akt1 cells as compared to parental F3 cells.

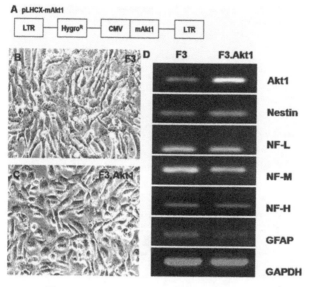

Figure 1. Characterization of human NSC lines. A: The retroviral vector encoding Akt1 (pLHCX.Akt1) used in the present study for the generation of HB1.F3.Akt1 (F3.Akt1) human neural stem cell (NSC) line. B and C: Phase contrast microscopy of F3 and F3.Akt1 human NSCs. Bar inciates 20 μm. D: Gene expression of cell type-specific markers as studied by RT-PCR in F3 and F3.Akt1 human NSCs. Both of F3 and F3.Akt1 NSC lines express cell type-specific markers Nestin (for neural stem cells), NF-L, NF-M and NF-H (for neurons), GFAP (for astrocytes) and Akt1.

Hydrogen Peroxide (H_2O_2)-Induced Cell Death

To determine the protective effects of Akt1 on the H_2O_2-induced cell death in hNSCs, F3 and F3.Akt1 hNSCs were exposed to varying concentration of H_2O_2. F3.Akt1 cells showed a higher survival rate to H_2O_2-induced cell death as compared to the F3 controls (Figures 2A–C). Whether H_2O_2 induces changes in phosphorylation status of Akt1 and caspase-3 cleavage in F3 and F3.Akt1 cells was investigated. When F3 and F3.Akt1 cells were exposed to 0.5 mM H_2O_2 for 6 hr, phosphorylated form of Akt1 and inactivated form of caspase-3 were found in F3.AKt1 cells (Figure 2D), while F3 cells showed the opposite expression pattern (Figure 2D). An increase in active fragment of caspase 3 (20 kDa) was found in F3 cells following H_2O_2 treatment. These results indicate that the enhanced cell survival of F3. Akt1 cells as compared to parental F3 cells following H_2O_2

exposure is due mainly from overexpression of Akt1 and its phosphorylation in F3.Akt1 cell line. And there might be some mediators from Akt1 to caspase-3 in signal pathway; we could consider that phospho-Akt1 finally inhibits caspase-3 cleavage and activation, thereby provides neuroprotection of the hNSCs.

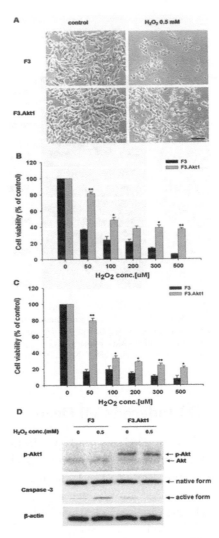

Figure 2. Cell viability increases to H2O2–induced oxydative stress conditions and Akt1 phosphorylation in F3.Akt1 human NSCs. A: Phase contrast microscopy of F3 and F3.Akt1 human NSCs following exposure to 0.5 mM H2O2 for 6 hr. F3.Akt1 NSCs survived well as compared to the parental F3 NSCs. B and C: F3.Akt1 NSCs were found to show resistance to H_2O_2-induced cell death as compared to control parental F3 NSCs at 6 hr (B) and 24 hr (C) respectively. D: Western blot analyses of protein levels of phopho-Akt1 and caspase-3 enzymes in F3 and F3.Akt1 NSCs following H2O2 treatment. F3.Akt1 NSCs showed an increased level of Akt1 phosphorylation, while the level in activation form of caspase-3 was reduced under the H2O2 treatment (* $p<0.05$, ** $p<0.001$).

Functional Recovery in ICH Animals by hNSC Transplantation

Motor performance of ICH animals receiving PBS, F3 Killed, F3 or F3 Akt1 hN-SCs was determined by the rotarod and neurology score (Figures 3A, B). The ICH mice receiving F3.Akt1 hNSCs showed improved behavioral performance on the rotarod compared with the PBS, F3.killed control groups, and the effect persisted for at least up to 8 weeks post-transplantation (PT, the point at which animals were sacrificed) (Figure 3A). Significant difference in rotarod test performance in F3.Akt1 vs F3 groups detected during the period of only 5 weeks PT (P<0.05). The F3.Akt1 transplantation group also showed marked improvement in the limb placement beginning 2 weeks PT and persisting for at least up to 8 weeks (Figure 3B). No significant difference in behavioral performance in F3.Akt1 vs F3 groups detected during the period of 1~8weeks PT.

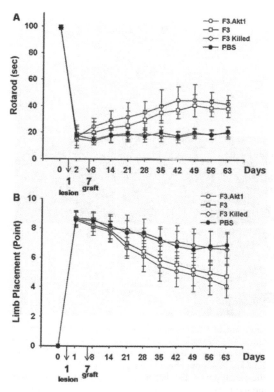

Figure 3. Behavioral improvement demonstrated in intracerebral hemorrhage (ICH) mice transplanted with F3 or F3.Akt1 human NSCs. A: Rotarod test. F3.Akt1-transplanted group showed better performance than PBS controls or F3 cell group, killed F3.Akt1 cell group 8 days onward, and these benefits continued up to 8 weeks post-transplantation (* P<0.05). B: In the modified limb placement test, F3.Akt1-transplanted group showed better performance than PBS, F3 or killed F3.Akt1 group (* P<0.05).

Transplanted NSCs Differentiate Into Neurons and Astrocytes

At 7 days after induction of experimental ICH, 2×10^5 cells/2 µl of F3 or F3.Akt1 hNSCs were transplanted into ICH mouse cerebral cortex overlying hemorrhage lesion site, 2 mm cranial to the hemorrhagic lesion. LacZ+ human NSCs migrated selectively to the hemorrhagic core and also located on the border of the hemorrhagic core and further away from the injection sites (Figures 4A–C). A large number of transplanted hNuMA (human specific nuclear matrix antigen)-positive F3.Akt1 cells (35–45%) differentiated into NF-H+ neurons in the perihematomal sites (Figures 4D–F). While only a small number of transplanted hNuMA+ F3.Akt1 cells (~4%) were GFAP+ astrocytes and the hNuMA+/GFAP+ double-positive cells were found along the border of hemorrhagic core (Figures 4G–I). These results indicate that a large portion of grafted F3.Akt1 cells differentiate into either neurons or astrocytes in response to signals from the local microenvironment provided by the hemorrhagic lesion.

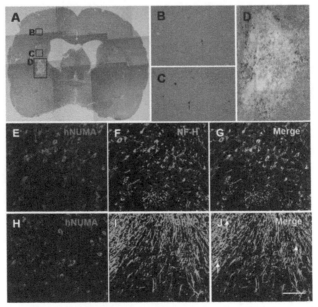

Figure 4. LacZ (beta-galactosidase)-labeled F3.Akt1 human NSCs in intracerebral hemorrhage (ICH) mouse brain at 2 weeks post-transplantation. A: One week after an ICH lesion (an intrastriatal injection of collagenase), LacZ -labeled F3.Akt1 NSCs were transplanted into the cortex overlying the ICH lesion. Two weeks post-transplantation, LacZ-positive F3.Akt1 NSCs were found to migrate extensively into the hemorrhage core and surrounding brain sites. Bar indicates 50 µm. B–D: Higher magnification of indicated areas. Bar indicates 20 µm. E–G: Double immunofluorescent staining of engrafted F3.Akt1 human NSCs in ICH mouse brain 8 weeks post-transplantation. F3.Akt1 NSCs in the lesion sites are found to differentiate into neurons as shown by double staining of human nuclear matrix antigen (hNuMA) and neurofilament protein (NF-H, a neuron specific marker). H–J: F3.Akt1 human NSCs in the lesion sites are found to differentiate into astrocytes (arrows) as shown by double staining of human nuclear matrix antigen (hNuMA) and glial fibrillary acidic protein (GFAP, an astrocyte specific marker). Bar indicates 20 µm.

Survival of Transplanted F3 and F3.Akt1 hNSCs in ICH Brain

Identification of grafted F3.Akt1 hNSCs in ICH brain was determined by immunostaining with hNuMA (Figure 5). Total numbers of hNuMA-positive F3 and F3.Akt1 hNSCs in the ICH brain sections were carried out using stereological estimation at 2- and 8-weeks post-transplantation (PT). Cell survival rate of F3.Akt1 hNSCs at 2-weeks PT is 107,770±2040 cells (54% of the initial population of 200,000 cells) and at 8 weeks PT the number is 64,890±1940 cells (33% of the initial population), while in control parental F3 hNSCs cell survival at 2 weeks PT is 78,320±1250 cells (39% of the initial population of 200,000 cells) and 32,540±4920 cells (16% of the initial population of 200,000 cells) at 8-weeks (p<0.05) (Figure 5). These results indicate that Akt1 overexpression in hNSCs resulted in a 40% increase in cell survival of transplanted hNSCs at 2 weeks PT and 100% increase at 8 weeks PT. In addition, F3.Akt1 human NSCs grafted in cortex overlying striatum were found to migrate extensively to hippocampus 8 weeks PT (Figure 6) indicating that F3.Akt1 hNSCs are capable of migrating to distant anatomical site.

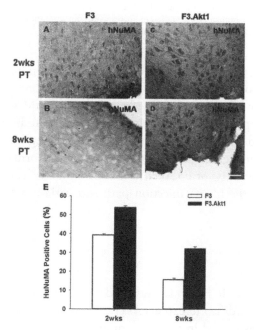

Figure 5. At 2 weeks post-transplantation (PT) in the hemorrhage core border areas, difference in number of human nuclear matrix antigen (hNuMA)-positive cells between ICH-F3 control group (A) and ICH-F3.Akt1 group (C) is not apparent, but at 8 weeks PT number of hNuMA-positive cells is much higher in F3.Akt1-ICH group (D) as compared to control F3-ICH group (B). Bar indicates 100 μm. E: Percentage of hNuMA-positive cells found in hemorrhage core border area is higher in ICH-F3.Akt1 group as compared to ICH-F3 group at both 2 and 8 weeks post-transplantation.

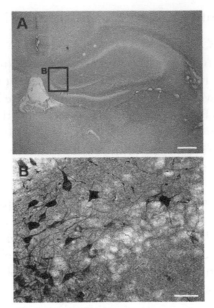

Figure 6. Survival of grafted F3.Akt1 human NSCs in hippocampus was demonstrated by immunoperoxidase microscopy of hNuMA at 8 weeks post-transplantation (PT). A: Lower magnification of hippocampus 8 weeks PT. Bar indicates 1 mm. B: F3.Akt1 human NSCs grafted in cortex overlying striatum were found to migrate extensively to hippocampus 8 weeks PT. Bar indicates 50 μm.

To determine whether Akt1 expression in F3.Akt1 cells induces their own proliferation, expression of cell proliferation marker Ki-67 was examined in ICH brain sections. Transplanted F3.Akt1 cells identified as hNuMA-positive cells were immunoreaction-negative for proliferation marker Ki-67 indicating that the grafted F3.Akt1 cells did not continue to proliferate following transplantation. Furthermore there was no sign of tissue distortion or tumor formation in brain of ICH animals grafted with F3 or F3.Akt1 hNSCs 6 months PT. Good survival of hNSCs was found in the hNSC injection path into striatum two days after transplantation.

Discussion

In the present study, mouse ICH model was used to provide proof-of-principle that hNSCs genetically modified to express Akt1, a general mediator of cell survival signals, can be transplanted in the brain of animal models of neurological diseases, and produces beneficial effects of increased survival of grafted NSCs and consequent functional recovery. We demonstrate that brain transplantation of hNSCs overexpressing Akt1 in the collagenase-induced ICH mice resulted in improvement in motor performance as determined by rotarod and limb placement tests, increased survival of grafted NSCs, and differentiation of grafted NSCs into neurons and astrocytes. These results are consistent with previous studies that Akt1, protein which is known as a general mediator of cell survival signals, promotes the survival of CNS neurons in vitro [23], [27]–[29], and provides favorable clinical outcome in animal models of ischemia [36].

In the present study we exposed F3 and F3.Akt1 hNSCs to hydrogen peroxide injury in vitro to study the neuroprotectiion provided by Akt1 under the conditions of oxidative stress, and the results indicate that F3.Akt1 cells showed a higher survival rate following H2O2 -induced injury as compared to the F3 controls (Figs. 2A–C). Following H2O2 treatment, phosphorylation of Akt1 and concomitant inactivation of caspase-3 were found in F3.AKt1 cells, while in control F3 cells non-phosphorylation of Akt1 and an increase in active fragment of caspase 3 were found (Fig. 2D). It is known that Akt1 phosphorylates caspase-9 at Ser-196, thereby blocking cytochrome c-mediated caspas-9 activation and inactivation of caspase-3 leading to inhibition of proapoptotic signals [37]. Recent studies have also shown that antioxidant enzymes such as glutathione peroxidase-1 (Gpx1), Cu/Zn-superoxide dismutase (Cu/Zn-SOD) and heme oxygenase-1 (HO-1) are target substrates of activated Akt and results in modulation of redox system and reduced toxic levels of reactive oxygen species (ROS) in various cell types [38]–[40].

From as early as 8-day post-transplantation (PT) to 8-weeks PT, F3.Akt1 hNSCs induced an increased survival of transplanted NSCs in the host brain. Survival of transplanted F3.Akt1 hNSCs in ICH mouse brain was identified by human nuclear matrix antigen (hNuMA)- or LacZ/β-gal-positive reaction. The cell count in ICH brain indicates that 40% increase in cell survival in F3.Akt1 group over control F3 group and 100% increase at 8-weeks PT. It should be noted that the majority of grafted F3.Akt1 cells differentiated into either neurons (β-tubulin III-, NF-L- or NF-H-positive) or astrocytes (GFAP-positive) along the border of lesion sites (Fig. 4). In addition, a majority of grafted F3.Akt1 hNSCs differentiated into either neurons or astrocytes in response to signals provided in the local microenvironment.

Previous studies have reported that the activation of PI3K/Akt signaling axis promotes growth and survival of tumor cells, and genetic perturbation of this pathway increases the survival of cancer cells [41]–[43]. For that reason, we were very concerned about possible tumorigenesis in the animals transplanted with F3.Akt1 hNSCs. Transplanted F3 or F3.Akt1 hNSCs in ICH lesion sites were immunoreaction-negative for cell proliferation marker Ki-67 indicating that hNSCs do not proliferate actively in vivo environment. In addition, none of the animals grafted with F3.Akt1 or F3 hNSCs showed tumor formation upon histological examination even in animals with 6-months PT.

Akt activity is induced following PI3K activation in various growth factor-mediated signaling cascades [23]. The PI3K-Akt signal pathway is well-known for the cell survival and it exhibits anti-apoptotic effects against a variety of apoptotic paradigms including withdrawal of extracellular signaling factors, oxidative and osmotic stress, irradiation and ischemic shock [24]–[27]. Akt regulates cell growth and survival mechanisms via phosphorylation of a large number of substrates such as Forkhead transcription factors (FOXO), GSK-3, BAD, caspase-9 and MDM2, and many of these proteins contribute to antiapoptotic signaling in various cell types [24]–[27]. Activation of PI3K/Akt signaling pathway promotes extended survival of neuronal cell types from cell death caused by injuries, including cerebellar granule cells [23] and hippocampal neurons [27]–[29].

Akt1 as well as its downstream molecules offer great promise for the development of therapeutics against a number of neurological disorders such as stroke, spinal cord injury and neurodegenerative diseases such as Alzheimer disease, Parkinson disease and ALS. Initially believed to have cellular functions directed primarily toward cell survival and growth, Akt1 is now seen as a potential broad cytoprotective agent. Akt1 can offer cellular protection not only through the modulation of intrinsic apoptotic machinery, but also through the activation of survival signal pathways. Akt1 drives cellular survival signals through a series of distinct pathways that involve the Forkhead family of transcription factors, GSK-3β, β-catenin, eIF2B, c-Jun, CREB, Bad, IKK, p53, and JIPs [25], [26]. Yet, it is evident that further work that clarifies the cellular environment controlled by Akt1 will be of exceptional value to refine our knowledge of Akt1 and to maximize the potential of this protein as a therapeutic agent.

Acknowledgements

We thank Dr. Min C. Lee for critical reading and comments of the manuscript.

Author Contributions

Conceived and designed the experiments: HJL MKK SUK. Performed the experiments: HJL MKK HJK SUK. Analyzed the data: HJL MKK HJK SUK. Wrote the paper: HJL SUK.

References

1. Gebel JM, Broderick JP (2000) Intracerebral hemorrhage. Neurol Clin 18: 419–438.

2. NINDS ICH Workshop Participants (2005) Priorities for clinical research in intracerebral hemorrhage: report from a National Institute of Neurological Disorders and Stroke workshop. Stroke 36: 23–41.

3. McKay R (1997) Stem cells in the central nervous system. Science 276: 66–71.

4. Gage FH (2000) Mammalian neural stem cells. Science 287: 1433–1438.

5. Kim SU (2004) Human neural stem cells genetically modified for brain repair in neurological disorders. Neuropathology 24: 159–171.

6. Lindvall O, Kokaia Z (2006) Stem cells for the treatment of neurological disorders. Nature 441: 1094–1096.

7. Muller FJ, Snyder EY, Loring JF (2006) Gene therapy: can neural stem cells deliver? Nat Rev Neurosci 7: 75–84.

8. Flax JD, Aurora S, Yang C, Simonin C, Wills AM, et al. (1998) Engraftable human neural stem cells respond to developmental cues, replace neurons, and express foreign genes. Nat Biotech 16: 1033–1039.

9. Chu K, Kim M, Jeong SW, Kim SU, Yoon BW (2003) Human neural stem cells can migrate, differentiate, and integrate after intravenous transplantation in adult rats with transient forebrain ischemia. Neurosci Lett 343: 129–133.

10. Chu K, Kim M, Park KI, Jeong SW, Park HK, et al. (2004) Human neural stem cells improve sensorimotor deficits in the rat brain with experimental focal ischemia. Brain Res 1016: 145–153.

11. Chu K, Kim M, Chae SH, Jeong SW, Kang KS, et al. (2004) Distribution and in situ proliferation patterns of intravenously injected immortalized human neural stem cells in rats with focal cerebral ischemia. Neurosci Res 50: 459–465.

12. Jeong SW, Chu K, Jung KH, Kim SU, Kim M, et al. (2003) Human neural stem cell transplantation promotes functional recovery in rats with experimental intracerebral hemorrhage. Stroke 34: 2258–2263.

13. Kelly S, Bliss TM, Shah AK, Sun GH, Ma M, et al. (2004) Transplanted human fetal neural stem cells survive, migrate and differentiate in ischemic rat cerebral cortex. Proc Nat Acad Sci USA 101: 11839–11844.

14. Ishibashi S, Sakaguchi M, Kuroiwa T, Yamasaki M, Kanemura Y, et al. (2004) Human neural stem/progenitor cells, expanded in long-term neurosphere culture, promote functional recovery after focal ischemia in Mongolian gerbils. J Neurosci Res 78: 215–223.

15. Lee HJ, Kim KS, Kim EJ, Choi HB, Lee KH, et al. (2007) Brain transplantation of immortalized human neural stem cells promotes functional recovery in mouse intracerebral hemorrhage stroke model. Stem Cells 25: 1204–1212.

16. Lee HJ, Kim KS, Park IH, Kim SU (2007) Human neural stem cells over-expressing VEGF provide neuroprotection, angiogenesis and functional recovery in mouse stroke model. PLoS ONE 2/e156: 1–14.

17. Lee ST, Chu K, Jung KH, Kim SJ, et al. (2008) Anti-inflammatory mechanism of intravascular neural stem cell transplantation in hemorrhagic stroke. Brain 131: 616–629.

18. Kim SU, Park IH, Kim TH, Kim KS, Choi HB, et al. (2006) Brain transplantation of human neural stem cells transduced with tyrosine hydroxylase and GTP cyclohydrolase 1 provides functional improvement in animal models of Parkinson disease. Neuropathology 26: 129–140.

19. Ryu JK, Kim J, Hong SH, Choi HB, Kim SU (2004) Proactive transplantation of human neural stem cells blocks neuronal cell death in rat model of Huntington disease. Neurobiol Disease 16: 68–77.

20. Lee ST, Chu K, Park JE, Lee K, Kang L, et al. (2005) Intravenous administration of human neural stem cells induces functional recovery in Huntington's disease rat model. Neurosci Res 52: 243–249.

21. Hwang DH, Lee HJ, Kim BG, Joo IS, Kim SU (2008) Intrathecal transplantation of human neural stem cells over-expressing VEGF provide behavioral improvement, disease onset delay and survival extension in transgenic ALS mice. Gene Ther. in press.

22. Meng XL, Shen JS, Ohashi T, Maeda H, Kim SU, et al. (2003) Brain transplantation of genetically engineered human neural stem cells globally corrects brain lesions in mucopolysacchridosis VII mouse. J Neurosci Res 74: 266–277.

23. Dudek H, Datta SR, Franke TF, Birnbaum MJ, Yao R, et al. (1997) Regulation of neuronal survival by the serine-threonine protein kinase Akt. Science 275: 661–665.

24. Kennedy SG, Kandel ES, Cross TK, Hay N (1999) Akt/protein kinase B inhibits cell death by preventing the release of cytochrome c from mitochondria. Mol Cell Biol 19: 5800–5810.

25. Brunet A, Datta SR, Greenberg ME (2001) Transcription-dependent and -independent control of neuronal survival by the PI3K-Akt signaling pathway. Curr Opin Neurobiol 11: 297–305.

26. Franke TF, Hornik CP, Segev L, Shostak GA, Sugimoto C (2003) PI3K/Akt and apoptosis: size matters. Oncogene 22: 8983–8998.

27. Matsuzaki H, Tamatani M, Mitsuda N, Namikawa K, Kiyama H, et al. (1999) Activation of Akt kinase inhibits apoptosis and changes in Bcl-2 and Bax

expression induced by nitric oxide in primary hippocampal neurons. J Neurochem 73: 2037–2046.

28. Yamaguchi A, Tamatani M, Matsuzaki H, Namikawa K, Kiyama H, et al. (2001) Akt activation protects hippocampal neurons from apoptosis by inhibiting transcriptional activity of p53. J Biol Chem 276: 5256–5264.

29. Chong ZZ, Kang JQ, Maiese K (2003) Erythropoietin fosters both intrinsic and extrinsic neuronal protection through modulation of microglia, Akt1, Bad, and caspase-mediated pathways. Br J Pharmacol 138: 1107–1118.

30. Kim SU (1985) Antigen expression by glial cells grown in culture. J Neuroimmunol 8: 255–282.

31. Kim SU, Moretto G, Lee V, Yu RK (1986) Neuroimmunology of gangliosides in human neurons and glial cells in culture. J Neurosci Res 15: 303–321.

32. Cho TS, Bae JH, Choi HB, Kim SS, Suh-Kim H, et al. (2002) Human neural stem cells: Electrophysiological properties of voltage gated ion channels. Neuroreport 13: 1447–1452.

33. Ryu JK, Choi HB, Hatori K, Heisel RL, Pelech SL, et al. (2003) Adenosine triphosphate induces proliferation of human neural stem cells: Role of calcium and p70 ribosomal protein S6 kinase. J Neurosci Res 72: 352–362.

34. Kim SU, Nagai A, Nakagawa E, Choi HB, Bang JH, et al. (2008) Production and characterization of immortal human neural stem cell line with multipotent differentiation property. Methods Mol Biol 438: 103–121.

35. West MJ, Slomianka L, Gundersen HJ (1991) Unbiased stereological estimation of the total number of neurons in the subdivisions of the rat hippocampus using the opost-transplantational fractionator. Anat Rec 231: 482–497.

36. Fukunaga K, Kawano T (2003) Akt is a molecular target for signal transduction therapy in brain ischemic insult. J Pharmacol Sci 92: 317–327.

37. Cardone MH, Roy N, Stennicke SM (1998) Regulation of cell death protease caspase-9 by phosphorylation. Science 282: 1318–1321.

38. Taylor JM, Ali U, Iannello RC, Hertzog P, Crack PC (2005) Diminished Akt phosphorylation in neurons lacking glutathione peroxidase-1 (Gpx1) leads to increased susceptibility to oxidative stress-induced cell death. J Neurochem 92: 283–293.

39. Rojo AI, Salinas M, Martin D, Perona R, Cuadrado A (2004) Regulation of Cu/Zn-superoxide dismutase expression via the phosphatidylinositol 3 kinase/Akt pathway and nuclear factor-kB. J Neurosci 24: 7324–7334.

40. Salinas M, Wang J, de Sagarra M, Martin D, Rojo A, et al. (2004) Protein kinase Akt/PKB phosphorylates heme oxygenase-1 in vitro and in vivo. FEBS Lett 578: 90–94.

41. Graff JR, Konicek BW, Manulty AM, Wang Z, Houck K, et al. (2000) Increased Akt activity contributes to prostate cancer progression by dramatically accelerating prostate tumor growth and diminishing p27Kip1 expression. J Biol Chem 275: 24500–24505.

42. Brognard J, Clark AS, Ni Y, Dennis PA (2001) Akt/protein kinase B is constitutively active in non-small cell lung cancer cells and promotes cellular survival and resistance to chemotherapy and radiation. Cancer Res 61: 3986–3997.

43. Roy HK, Olusola BF, Clemens DL, Karolski WJ, Rotashak A, et al. (2002) Akt proto-oncogene overexpression is an early event during sporadic colon carcinogenesis. Carcinogenesis 23: 201–205.

Copyrights

Index

Printed and bound by CPI Group (UK) Ltd, Croydon, CR0 4YY

23/10/2024

01777692-0003